D. I. BOLEF   TRADE 13.951

AN INTRODUCTION TO
QUANTUM PHYSICS

AN INTRODUCTION TO QUANTUM PHYSICS

GARRISON SPOSITO

Department of Physics
Sonoma State College

John Wiley & Sons, Inc.
New York   London   Sydney   Toronto

Copyright © 1970 by John Wiley & Sons, Inc.

Library of Congress Catalog Card Number: 72–110175
ISBN 0–471–81755–4

Printed in the United States of America

10 9 8 7 6 5 4 3 2 1

To Volney

I drew one conclusion which I believe to be correct: that is, though there is no system, there really is a sort of order in the sequence of casual chances— and that, of course, is very strange.

Fyodor Dostoevsky, *The Gambler*

Preface

Things must change, so they can stay as they are.

Sometimes it is hard to remember how much the ways of doing physics have changed during the past decade. Walking about the modern university laboratory or browsing through its library, we see little to remind us of the time when "radiation counter" referred to a large blue box filled with vacuum tubes, or when "coherent light" meant an angstrom linewidth and "elementary particles" meant three dozen corpuscles. Physics today is truly one of the most modern of the sciences, and it is only in consonance with the rapid changes it has undergone that many earnest attempts have been made recently to significantly alter the way it is taught, particularly at the introductory level. The basic motivation behind these attempts is obvious: if the physics taught is to preserve its traditional relevance to the physics done, it must keep in stride with the latter. This premise seems unassailable. What seems to be less certain is how we can maintain relevance without becoming faddish, or how we can nourish and yet not be force-feeding .

The result of this uncertainty has been a great deal of controversy over what should be done to introductory courses and a concomitant neglect of the intermediate curriculum. This is a little ironical, since a very good case can be made that an appreciable chasm *has* developed between what is done at the graduate level, primarily in courses on quantum theory, and what is done at the intermediate level. The year or so of quantum physics to which advanced undergraduates are exposed is often appreciably different in form and content from what is found in a text at the level of, say, Messiah's *Quantum Mechanics.* Part of the reason for this lapse may be that there simply is not enough time to discuss thoroughly the more theoretical aspects of quantum physics and still do justice to the empirical portions. More than this, however, the reason may be that no intermediate textbook presently available seriously attempts to describe quantum phenomena in a truly quantum-theoretical language. Rather, the usual approach has been one wherein either a text discusses just the phenomena, insulated from theory by a host of questionable semiclassical models ("Modern Physics"), or develops only the theory as a kind of applied mathematics in isolation from the facts which prompted it ("Quantum Mechanics").

The method of the present volume, which appears to be replacing the older approach, is to avoid this rather inefficient schism by giving an introduction to quantum physics with experiment and theory completely interwoven. In this way the theory is developed as the natural language with which to interpret experiment. It can certainly be argued that such a method as this requires a somewhat more mathematical treatment than does its alternative. But we must also bear in mind the not inconceivable possibility that further progress in microscopic physics may someday depend on our having produced a generation of physicists which attributes to operators and wavefunctions the same degree of intuitive reality its predecessors attributed to vectors in three-space. It is this possibility which motivates the present work.

The readers of this book are presumed to have observed basic quantum phenomena, such as line spectra and radioactivity, in the laboratory. They also should understand quantum theory at the elementary level and know classical mechanics, electrodynamics, and differential equations at the intermediate level. A working knowledge of linear algebra is helpful, but not necessary.

It is also the premise that this book is best used as an organon for self-study. The text has been written to be read, not referred to. As an aid to this kind of enterprise the pages of each chapter have been generously sprinkled with statements which are not necessarily self-evident and which require the reader to *instantly* verify them in order that he completely understand what is being said. Many of these verifications are, in fact, requested in the problem sets which follow each chapter. Some, however, challenge in silence, without hints, leaving only the reader's curiosity (or his professor's goading) to move him. The majority of the 180 problems in this book are intended to be non-trivial extensions of the main points discussed *and thus are to be regarded as an integral part of the text*. Several of the problems are worth lecture time in their own right under the aegis of Special Applications. Others provide a counterpoint to the material of the main text which may emerge in a later chapter as a principal theme. (An example of this approach occurs with the Pauli matrices, which are introduced and developed solely in the problems for Chapters 6 and 7, but then appear as an important part of the Dirac theory of the electron in Chapter 12.) All in all, it is expected that the problems will be employed as a significant teaching device in any course of lectures supported by this book.

Often it is useful to know of supplementary reading material which may further illuminate the main points of a text. This information has been provided here primarily through footnotes punctuating the discussion rather than through the common end-of-the-chapter list of references. In this way the relevance of the cited work should be more obvious and its possibilities as a guide to deeper understanding less easy to ignore. For those wishing more than a single suggested reference, the Bibliography following the Appendix presents a number of texts grouped according to the chapters they best amplify.

Quantum Physics is for a one-year course as written, but can be used as the textbook for one-semester or two-quarter courses on quantum mechanics as well. These options are fully detailed in a table following the Teaching Notes.

I should like to thank the several anonymous reviewers, enlisted by Donald H. Deneck, the physics editor at John Wiley and Sons, for their many helpful criticisms of the manuscript. Gratitude is also expressed to C. E. Falbo, for his comments on an early version of Chapter 3; to Leo Falicov, for his review of Chapter 9; and to Eugen Merzbacher, for his critical reading of the entire manuscript and his dedication to keeping the number of errors and obscurities I have committed down to a relative minimum.

I of course accept all responsibility for any remaining errors and will greatly appreciate being informed of them by interested readers.

January, 1970   GARRISON SPOSITO

Teaching Notes

The following remarks are based on my experience in using the material in this textbook during the last few years. The suggestions may prove helpful in arranging lectures or assigning reading.

Chapter 1. The Empirical Foundation of the Quantum Theory. This chapter serves both as a reminder of the quantum phenomena usually discussed in the introductory sequence and as an apology for the term *quantum*. It is very important that students get an early appreciation of the discrete character of energy and that they realize the radical nature of Planck's Postulate. It may be disturbing to some that the Stern-Gerlach experiment is discussed in this chapter as if it were evidence for quantized atomic orbits. The reason for doing this is pedagogical: the early (incorrect) interpretation of the experiment is first used to introduce the quantal atom and then, in Chapter 7, to bring home the need for a spin variable in quantum theory.

Chapter 2. The Non-Relativistic Electron. At first glance this chapter seems trivial, but first impressions can be misleading. The purpose here is to develop the well-known particle properties of the electron, then show that, in fact, light can also have such properties; but the particles so conceived can no longer be described deterministically. The latter point is most important and should be given plenty of lecture time: it should be made clear that photon trajectories and interference patterns are mutually exclusive. This paves the way for a non-deterministic electron as soon as *its* wave properties are made known. The student should leave this chapter as convinced as purely conceptual arguments can make him that the correct theory of the electron cannot be deterministic.

Chapter 3. Fundamentals of Quantum Mechanics. This is a long and difficult chapter, to be read and reread with care and patience. The first two sections are not intended to be mathematically rigorous and should be taught informally. They are likely to be most palatable if the formal analogies with ordinary three-space are stressed. But there is no getting around the fact that learning a new language is a strain. Physical analogs, such as the modes of a vibrating string, may be useful in making the eigenvalue problem seem more friendly. The section on the free electron is worth whatever time it takes to make students at ease with operators and basis states. This is one chapter where it would not be unreasonable to assign every problem as homework.

Chapter 4. The Hydrogen Atom. The presentation in this chapter is standard except for the important *conceptual* difference that we are now discussing orthonormal basis states, not the solutions of a partial differential equation. The fact that the spherical harmonics span finite-dimensional subspaces of a Hilbert space should be exploited both mathematically and physically. One technique I have found useful in doing the former is to consider the hydrogenic *p*-state geometrically. Draw the basis states Υ_{+1}^{-1}, Υ_{+1}^{0}, and Υ_{+1}^{1} as three mutually perpendicular unit vectors in three-space, then draw in some possible arbitrary *p*-states and interpret the direction cosines physically. It is very important that the language of quantum theory be made to work here; otherwise, it will become artificial. Problem 21 is a must.

Chapter 5. Quantum Mechanical Approximation Methods. This chapter should present no great difficulty so long as every mathematical section assigned is accompanied by an application section. A previous knowledge of linear algebra is helpful here.

Chapter 6. Matrix Quantum Mechanics. This chapter is optional in a minimum program. Its greatest value to an undergraduate lies with the section on the harmonic oscillator. With the second quantization formulation becoming so important in nuclear, solid-state, and statistical physics, it is no longer just a matter of elegance to consider construction operators. The section on measurement may be taught seriously or *cum granis salis*, depending upon how good one feels about this aspect of quantum theory. Problem seven may be worth lecture time to discuss its application to two-state systems (for example, the ammonia molecule or the K-mesons). Problem 14 is a must.

Chapter 7. Many-Electron Atoms. A. This chapter introduces the concept of spin. It is here that the language of quantum theory begins to pay dividends, since spin-space is not like ordinary three-space, but spin operators and basis states are quite like other angular momentum operators and basis states. The vector model as presented in section three should be considered tentative if Chapter 9 is to be read. Problems 5, 7, 10, and 14 are the most important in Chapter 7, the last one because it can yield great confidence in the applicability of the vector model to real phenomena.

Chapter 8. Many-Electron Atoms. B. This is another chapter which can be omitted if necessary. Even if everything else is passed over, section 8.1 is worth lecture time as a good application of the variational principle. Hartree-Fock theory might seem a little specialized for an undergraduate text, but it is hard to deny its value in creating a viable picture of a many-electron atom.

The approach taken here is to stress the physical aspects and to nearly ignore the technical details. This method can be extended to the lecture through descriptions of the use of Hartree-Fock theory in nuclear and solid-state physics. The theory of superconductivity affords an especially interesting illustration.

Chapter 9. Many-Electron Atoms. C. This chapter introduces some of the concepts of the theory of groups and their representations. Once again, it is a matter of learning a somewhat foreign language. The first two sections require a careful reading aided by all the classroom example one can think of. It is a good idea to bring in geometrical figures—equilateral triangles, squares, tetrahedra—to illustrate finite groups. Alexandroff's little high school text, cited in the Bibliography, is very useful for supplementary reading. When one considers the beauty and significance of the applications in section four and five, it is clear that symmetry should become part of the analytic repertoire of every serious student of physics. If time does not permit group theory to appear in the lectures, sections six and seven may still be read, as they have been written to make use only of the material in Chapter 7.

Chapter 10. Molecular Spectra. The study of molecular phenomena is optional in a short course for physicists, but necessary in the course for chemists. Only the basic ideas are given; these should be illustrated in lecture by discussions of laboratory data on rotational and vibrational spectra.

Chapter 11. Scattering Theory. This is one of the more important applications chapters. Everything has been done in terms of partial waves as this formalism seems to balance physics and logical consistency better than alternative introductory discussions. The advantages are a picture of scattering that "holds together" and a satisfying milieu in which to present the first Born approximation. The disadvantages lie with the need to bring in spherical Bessel and Neumann functions from out of the blue. It can be helpful in lecture to distribute summary sheets on the properties of these functions.

Chapter 12. The Relativistic Electron. What this chapter may lack in mathematical simplicity it always seems to make up in physical sensationalism. Students have yet to be left unaffected by the discovery of negative-mass electrons. Moreover, the idea of energy-mass levels generally strikes their fancy and refreshes their appreciation of the conservation of energy in electronic transitions. One need not approve of the method used in section one for bringing in Lorentz invariance; any reasonable path to the Dirac Postulate suffices. It is worth the time to go over the Lorentz covariance of the Dirac equation. Problem 9 may be used as a substitute for the part of section 12.2 discussing the spin of a Dirac particle. If the Pauli matrices have not been discussed yet, the fourteenth problem in Chapter 6 and the seventh one in Chapter 7 should be assigned as a classroom exercise.

Chapter 13. Nuclear Phenomena. At least part of this chapter should be covered in a minimal program. The discussion in sections 13.1 and 13.2 requires only Chapters 3 and 4 as background. One might wish for more on nuclear structure and spectra than is given here, but to do so is to overlook the fact that most undergraduate physics programs include courses on nuclear physics. Chapter 13 is written to serve as a jumping-off point to such courses. The last section of the chapter, on beta decay, is not completely comprehensible unless the Dirac equation has been studied. In my view, this fact alone makes Chapter 12 mandatory. The first sixteen problems following Chapter 13 can be done without Chapter 12, however, and they should be assigned to round out the material. Problems 14 and 15 may be worth some time in lecture.

Shorter Courses using this Book. With a judicious choice of reading assignments, it is possible to use this textbook in one-semester or two-quarter courses on quantum *mechanics* (that is, primarily theoretical quantum physics). Following is a table which suggests courses that might be taught either in physics or in chemistry.

	PHYSICS	CHEMISTRY
One Semester	Chapters 3–7, 11 and 12 Chapter 13, (Sections 13.1 and 13.2)	Chapters 3 and 4 Chapter 5, (Sections 5.1, 5.3 and 5.6) Chapters 6 and 7 Chapter 9, (Sections 9.6 and 9.7) Chapter 10
Two Quarters	Chapters 3–7 Chapters 9, 11–13	Chapters 3–7, 9 and 10

Contents

2 lectures

Finished Sept. 13

Sept. 24

Contents

xvi

Contents

AN INTRODUCTION TO
QUANTUM PHYSICS

PART I

PRIMEVAL QUANTUM CONCEPTS

CHAPTER 1

The Empirical Foundation of the Quantum Theory

Nature is, after all, discrete.

1.1. ATOMIC LINE SPECTRA

Every gas can be made to radiate light. As a case in point, let us suppose that we are in a laboratory and have at hand a small quantity of helium gas sealed into a glass tube whose center is drawn fine, to be of capillary diameter. Mounted at each end of the tube is an electrode which is connected to one terminal of a high-voltage power supply capable of delivering a current of several milliamperes. Now the power supply is switched on: the current is sent through the gas, and the laboratory is filled with a brilliant light the color of pink carnations.

Because the light from the glowing tube is not the purest red, we suspect that it is not monochromatic, and would seek to resolve it into component colors, much as we might a beam of ordinary sunlight. However, upon passing the light through a prism, we discover that the spectrum cast is a great deal different from the familiar multicolored swath resolved from sunlight. For, superimposed on a dark background, we see only narrow, brightly colored lines. (See Figure 1a.)

As is well known, our experience is not unique. Other elemental gases, such as hydrogen or neon, and the vapors of metals give rise to a line spectrum just as does helium. To be sure, the ordering and intensities of the spectral lines are different for each elemental gas, but for each the spectrum is discrete —composed of narrow lines—rather than continuous. Moreover, the spectral lines are never limited to the visible light region, but are to be found in the X-ray, ultraviolet, and infrared regions as well.

Experimental studies of atomic line spectra have shown quite conclusively

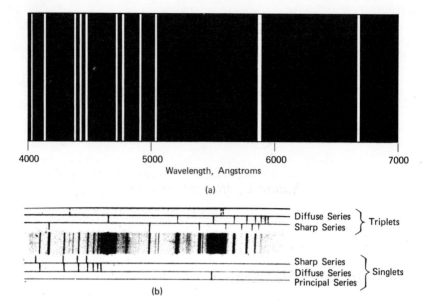

Figure 1. Atomic line spectra. (a) Helium gas. (b) Magnesium vapor, showing two series of triplets and three series of singlets. (Reprinted, with permission of the publisher, from G. Herzberg, *Atomic Spectra and Atomic Structure*, Dover Publications, New York, 1944.)

that once a spectrum has been recorded, a means for unambiguously identifying the element producing it is at hand. This means of "fingerprinting" the gas is made feasible by the interesting fact that the lines in any spectrum appear to fall into series whose members show a regular decrease in separation and intensity as their wavelengths grow shorter. As an example, consider the portion of the line spectrum of magnesium vapor shown in Figure 1b. Three series are evident here. One is composed of very intense lines and is called the Principal Series; one is composed of intense narrow lines and is therefore called the Sharp Series; and another, whose lines are diffuse in character, is called, predictably, the Diffuse Series. (Another series, not shown in the illustration, lies in the infrared region and is called the Fundamental Series.) We notice that the sharp and diffuse series show a definite bunching together of their constituent lines as the wavelengths of the latter decrease. Also, the series for magnesium, upon close scrutiny, appear to be composed of both single lines, called singlets, and trios of closely spaced lines, called triplets. This property, however, is not a very general one, as other line spectra are found which contain doublets, quartets, and so on. In any case, the series of the different kinds of line usually overlap in a complicated way throughout a given spectrum. But this fact does not diminish the great utility of these series in identifying the gas they represent, for it turns out that the wavelengths corresponding to the component lines may be arranged and calculated in many cases with the help of the empirically deduced expression[1]

$$\frac{1}{\lambda_{pk}} = R\left[\frac{1}{(p + \alpha)^2} - \frac{1}{(k + \beta)^2}\right], \qquad (p < k, k = 2, 3, 4, \ldots) \qquad (1.1)$$

where λ is the wavelength, R is the "Rydberg constant," a number dependent on the atomic number of the elemental gas, p is a positive integer less than the smallest value taken on by k, another positive integer, and α and β are constants for the spectral series in question. Thus, to each element we may assign a value of R; to each series, a value of p; and to each component, whether singlet, doublet, triplet, or whatever, a value of the integer k. That whole numbers appear in the expression for the wavelengths of spectral lines is remarkable, but entirely consistent with the fact that the spectra are discrete and not continuous.

Table 1.1. A comparison between equation (1.1) and experiment for certain of the lines in the spectrum of magnesium. The constants are $\alpha = -1.0673913$, $k = \infty$, $R = 4.3893932 \times 10^5 \text{ cm}^{-1}$. [The data are from ref. 1 and C. E. Moore, *Atomic Energy Levels*, U.S. Bureau Stand Circ. 467 (1949), p. 109.]

$1/\lambda_{p\infty}$ (obsd.) cm^{-1}	$1/\lambda_{p\infty}$ (calcd.) cm^{-1}	p
121,267	117,524	3
51,462	51,039	4
28,481	28,383	5
18,069	18,041	6
12,483	12,472	7
9,138	9,133	8

Given the experimental fact of line spectra and the ostensible correctness of equation (1.1), we are led naturally to wonder whether or not these things can be related within the context of some theoretical model of the atom. This question can be answered only by trying. The simplest model we might imagine is that which conceives of the atom as a microscopic solar system

[1] For a discussion of empirical spectroscopic formulas, see B. Edlén, Atomic spectra, *Handbuch der Physik* **XXVII**: 80–90 (1964).

wherein the role of gravitational forces is played by coulomb forces. Thus, particles of negative charge (electrons) are made to move about the positively charged central core (nucleus) of the atom in well-defined orbits while acted upon by a centripetal force. This structure is sufficient to insure that the electrons will radiate energy during their motions, as can be shown through a mild recasting of Maxwell's equations.

Consider the Maxwell curl equations (in the usual gaussian notation):

$$\nabla \times \mathbf{E} = -\frac{1}{c} \frac{\partial \mathbf{H}}{\partial t} \tag{1.2}$$

$$\nabla \times \mathbf{H} = \frac{1}{c} \frac{\partial \mathbf{E}}{\partial t} \tag{1.3}$$

appropriate to a finite region of empty space V bounded by a surface S. If the scalar product of equation (1.2) with $\mathbf{H}$ and that of equation (1.3) with $\mathbf{E}$ be formed, there results

$$\mathbf{H} \cdot (\nabla \times \mathbf{E}) = -\frac{1}{2c} \frac{\partial}{\partial t} (\mathbf{H} \cdot \mathbf{H}) \tag{1.4}$$

$$\mathbf{E} \cdot (\nabla \times \mathbf{H}) = \frac{1}{2c} \frac{\partial}{\partial t} (\mathbf{E} \cdot \mathbf{E}). \tag{1.5}$$

Now equation (1.5) is subtracted from equation (1.4):

$$\nabla \cdot (\mathbf{E} \times \mathbf{H}) = -\frac{1}{2c} \frac{\partial}{\partial t} (\mathbf{E} \cdot \mathbf{E} + \mathbf{H} \cdot \mathbf{H}) \tag{1.6}$$

where we have employed the vector identity

$$\mathbf{H} \cdot (\nabla \times \mathbf{E}) - \mathbf{E} \cdot (\nabla \times \mathbf{H}) = \nabla \cdot (\mathbf{E} \times \mathbf{H}).$$

Finally, equation (1.6) is integrated over V and the divergence theorem is applied where appropriate to obtain

$$\frac{c}{4\pi} \int_S (\mathbf{E} \times \mathbf{H}) \cdot d\mathbf{S} = -\frac{1}{8\pi} \frac{\partial}{\partial t} \int_V (\mathbf{E} \cdot \mathbf{E} + \mathbf{H} \cdot \mathbf{H}) \, dV. \tag{1.7}$$

Equation (1.7) represents the energy balance in the region V. The right-hand side is the rate of decrease in electric and magnetic field energies within V. To preserve the conservation of energy, the left-hand side must be interpreted as the rate of energy loss from V *as radiation*. In general, this side may be made to vanish by taking S to be arbitrarily large in extent. However, in the case of an electron of charge e moving in a circular orbit with a constant

6

angular speed, it can be shown[2] that the contributions of **E** and **H** to the left-hand side of equation (1.7) lead to

$$\frac{c}{4\pi} \int_{S} (\mathbf{E} \times \mathbf{H}) \cdot d\mathbf{S} = \frac{2}{3} \frac{e^2}{c^3} (R\omega^2)^2 \left(1 - \left(\frac{R\omega}{c}\right)^2\right)^{-2} \tag{1.8}$$

where R is the radius of the circle and ω is the angular speed. We see from this result that, according to classical electromagnetic theory, *an electron moving uniformly in a circular orbit will constantly radiate energy*. Indeed, fast electrons will rapidly lose energy.

The question remains as to what will be the nature of the radiation from our classical atom. Evidently the total energy of the electron is decreased by the radiative process, so that the particle is obliged to decrease continually its potential energy. This action, in turn, must diminish the orbital radius in a continuous manner. Thus the atom passes through all imaginable energy states, radiating a *continuous spectrum*, in sharp disagreement with what is observed experimentally. Moreover, our atom is unstable: its orbit diminishes as the radiating electron spirals into the nucleus. We are left to conclude that the classical theory of the electron is consistent with the observation of a radiating atom,[3] but can explain neither the occurrence of line spectra nor, unfortunately, even the existence of stable atoms composed of electrons and nuclei.

1.2. THE FRANCK-HERTZ EXPERIMENT

Classical electromagnetic theory has led us quite rapidly to an unsuccessful model of the atom and its radiative process. It is wisest, when confronted by such a difficulty, to appeal to experiment once again in the hope that new information will point the way to better theory. But what kind of experiment? A close examination of our incorrect theory may help answer this question. If we adopt the position that it is still reasonable to consider an atom as a system of electrons orbiting about a central nucleus, then it must be that our analysis of the radiative process itself is faulty. Evidently the transfer of electronic energy does not occur in the way we have imagined. It seems, therefore, that an experiment on the manner in which electrons in atoms lose and gain energy is required. Such an experiment was reported by James Franck and Gustav Hertz in 1913.[4]

---

[2] See, for details, J. B. Marion, *Classical Electromagnetic Radiation*, Academic Press, New York, 1965, Chapter 7.

[3] Notice, however, that the atom radiates merely by virtue of the electron-nucleus interaction. This result is certainly in contradiction with the facts and leads directly to the instability of the classical planetary atom.

[4] J. Franck and G. Hertz, Uber Zusammenstösse zwischen langsamen Elektronen und Gasmolekülen II, *Verhandlungen der Deutschen Physikalischen Gesellschaft* **15**: 613–620 (1913).

7

These investigators constructed the apparatus shown diagrammatically in Figure 2a. An incandescent filament inserted through the plate a is the source of thermally emitted electrons which move toward the grid b through a potential difference V. A retarding potential difference (that is to say, one opposite in sign to V) is established between the grid and the collector plate c. In this way the number of electrons reaching the galvanometer wired to c may be controlled. When a gas at low pressure is introduced between the filament and the collector plate, it is to be expected that the electrons leaving the

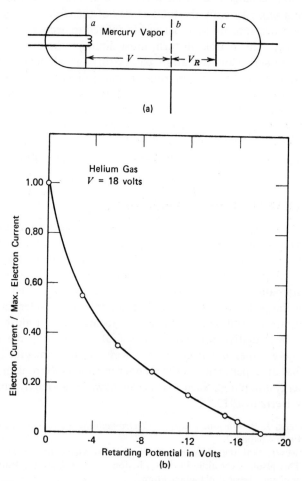

Figure 2. The Franck-Hertz experiment. (a) A diagram of the apparatus. (b) The dependence of the 18-volt electron current upon the retarding potential.

former to accelerate toward the latter will suffer collisions. During these collisions, which, according to our model, are free electron-atomic electron interactions, energy and momentum transfer will occur. However, the amount of energy exchange is expected to be small (because the atomic mass is so large), provided that no alterations in the *internal energy* of the atom (inelastic collisions) have taken place. Since the accelerating

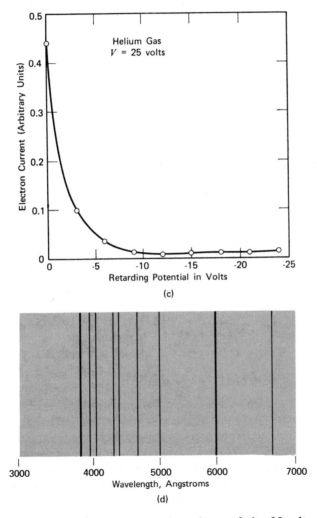

Figure 2 (*continued*). (c) The dependence of the 25-volt electron current upon the retarding potential. (d) The spectrum of the radiation produced by 25-volt electrons traversing helium gas.

9

potential V may be fixed at will, the galvanometer current for a given retarding potential should provide a measure of the number of free electrons not exchanging kinetic energy in their trips through the gas. In general, it is expected that a plot of galvanometer current against retarding voltage should be a regularly descending curve, the current going to zero when the retarding voltage is equal and opposite to the accelerating voltage. In Figure 2b are shown the data of Franck and Hertz plotted for electron motion through helium gas at a pressure of 1.3 torr. The curve is sufficiently regular in its descent to warrant the conclusion that, when the accelerating potential is 18 volts, the collisions between free and helium electrons are virtually elastic. Now we turn to the data plotted in Figure 2c. The conditions of the experiment remained the same as for that represented by Figure 2b, with the exception of an increase in the accelerating potential from 18 to 25 volts. The striking difference between the two figures leads us immediately to state that, when the accelerating potential is 25 volts, a retarding potential of about 6 volts is quite sufficient to prevent electrons traveling through helium gas from ever reaching the collector plate. With reference to energy transfer, we can say alternatively that free electron-helium electron collisions have suddenly become inelastic. It can only be that the helium electron is for some reason especially sensitive to collisions with 25- to 19-volt electrons; that is, the energy transfer is a selective, *discontinuous* process. This remarkable phenomenon is made all the more so by the fact that when the gas is subjected to 25-volt electrons it begins to radiate light which can be resolved as part of the line spectrum of helium (Figure 2d). When accelerating potentials less than 19 volts are used, no light is observed. Therefore, we may conclude from the experiment of Franck and Hertz that *the line spectrum of a gaseous element is in some way connected with electronic energy transfers that are discrete rather than continuous processes.*

1.3. THE STERN-GERLACH EXPERIMENT

The investigation thus far has yielded the impression that atomic electrons are not free to exchange all imaginable energies, but only those of certain, fixed magnitudes. This exchange process, being selective, gives rise directly to the atomic line spectrum. We have as yet no information on the physical criteria prompting the radiative process, however, nor on the manner in which the electronic energy spectrum imposes conditions on the structure of the atom. We suspect that these two questions are closely related, being but complementary expressions of the same problem. Therefore, a return to the analysis of our faulty classical atomic model should continue to produce insight as to the physical basis of the emission spectrum. Our atom, in light of the Franck-Hertz experiment, is now conceived as a miniature solar system wherein the

orbiting electrons are not disturbed by external influences unless the perturbations impart energy to the system in quantities greater than some
minimum amount—at least this is so when the perturbation takes the form
of a *particle* such as an electron. But what happens when the disturbance is
created by a *field*? Perhaps the discontinuous energy transfer in the atom is
conditioned by the use of a discrete probe and would comply with classical
expectations if a continuous stimulus were employed! This hypothesis should
not be hard to test, for the atom, according to our model, should respond to
the application of an inhomogeneous magnetic field.

To see this, we can imagine the electron circling in its orbit to be the same
as a microscopic, stationary current loop. A current loop will in general give
rise to a magnetic field equal in magnitude to that produced by a magnetic
dipole of moment[5]

$$\mu = \frac{I}{c}\mathbf{S}$$

where μ is the magnetic moment vector, I is the magnitude of the current, and
$\mathbf{S}$ is the vector area of the loop. If the dipole is placed in a space-dependent
magnetic field $\mathbf{H}$, it is subjected to the force

$$\mathbf{F} = (\mu \cdot \nabla)\mathbf{H}. \qquad (1.9)$$

Now suppose a beam of atoms (magnetic dipoles) is projected in, say, the
positive x-direction perpendicularly to the direction of an inhomogeneous
magnetic field along the z-axis. In this case, the x- and y-components of $\mathbf{H}$
are zero and equation (1.9) reduces to

$$F_z = \mu_z \frac{\partial H_z}{\partial z}.$$

The only effective force acting upon the atom in this situation, then, is
directed along the z-axis. Accordingly, this force will cause a deviation of the
atomic beam, as it travels through the field, in the z-axis direction. The
magnitude of the deflection will depend on how long is the path of the beam
in the field, the initial velocity of the atoms, and of course, the strength of
the deflecting force.

In our classical atom there is no provision for any particular size or
orientation of the electronic orbits, with the result that all imaginable values
of the magnetic moment and, therefore, the deflecting force are permitted.
Consequently, a *continuous* deflection of the beam in magnitude from zero to
a maximum consistent with the physical criteria mentioned above is expected
from an experimental test. The first test came in 1922, when Otto Stern and

[5] See R. Becker, *Electromagnetic Fields and Interactions*, Ginn and Co., New York,
1964, Vol. I, §47.

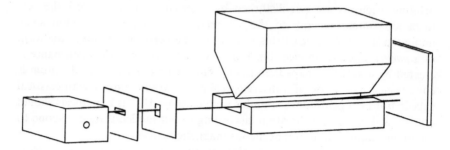

Figure 3. A diagram of Stern and Gerlach's apparatus for observing the deflection of silver atoms in an inhomogeneous magnetic field.

Walther Gerlach published the result of their investigation on the behavior of silver atoms in an inhomogeneous magnetic field.[6] The experimental apparatus is diagrammed in Figure 3. The essential parts are an oven O for evaporating silver at a known temperature, a pair of slits for collimating the atomic beam, and a pair of magnet pole-pieces designed to produce a strong, very inhomogeneous magnetic field close (about 0.2 mm) to the knife-edge pole, which was about 3.5 cm long. The maximum deflection expected with this arrangement was about one-tenth of a millimeter. In Figure 4 is shown the

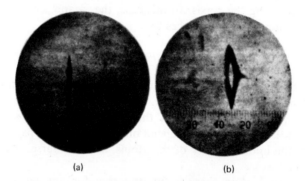

(a) (b)

Figure 4. The result of the Stern-Gerlach experiment.
(a) The beam deposit in the absence of a magnetic field.
(b) The beam deposit in the presence of an inhomogeneous magnetic field.

[6] W. Gerlach and O. Stern, Der experimentelle Nachweis der Richtungsquantelung im Magnetfeld, *Zeitschrift für Physik* **9**: 349–352 (1922). Our interpretation of this experiment is a little too much like that of classical mechanics to be correct, but it is helpful as a guide at this point. We shall have more to say about this matter in Chapter 7.

result of the Stern-Gerlach experiment. The atomic beam is *not* observed to deflect in a continuous manner; instead, the beam splits into a pair of components about 0.1 mm apart. (To be sure, the deflected beams are not sharply centered on single points; but this is to be expected since the atoms emerging from the oven have velocities distributed regularly about an average value, depending on the oven temperature.) We are forced to conclude that the magnetic moment of the silver atom is restricted to only two values, rather than a multitude of them, and, accordingly, that the electronic orbits in the silver atom are *not* capable of taking on every orientation and dimension conceivable, but that the structure of the atom is ordered; that is to say, *spacially discrete.*[6]

1.4. BLACKBODY RADIATION

A significant refining of our conception of the atom has come from the results of two ingenious experiments. We know now that if the planetary model is retained, the electronic orbits must be of certain dimensions and orientations in space, and that the electrons must be required to exchange energy in discrete quantities greater than some minimum amount. This information, however, is the limit of our insight; we have no way of quantitatively relating the discrete aspects of atomic structure to the discrete aspects of atomic emission spectra. For that we must turn to a study of radiation itself.

Now, a solid at any temperature above absolute zero will radiate. The spectrum, unlike that of the light emitted from elemental gases, is continuous, there being some finite intensity of radiation for every imaginable wavelength. (See Figure 5.) If the solid is a very efficient emitter of radiation—that is, if the intensity of its radiation is virtually the greatest possible—it is said to approximate closely the behavior of a *black body*, the perfect emitter and absorber of radiation. As an example of a black body, we may consider a solid containing a rough-walled cavity whose only contact with its surroundings is through a very small hole. Light incident upon the hole fills the cavity with radiation which, after some reflection, is absorbed entirely (or essentially so, if the area of the hole is much smaller than the area of the cavity walls). Conversely, if the solid is at some non-zero temperature, the radiation filling the cavity and emitted through the hole will be blackbody radiation. An important theorem, first proved by Kirchhoff,[7] states that the frequency distribution of the spectral energy density (the energy per unit volume of space per unit frequency) for blackbody radiation does not depend on the

[7] See R. Becker, *Electromagnetic Fields and Interactions*, Ginn and Co., New York, 1964, Vol. II, p. 277*f.*

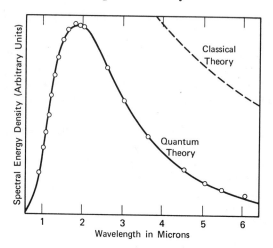

Figure 5. The spectral energy density of black-body radiation at 1956°K as a function of wavelength. The solid line represents equation (1.17) while the dashed line represents equation (1.15). (After W. W. Coblentz, Constants of spectral radiation of a uniformly heated inclosure or socalled black body. II. *Bull. Bureau of Standards* **13**:459–477 (1917).)

material nature of the emitting solid, but only on its absolute temperature. Hence the very great significance of this phenomenon for our study of the properties of radiation.

We wish first to consider the problem from a thermodynamic point of view so as to gain the most information possible without recourse to a microscopic model. From experiment it is found that the total energy of blackbody radiation per unit volume is related to the absolute temperature by[8]

$$u(T) = aT^4 \tag{1.10}$$

where a is a constant. This observation implies that the entropy of the radiation is proportional to the product of the volume and the cube of the absolute temperature:

$$T \equiv \left(\frac{\partial U}{\partial S}\right)_V = V\left(\frac{\partial u}{\partial S}\right)_V = 4aVT^3\left(\frac{\partial T}{\partial S}\right)_V$$

[8] See, for example, M. Born, *Atomic Physics*, Hafner Publ. Co., New York, 1962, pp. 251; 422*ff.*

and

$$S - S_0(V) = \int_0^T 4aVt^2 \, dt = \tfrac{4}{3}aVT^3. \tag{1.11}$$

Now suppose an adiabatic compression of blackbody radiation enclosed by perfectly reflecting walls takes place. The manifest result, according to equation (1.11), is an increase in the temperature; there will also be a change in the energy of radiation per unit frequency per unit volume (the spectral energy density) because the frequency of the radiation impinging upon the compressing piston will be altered by the piston motion (the Doppler effect). A straightforward but tedious argument[8] shows that the change in the spectral energy density is given by

$$d(u(\nu, V)V) = \frac{1}{3}\left(\frac{\partial u}{\partial \nu}\right)_V dV \tag{1.12}$$

where $u(\nu, V)$ is the spectral energy density and ν is the frequency of the radiation impinging upon the piston. By expanding equation (1.13), we get a partial differential equation for $u(\nu, V)$:

$$V\frac{\partial u}{\partial V} + u(\nu, V) = \tfrac{1}{3}\nu\frac{\partial u}{\partial \nu}.$$

If $u(\nu, V) \equiv \nu^3\phi(\nu^3 V)$ is introduced into this equation, it reduces to

$$\nu^3 V\frac{\partial \phi}{\partial V} = \tfrac{1}{3}\nu^4\frac{\partial \phi}{\partial \nu}$$

which is seen to be an identity, upon making the change of variable $x \equiv \nu^3 V$. Therefore, $u(\nu, V) = \nu^3\phi(\nu^3 V)$ is a solution of equation (1.12). Because the process under consideration is adiabatic, we have from equation (1.11)

$$u(\nu, T) = \nu^3 F\left(\frac{\nu}{T}\right)$$

an expression first derived by W. Wien in 1893. We notice that equation (1.12) implies equation (1.10), since

$$u(T) \equiv \int_0^\infty u(\nu, T) \, d\nu = \int_0^\infty \nu^3 F\left(\frac{\nu}{T}\right) d\nu = T^4 \int_0^\infty z^3 F(z) \, dz \tag{1.10'}$$

where $z = \nu/T$.

Our thermodynamic analysis has led us to what must be the general form of the spectral energy density. Any further elucidation of this function must come from calculations based upon some microscopic model of radiation by a solid. The most intuitive picture of a radiating solid is that which portrays the solid atoms as microscopic harmonic oscillators. Radiation occurs

15

because of the relative displacements of the centers of charge and mass of the atom during its vibratory motion (dipole radiation). The spectral energy density in this case can be shown to be proportional to the statistical average of the oscillator's energy:[9]

$$u(v, T)\, dv = \frac{8\pi v^2}{c^3} \langle\langle E(v, T)\rangle\rangle\, dv, \tag{1.13}$$

where $\langle\langle\ \ \rangle\rangle$ refers to a statistical average. The problem has thus been reduced to deducing the form of the average energy. If we assume the energy of the oscillator can take on all values from zero upward, then the average energy follows directly from Boltzmann's theorem[10]

$$\langle\langle E(v, T)\rangle\rangle = \frac{\int_0^\infty E(v) \exp\left(-E/kT\right) dE}{\int_0^\infty \exp\left(-E/kT\right) dE} = \frac{(kT)^2}{kT} = kT \tag{1.14}$$

where k is Boltzmann's constant. In accordance with this result, equation (1.13) becomes

$$u(v, T)\, dv = \frac{8\pi v^2}{c^3} kT\, dv. \tag{1.15}$$

A glance at Figure 5 shows that equation (1.15) is in violent disagreement with the facts save for the smallest values of the frequency. Moreover, although it does not contradict Wien's expression, it does not lead to equation (1.10) because the integral

$$\int_0^\infty v^2\, dv$$

does not exist. Once again our assumption of continuity has failed us: we have been led astray by assuming that the energy of the oscillator can vary *continuously* from zero upward. Evidently the contrary is true, but it is not at all clear in what manner blackbody radiation exhibits discrete characteristics. Indeed, to impart such a property to a wave phenomenon seems a contradiction in terms. We shall see that the price for a profound, new understanding of radiation is the adoption of this curious duality notion and the concomitant difficulties with the interpretation of microscopic phenomena.

1.5. PLANCK'S POSTULATE

From these considerations we see that, if we wish to calculate the way in which a thermodynamic process takes place in time, such a formula-

[9] See, for example, R. M. Eisberg, *Fundamentals of Modern Physics*, John Wiley and Sons, New York, 1961, pp. 51–62.
[10] E. A. Desloge, *Statistical Physics*, Holt, Rinehart, and Winston, New York, 1966, Chapter 12. This is one of many discussions on Boltzmann's theorem.

tion of initial and boundary conditions as is perfectly sufficient for a unique determination of the process in thermodynamics, does not suffice for the mechanical theory of heat or for the electrodynamical theory of heat radiation. On the contrary, from the standpoint of pure mechanics or electrodynamics the solutions of the problem are infinite in number. Hence, unless we wish to renounce entirely the possibility of representing the thermodynamic process mechanically or electrodynamically, there remains only one way out of the difficulty, namely, to supplement the initial and boundary conditions by special hypotheses of such a nature that the mechanical or electrodynamical equations will lead to an unambiguous result in agreement with experience.[11]

These are the words of Max Planck, who, in 1900, had turned to a consideration of the thermodynamic properties of electromagnetic radiation. By this time he had through empirical means found an expression for $\langle\langle E(v, T)\rangle\rangle$ which, when introduced into equation (1.13), produced a function which fit experimental data such as those shown in Figure 5. One year later he published a theoretical derivation of $\langle\langle E(v, T)\rangle\rangle$ for the harmonic oscillator.[12] In doing so he postulated that the energy of the oscillator does not vary upward from zero in a continuous manner. Instead, on the average, the energy is composed of a whole number of discrete elements, each having a value directly proportional to the vibration frequency; thus,

$$E(v, N) = Nhv \qquad (N = 0, 1, 2, \ldots), \qquad (1.16)$$

where h is called *Planck's constant*. The integrals in equation (1.14) must be accordingly replaced by summations:

$$\int_0^\infty \exp(-E/kT)\, dE \rightarrow \sum_{N=0}^\infty \exp(-Nhv/kT)$$

$$= \sum_{N=0}^\infty [\exp(-hv/kT)]^N = [1 - \exp(hv/kT)]^{-1},$$

$$\int_0^\infty E \exp(-E/kT)\, dE = -\frac{\partial}{\partial(1/kT)} \int_0^\infty \exp(-E/kT)\, dE \rightarrow$$

$$-\frac{\partial}{(\partial 1/kT)} [1 - \exp(-hv/kT)]^{-1} = [1 - \exp(-hv/kT)]^{-2} hv \exp(-hv/kT)$$

---

[11] M. Planck, *The Theory of Heat Radiation*, Dover Publications, New York, 1959, p. 115. Reprinted with the permission of the publisher.
[12] M. Planck, Über das Gesetz der Energieverteilung im Normalspectrum, *Annalen der Physik* **4**: 553–563 (1901).

upon summing the geometric series in $exp\,(-hv/kT)$, so that

$$\langle\langle E(v, T)\rangle\rangle = \frac{[1 - \exp\,(-hv/kT)]^{-2}hv\,\exp\,(-hv/kT)}{[1 - \exp\,(-hv/kT)]^{-1}}$$

$$= \frac{hv}{[\exp\,(hv/kT) - 1]}.$$

The spectral energy density becomes now

$$u(v, T)\,dv = \frac{8\pi hv^3/c^3}{[\exp\,(hv/kT) - 1]}\,dv \tag{1.17}$$

which, if we put $h = 6.63 \times 10^{-27}$ erg-sec, is in excellent agreement with experiment. (See Figure 5.) From equation (1.10') we find

$$u(T) = \frac{8\pi h}{c^3}\int_0^\infty \frac{v^3\,dv}{[\exp\,(hv/kT) - 1]} = \left(\frac{8\pi^5 k^4}{15(hc)^3}\right)T^4. \tag{1.18}$$

This expression is also in agreement with experiment, both qualitatively and quantitatively. We notice that the expression (1.17) is in agreement with Wien's relation. However, equation (1.16) is not unique. If an arbitrary function of v were added to this expression, the subsequent equations would be unaffected since $E(v, N)$ could be redefined relative to the arbitrary function as zero.

Let us reflect a moment on the significance of Planck's postulate. It states that the energy of an harmonic oscillator, expressed in terms of its frequency of vibration, may *on the average* occur in units no smaller than hv. It does not say what is the energy of the oscillator at any instant, nor does it say how the value of h is to be calculated. From this we may conclude three things. First, Planck's postulate is *empirical*; that is, it contains a parameter h whose value cannot be calculated except through the comparison of some expression in which it appears with the appropriate experimental result. There is no theory of Planck's constant. The only physical reason we can give for its existence is that it is invaluable for the elucidation of microscopic phenomena. Secondly, the postulate is *statistical* in that it refers only to the time-averaged behavior of the microscopic oscillator. Nothing is said about energy except that it varies directly with the frequency if evaluated over periods of time long compared with v^{-1}. Therefore, it appears we have a less than complete picture of the radiative process in a solid. The laws of Newtonian mechanics tell us that the time-averaged oscillator energy is

$$E(v) = 2m(\pi A)^2 v^2$$

where A is the amplitude of the vibration. This expression certainly does not agree with equation (1.16). We are therefore left with a mistrust of Newtonian mechanics in its description of the details of oscillator motion; but Planck's postulate does not seem to provide enough information to fill the gap in our knowledge.

Finally, the postulate does not in itself suggest that a flaw exists in the wave theory of light. It states only what energy an oscillator must have on the average if it is to remain in thermodynamic equilibrium with a radiation field. We should not conclude from this that the latter must necessarily exhibit discrete energies when interacting with matter—we have verified the Quantum Hypothesis only for the emission of radiation by a heated solid! Nonetheless, it would seem to contradict the notion of dynamical equilibrium if absorption processes did not in some way also show discrete characteristics.

PROBLEMS

1. In the table below are listed the wavelengths of the first few lines in the principal series for lithium. For this series, $p = 1$ and $\alpha = 0.5884$.
   (a) Calculate the value of R and β in the series expression.
   (b) Calculate the wavelength of the last line in the series.

k	λ_{1k}, Å
2	6706.84
3	3232.61
4	2741.31
5	2562.54
6	2475.29

2. These are some of the wavelengths of the lines in one of the series for hydrogen. This series has $p = 2$ and $\alpha = 0$. Calculate the value of R and β in the series expression.

k	λ_{2k}, Å
3	6562.79
4	4861.33
5	4340.47
6	4101.74
7	3970.07

3. Consider a repetition of the Franck-Hertz experiment in which the *retarding potential* is held constant while the *accelerating potential* is increased. Draw a graph of plate current *versus* accelerating potential for helium gas with the retarding potential fixed at 0.5 volt. Interpret the drawing physically.

4. The mean deflection of the silver atoms used in the Stern-Gerlach experiment should be

$$S = \frac{1}{2}\left(\frac{F_z}{m}\right)t^2$$

where m is the mass of a silver atom and t is the time spent by the beam in the inhomogeneous magnetic field. We can estimate t by

$$t = \frac{l}{v_x}$$

where l is the length of the beam path and

$$v_x^2 = 3kT/m$$

T being the absolute temperature and k, Boltzmann's constant. Show that

$$S = \frac{\mu_z(\partial H_z/\partial z)l^2}{6kT}.$$

5. Show that the wavelength for which $u(\lambda, T)$ is a maximum is inversely proportional to the absolute temperature. Note that

$$d\nu = -c\, d\lambda/\lambda^2$$

and

$$u(\nu, T)\, d\nu = u(\lambda, T)(-d\lambda).$$

6. Show that if Planck's postulate is expressed

$$E(\nu, N) = (N + \tfrac{1}{2})h\nu \qquad (N = 0, 1, 2, \ldots),$$

an infinite total energy density will result unless the energy is renormalized relative to $\tfrac{1}{2}h\nu$.

7. Why is it not in contradiction with Planck's postulate to express the total energy density as an integral over the spectral energy density, rather than a sum?

8. Show that, in the limit of very small frequency, the spectral energy density is independent of Planck's constant and is of the same form as the expression obtained in section 1.4. (*Hint*: Use the Taylor series for e^x in your argument.)

CHAPTER 2

The Non-Relativistic Electron

Determinism to uncertainty—a quantum jump.

2.1. THE PHYSICAL PROPERTIES OF CATHODE RAYS

We have laid the foundation for the quantum theory. Through experiment and the postulate of Planck, our conception of the atom and radiation now has no place for a continuous microscopic model of matter and light. We require that a planetary atom consist of orbits whose dimensions and inclinations are restricted to certain values, and that the radiation from this atom represent no smaller energy release than that stipulated by the product of Planck's constant and the emission frequency. Moreover, the negative charges in the atom may exchange energy with other particles only in amounts larger than a certain characteristic value. A parallelism has clearly developed here: the atomic electrons can receive energy from collisions only in discrete quantities and the radiation they emit can be released only in discrete quantities. But what is the relation between the two phenonema? To answer this question we must know more about the physical properties of the electron and the result of the direct interaction between this particle and radiation. Let us first consider the electron. We restrict the discussion to the particle having low, or non-relativistic, kinetic energies.

Suppose a glass tube fitted with electrodes like the one in which was observed the emission spectrum of helium is filled with air at 1 atmosphere pressure. Care is taken not to seal the hole through which the gas was introduced, but to provide it with a capillary outlet and stopcock so that the air pressure can be controlled. A high-voltage power supply is again connected to the electrodes and a vacuum pump lead is attached to the gas outlet. The pump is switched on and used to reduce the air pressure to about 1 per cent

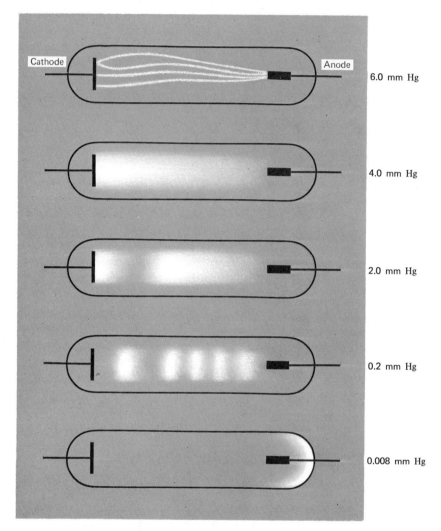

Cathode   Anode   6.0 mm Hg

4.0 mm Hg

2.0 mm Hg

0.2 mm Hg

0.008 mm Hg

Figure 6. The appearance of electric discharge in air at low pressures.

of its original value. Then the power supply is turned on. We observe a discharge between the electrodes that consists of blue streaming rays traveling from the negative electrode (cathode) to the positive one (anode). Reduce the air pressure again by 50 per cent. The entire tube is now filled with the discharge which has taken on a pink color reminiscent of our experience with helium. Another decrease in the pressure by half brings about the appearance of a dark region in the vicinity of the cathode. The dark space separates from the pink discharge a blue discharge existing immediately adjacent to the cathode. A further decrease in the pressure induces another dark space between the cathode and the blue discharge, the widening of the first-observed dark space, and the breakup of the pink discharge into uniformly arranged vertical bands. An increase in the vacuum, until the air pressure is 1/100,000 of its original magnitude, causes the dark space to fill the tube and the glass at the anode end to glow with the color of limes.

This spectacular and beautiful phenomenon was the subject of intense scientific inquiry during the last half of the nineteenth century. Of most interest was the behavior in the discharge tube at the lowest pressure. Not long after the first observations were made, it was hypothesized that the green glow was caused by invisible rays emitted by the cathode. The next decade of experimentation brought to light the following facts about these "cathode rays":

(a) *The cathode rays travel in straight lines.* This fact was demonstrated by placing an object between the cathode and the glowing spot on the tube. The result was a shadow of the object cast on the glass wall.

(b) *A magnet placed near the discharge tube will change the position of the glowing spot.*

(c) *The rays can heat a thin metal foil until it is bright red.*

(d) *The rays can exert a force on objects placed in their path.* This was shown by suspending a thin glass filament in front of the cathode of a discharge tube, with the ensuing deflection of the object.

(e) *The rays carry a negative charge.* This property was not easily discovered. It finally was demonstrated by allowing cathode rays to strike a charge-collecting electrode housed inside a cylindrical anode. The electrode invariably became negatively charged when struck by the rays, but remained neutral if property (b) were employed to prevent bombardment.

(f) *The physical properties of the rays are independent of what material is used to make the cathode.*

The interpretation of these facts was beset with not a little controversy.[1] Scientists working on the continent of Europe favored a model which portrayed the cathode rays as a form of electromagnetic radiation, while those in Great Britain preferred to think of the rays as beams of particles. The latter group, largely because of the work of J. J. Thomson at Cambridge, eventually was to win out. Thomson believed the cathode rays to consist of

---

[1] A very good historical account is given by D. L. Anderson in *The Discovery of the Electron*, D. Van Nostrand Co., Princeton, 1964, Chapter 2.

particles having definite charge and mass. He sought to prove it by making use of the behavior of cathode rays in electric and magnetic fields to measure the charge-to-mass ratio.

Let us consider these phenomena a moment from the point of view of Newtonian mechanics. Suppose a particle of charge e and mass m moves, in a direction we shall designate as along the x-axis, with a speed v_x and then enters an *electric field* directed along the positive y-axis. The equations of motion for the non-relativistic particle are

$$m\frac{d^2x}{dt^2} = 0 \tag{2.1}$$

$$m\frac{d^2y}{dt^2} = eE \tag{2.2}$$

there being no motion in the z-direction. The solutions of equations (2.1) and (2.2) are

$$x(t) = x_0 + v_x t$$

$$y(t) = y_0 + \frac{e}{m}E\frac{t^2}{2}$$

where (x_0, y_0) are the position coordinates of the particle just as it enters the transverse field E. The deviation of the particle from its initial path may be expressed in terms of the deflection angle θ as

$$\tan\theta \equiv \frac{v_y(t)}{v_x(t)} = \frac{eE}{mv_x}t$$

so that

$$\theta = \tan^{-1}\frac{eE}{mv_x}t = \left(\frac{eE}{mv_x}\right)t - \frac{1}{3}\left(\frac{eE}{mv_x}\right)^3 t^3 + \cdots. \tag{2.3}$$

Equation (2.3) is an expression for the deflection of a charged particle by an electric field as a function of the charge-to-mass ratio of the particle, its initial speed, the field strength, and the time the particle is subjected to the field. We notice that for short times the deflection angle is a linear function of the time.

Now consider the same particle entering a *magnetic field* directed along the positive z-axis. The equations of motion are, in the low-speed limit,

$$\frac{d^2x}{dt^2} = \frac{eH}{mc}\frac{dy}{dt} \tag{2.4}$$

$$\frac{d^2y}{dt^2} = -\frac{eH}{mc}\frac{dx}{dt}, \tag{2.5}$$

again ignoring the motion in the z-direction. By differentiating equation (2.4) once with respect to time, and substituting into the result equation (2.5), we get

$$\frac{d^3x}{dt^3} + \left(\frac{eH}{mc}\right)^2 \frac{dx}{dt} = 0$$

which has the solution

$$x(t) = x_0 - \rho \cos \left(\frac{eH}{mc}\right)t$$

where x_0 has its previous meaning and ρ is defined by

$$\rho^2 \equiv (x(t) - x_0)^2 + (y(t) - y_0)^2 ;$$

that is, the path of the deflected particle is a circle of radius ρ. In a similar manner the solution of equation (2.5) is found to be

$$y(t) = y_0 + \rho \sin \left(\frac{eH}{mc}\right)t.$$

We can express the deviation of the particle from its original trajectory by

$$\tan \phi \equiv \frac{(y(t) - y_0)}{(x(t) - x_0)} = -\tan \left(\frac{eH}{mc}\right)t$$

or

$$\phi = -\frac{eHt}{mc}. \tag{2.6}$$

In this instance the deflection angle is always of the first degree in the time.

The apparatus shown schematically in Figure 7 was used by Thomson, in 1897, to measure the ratio e/m for cathode-ray particles. The beam from the

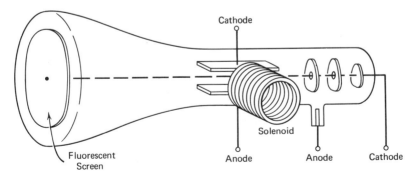

Figure 7. A diagram of Thomson's apparatus for determining the charge-to-mass ratio of cathode rays.

cathode is collimated by the tiny holes in the diaphragms and then passes between the metal plates and the solenoids on its way to the fluorescent screen. An electric field established across the metal plates is used to deflect the beam downward; a magnetic field induced by passing a current through the solenoids is used to deflect the beam upward. Thomson's use of these involved two measurements. In the first, he adjusted the electric and magnetic fields so that the deflections of the cathode-ray beam produced by them were mutually cancelling:

$$\theta = \frac{eE_0}{mv_x} t = -\phi = \frac{eH_0}{mc} t$$

where E_0 and H_0 are the field magnitudes necessary to produce no resultant deflection of the beam. (We have assumed the deflection to be small.) The initial speed of the beam particles under this condition is, therefore,

$$v_x = \frac{E_0}{H_0} c.$$

In the second measurement the magnetic field was switched off and the deflection angle θ was measured for an applied field of magnitude E_0:

$$\theta = \frac{eE_0}{mv_x} t = \frac{eH_0}{mc} t.$$

The time τ spent by the beam in traversing the field between the deflection plates may be estimated as l/v_x, for small θ, where l is the length of one of the plates. Thus,

$$\theta = \frac{eH_0{}^2}{mE_0c^2}$$

and

$$\frac{e}{m} = \frac{\theta E_0 c^2}{H_0{}^2}. \tag{2.7}$$

A downward deflection signifies a negative value of e. (θ would be negative.) Thomson found from his measurements the average value

$$\frac{|e|}{m} \equiv \frac{|e|}{m_e} = 2.3 \times 10^{17} \text{ esu/gm.}$$

The value accepted at present is

$$\frac{|e|}{m_e} = 5.27274 \pm 0.00006 \times 10^{17} \text{ esu/gm.}$$

Thomson concluded from his measurements that the cathode rays were indeed composed of particles. He argued further that, in view of the (then) lately reported data on the great penetrability into matter of the cathode rays, the largeness of the charge-to-mass ratio he measured, relative to that for hydrogen ions (about 2.88×10^{14} esu/gm), was a result of the *smaller mass* of the cathode-ray particle. The mass would, in fact, have to be about 2,000 times as small, if the magnitudes of the charges on the two were equal. Thomson speculated that[2]

> ... we have in the cathode rays matter in a new state, a state in which the subdivision of matter is carried very much further than in the ordinary gaseous state: a state in which all matter—that is, matter derived from different sources such as Hydrogen, Oxygen, etc.—is of one and the same kind; this matter being the substance from which all the chemical elements are built up.

The statement is, as far as it goes, in agreement with our planetary model of the atom. We shall find, however, that Thomson's view of the electron as a particle is far too naive for more than a semiquantative theory of atomic structure.

2.2. THE CHARGE ON THE ELECTRON

Having established that cathode rays are composed of particles, we must seek an answer to whether or not the properties of the particles are universal. Specifically, does only the charge-to-mass *ratio* remain constant for the electron, while the quantities charge and mass vary widely, or are the charge and mass themselves constant properties? The most certain way to decide the question is to measure charge or mass separately; if one quantity varies, so must surely the other.

The solution was provided by Robert Millikan in a paper published in 1913. Millikan had perfected a clever method for measuring the charge on the electron. Droplets of mineral oil were introduced by means of an atomizer between the plates of a capacitor otherwise separated by stagnant air. (See Figure 8.) Certain of the oil drops would become charged during their formation or by the radiation from a specially constructed cathode-ray tube. The charged drops then must proceed toward the appropriate metal plate of the capacitor, once the latter is itself charged. Now, if the upper plate is positively charged, negatively charged drops will be subjected to opposing forces: one from the gravitational field of the earth and one from the electric field in the capacitor. In principle, these two forces can be mutually balanced; this fact forms the basis of Millikan's experiment.

---

[2] D. L. Anderson, *The Discovery of the Electron*, D. Van Nostrand, Co., Princeton, 1964, p. 47. Reprinted by permission of the publisher.

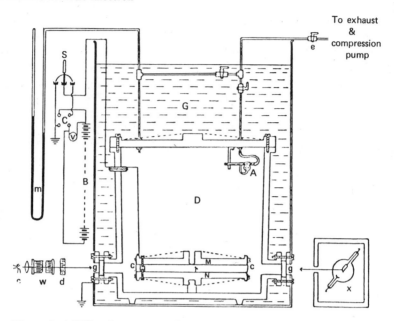

Figure 8. Millikan's apparatus for measuring the charge on the electron, consisting of (A) an atomizer, (B) a bank of batteries, (C) a double-pole-double-throw switch, (D) a brass cylinder, (e) a stopcock, (G) a constant-temperature oil bath, (c, w, d) an arc light with heat absorbers, (g) a trio of glass windows, (M, N) a capacitor, (m) a manometer, and (X) an X-ray tube. (Redrawn with permission from R. A. Millikan, *The Electron*, University of Chicago Press, Chicago, 1917.)

Suppose a negatively charged, spherical oil drop is in the capacitor and rises toward the positively charged plate with a speed v_1. The laws of Newtonian mechanics require that, if the speed is constant, the forces on the droplet must each cancel one another:

$$\mathbf{F}_E + \mathbf{F}_G + \mathbf{F}_B + \mathbf{F}_V = 0$$

where $\mathbf{F}_E$ refers to the force from the electric field, $\mathbf{F}_G$, that from the gravitational field, $\mathbf{F}_B$, that from the droplet's buoyancy, and $\mathbf{F}_V$, that from friction. Explicitly, we have

$$\frac{qV}{d} - mg + \frac{4\pi r^3}{3}\rho_a g - 6\pi\eta r v_1 = 0 \qquad (2.8)$$

where q is the charge on the drop, V is the potential difference across the capacitor, d is the width of the capacitor, m is the mass of the drop, r is its

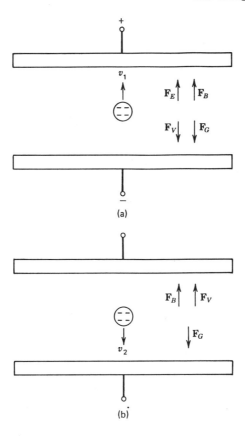

Figure 9. The dynamical aspects of Millikan's experiment. (a) the oil drop rising at constant speed in an electric field and a gravitational field. (b) The oil drop falling at constant speed in a gravitational field.

radius, ρ_a is the density of air, η is its viscosity, and π and g have their usual meanings. Equation (2.8) may be solved for v_1, to yield

$$v_1 = \frac{(qV/d) - (4\pi r^3/3)(\rho_d - \rho_a)g}{6\pi\eta r} \tag{2.9}$$

where ρ_d is the density of the oil from which the drop was made. Now suppose the field is switched off. The oil drop will fall in the gravitational field until it

29

achieves the terminal speed and the concomitant equilibrium condition. By our foregoing analysis, its speed sould be

$$v_2 = \frac{2r^2(\rho_d - \rho_a)}{9\eta} g. \tag{2.10}$$

The quotient of equations (2.9) and (2.10) provides an expression for the charge on the droplet:

$$|q| = \frac{d}{|V|} \left(1 + \frac{v_1}{v_2}\right) \frac{4\pi r^3}{3} (\rho_d - \rho_a)g. \tag{2.11}$$

This equation, however, is experimentally impractical because it contains the droplet radius, which is difficult to measure directly. We can alleviate the problem by solving equation (2.10) for the radius and substituting the result into equation (2.11). The expression for the charge then becomes

$$|q| = \frac{18\pi}{\sqrt{2}} \frac{(d\eta^{3/2}/|V|)}{[(\rho_d - \rho_a)g]^{1/2}} (v_1 + v_2)v_2^{1/2}. \tag{2.12}$$

Millikan's experiment consisted of measurements of v_1 and v_2 made by timing the oil drops as they fell or rose through a prescribed distance. A measurement of d, V, and η, along with the tabulated values of the two densities is then sufficient to calculate the charge. One other point should be mentioned. Millikan discovered[3] that the viscous force was not adequately represented by $6\pi\eta rv$ in his experiment, but by

$$F_V = \frac{6\pi\eta rv}{(1 + b/pr)}$$

where b is an empirical constant, p is the atmospheric pressure, and r is the radius of the drop. Thus, the expression for the droplet charge was actually divided by the factor $(1 + b/pr)^{3/2}$. When the charges were determined on a large number of oil drops, Millikan discovered that each was a virtually integral multiple of the number

$$4.77 \times 10^{-10} \text{ esu;}$$

that is, that *the charge on the electron does not vary continuously over all imaginable values from zero upwards, but is fixed* at the value just given. Once again the discrete aspect of the electron has made itself known: charge transfer, like energy transfer, is a discontinuous process.

Curiously enough, Millikan's experiment suffered from a hidden flaw. The value determined by him and his colleagues for the viscosity of air was too

---

[3] See R. A. Millikan, *The Electron*, University of Chicago Press, Chicago, 1917, pp. 88–100. This book contains much information on Millikan's measurement of the electronic charge.

small, making, in turn, the value of the electronic charge too small. The error was finally rectified some twenty years after, but not without some controversy. The presently accepted value is

$$|e| = 4.80298 \pm 0.00020 \times 10^{-10} \text{ esu.}$$

We can immediately deduce the mass of the cathode-ray particle from this result and that of Thomson's experiment. The mass turns out to be

$$m_e = 9.1091 \pm 0.0004 \times 10^{-28} \text{ gm,}$$

a result which firmly establishes the uniqueness, as a particle, of the non-relativistic electron.

2.3. THE PHOTOELECTRIC EFFECT: THEORY

We can now turn to the second part of our inquiry, armed with fresh knowledge of the properties of the electron, and ask: What is the nature of the direct interaction of light with this particle? As done several times before, we shall consider an experiment whose result should be decisive in this matter. Let us suppose that a clean metal plate, the surface of which we assume to be composed of atoms and, therefore, atomic electrons, is subjected to a beam of ultraviolet light. The result of this perturbation is made manifest easily if we juxtapose the plate with another just like it and use the pair to form a circuit, as shown in Figure 10. Now, with no potential difference across the

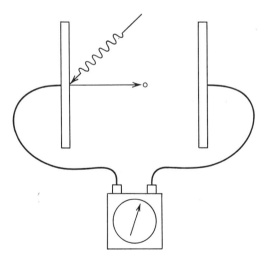

Figure 10. An idealized photoelectric circuit.

plates, we observe that current flows in the circuit when the light strikes the first metal plate. This phenomenon is called the *photoelectric effect*. The first reported observation of the photoelectric effect was that made by Heinrich Hertz in 1887, who noticed that the maximum distance traversed by an electrical discharge between two metal electrodes could be increased if ultraviolet light impinged upon the cathode. Throughout the next one and one-half decades the experimental description of the photoelectric effect was completed. The facts may be summarized in terms of the circuit diagrammed in Figure 10:

(a) Electrons are ejected when light of the appropriate frequency strikes a clean metal surface.
(b) The current between the plates depends linearly on the intensity of the light beam.
(c) The speed of the photoelectrons is not dependent on the light *intensity*, but is dependent on the light *frequency*.
(d) A current will not flow between the plates unless the frequency of the impinging light is greater than a certain value characteristic of the metal producing the photoelectrons.

If we assume, contrary to the essence of our experience with atomic radiation, that the electrons in the metal absorb light in a *continuous* manner, then we should encounter grave problems in attempting an explanation of the photoelectric effect. For, if the ejected electrons each may have absorbed any imaginable quantity of energy from the bombarding radiation, the photoelectric current should appear at all frequencies, but vanish when the light *intensity* falls below some characteristic value needed to free the electrons from the metal. But this does not happen: no current is detected if the light is of too low a frequency and the relation between current and light intensity stated above holds at the lowest measurable values of each. Again, if the energy exchange between light and electrons is perfectly continuous, an increased light intensity should serve to increase the kinetic energy and thus the speed of the photoelectrons. This prediction contradicts the facts, too. It seems once again we are forced to accept a discrete, rather than continuous, transfer of energy between radiation and the electron.

The problem was solved by Albert Einstein in 1905. He simply adopted the conclusion just drawn and made it the foundation of his theory of the photoelectric effect. The reasoning goes as follows. We shall assume that the incident radiation may be viewed as an assembly of *particles*, each bearing the kinetic energy $h\nu$, where h is the constant introduced by Planck and ν is the frequency of the light. When these "particles," or *light quanta*, are of the appropriate energy and strike electrons bound into the metal surface, they are completely absorbed. The energy so transferred is in part used to free the electron from the metal atom, while the rest becomes its kinetic energy. Therefore, we may write the conservation-of-energy condition

$$hv = \tfrac{1}{2}m_e v^2 + hv_0 \qquad (2.13)$$

where v is the speed of the photoelectron and v_0 is a constant to be determined by experiment. Equation (2.13) and its quantum-physical interpretation contain the full explanation of the photoelectric effect:

(a) Electrons are ejected from the irradiated metal because each absorbs a light quantum and its concomitant energy.

(b) The number of photoelectrons produced per unit time is proportional to the number of light quanta absorbed per unit time.

(c) The speed of the photoelectrons will depend on the magnitude of v (and that of v_0).

(d) When $v \leqslant v_0$ no kinetic energy is imparted to the irradiated electrons, and so no photoelectrons appear.[4]

The theory of the photoelectric effect has been obtained at the price of a radical new conception of the nature of electromagnetic radiation: when the radiation interacts directly with atomic electrons, it must be considered as an assembly of particles. But such a curious particle! We have to do with a corpuscle which moves with very high speed, does not obey the laws of Newtonian mechanics, and, under most conditions, exhibits wave behavior. How are we to explain interference and diffraction phenomena in terms of light quanta? Or is this extending the model too far? If so, how do we reconcile the quantum model with the wave model?

These serious questions make it seem as if Einstein's solution has created more problems than it has solved. Indeed, it has! But for the moment let us put them by and consider the light quantum for what it is—a notion whose intrinsic significance must be fully developed if we are to proceed further in the quantum theory.

We can deepen our understanding of the concept of light quanta by returning for a moment to the problem of blackbody radiation. Consider a rough-walled cavity in a solid at finite temperature. This time we shall adopt a dynamic approach to the problem and calculate the number of absorptions and emissions per unit time by the atoms of the solid in equilibrium with the radiation. The argument we give was first put forth by Albert Einstein in 1917.

---

[4] It should be noted that equation (2.13) affords a means for calculating only the *maximum* kinetic energy of the photoelectrons. If these particles are ejected from atoms beneath those in the metal surface, or if they are propelled at other than right angles to the latter, their kinetic energies will be smaller than the theoretical maximum.

The Non-Relativistic Electron

The number of absorptions taking place in the time interval dt must be directly proportional to the number of atoms receiving quanta, to the spectral energy density of the radiation, and to the time interval dt itself:

$$Z_a = BN_a u(v, T) \, dt \qquad (2.14)$$

where B is a proportionality constant characteristic of the absorption process, N_a is the number of atoms receiving quanta, and $u(v, T)$ is the already-familiar spectral energy density. Moreover, as mentioned above, it is reasonable to expect that during the time interval dt certain of the atoms will be *emitting* light. These may be divided into two classes: those which emit spontaneously a radiation of the appropriate frequency and those which emit radiation because they have been stimulated to do so by the incident light quanta. We may call the two kinds of emission *spontaneous emission* and *stimulated emission*, respectively. It follows that the number of emissions in the time interval dt always should be proportional to the number of stimulated atoms, but not always to the spectral energy density:

$$Z_e = N_e(A + B'u(v, T)) \, dt \qquad (2.15)$$

where A and B' are constants characteristic of the emission process. Equation (2.15) states, in accordance with the remarks preceding it, that emission may occur during dt even when there is no radiation present. This is what is meant by spontaneous emission.

We shall now invoke three physical requirements:

(a) The atoms are in thermodynamic equilibrium, necessitating that $Z_a = Z_e$.

(b) The form of $u(v, T)$ when hv/kT is very small has to be that obtained by means of the classical theory of electromagnetic radiation. [See equation (1.15).] This, it turns out, requires that $B = B'$.

(c) Boltzmann's theorem on the equilibrium distribution of two assemblies having different energies shall be valid. (See the reference cited in footnote 9 of Chapter 1.) Then we have

$$N_e = N_a \exp\left(-hv/kT\right).$$

By imposing upon equations (2.14) and (2.15) the foregoing restrictions, we get

$$BN_a u(v, T) \, dt = N_e(A + B'u(v, T)) \, dt$$
$$= N_a \exp\left(-hv/kT\right)(A + Bu(v, T)) \, dt$$

or

$$u(v, T) = \frac{A/B}{(\exp\left(hv/kT\right) - 1)} \qquad (2.16)$$

The proportionality constants A and B are arbitrary to the extent that their ratio may be fixed at will. If we put

$$A/B = 8\pi h\nu^3/c^3,$$

then equation (2.16) becomes identical with the expression for the spectral energy density derived from Planck's postulate. Therefore, it may be inferred that *the existence of light quanta is fully compatible with the phenomenon of thermal equilibrium between matter and radiation*; indeed, an hypothesis such as Einstein's is absolutely necessary for an adequate microscopic description of this condition. The particle model for light is now seen to be an integral part of the foundation of the quantum theory. Notwithstanding the very great problems of interpretation it engenders, we cannot extricate this view from physical theory and still retain an understanding of the behavior of electrons interacting with radiation.

2.4. THE PHOTOELECTRIC EFFECT: APPLICATION

Of course, Einstein's equation for the photoelectric effect did not have the complete support of experiment when it was first proposed. For example, it was not certain at the time that the kinetic energy of the photoelectrons is a linear function of the radiation frequency. Even if it were, the proportionality constant would not of necessity have to be that introduced by Planck. Thus some doubt still could be had concerning the suitability of equation (2.13), and this skepticism was all the more whetted by the fundamental disruption of the theory of electromagnetic radiation concomitant with the expression. What was needed, then, was an experiment—one which would permit the overall verification of the photoelectric theory as well as result in a precise determination of the constant h. The call was answered admirably by Millikan in 1916. He devised what must be regarded as one of the most ingenious of experiments, making use of the photoelectric emission by clean alkali metals. A diagram of the apparatus is given in Figure 11. An electric arc was used to produce visible and ultraviolet light which, in turn, was collimated and focused, then put through a prism. The optical system was made from quartz so as to insure the transmittance of the high-frequency radiation. The dispersed light was focused on a screen that possessed a small hole at an appropriate spot; in this way, only a narrow range of frequency was permitted the light that finally struck the photoelectric tube. The beam passed through a quartz window and impinged directly upon the surface of either lithium, sodium, or potassium metal. The choice of metal was abetted by mounting targets on a wheel that could be rotated freely. Surface cleanliness—an absolute prerequisite for a precise measurement of the photoelectron speed—was achieved by rotating the target first

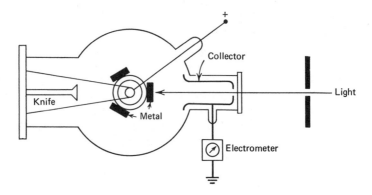

Figure 11. A diagram of Millikan's apparatus for the determination of Planck's constant.

to a position before a knife, which then shaved the metallic surface. After this operation, the sample was rotated through 180° to be in line with the bombarding radiation. Photoelectrons ejected by the metal were collected at a beaker-shaped electrode which filled the front of the tube; the charge transfer was measured by a grounded electrometer in series with the collecting electrode.

The ingenious aspect of all of this work was that Millikan had constructed the photoelectric tube so that the metal targets could be maintained at a negative potential difference relative to the collecting electrode. For a sufficiently high value of this potential the production of photoelectrons could be stopped. In that case,

$$\tfrac{1}{2}m_e v^2 = eV_R \tag{2.17}$$

where the left-hand side represents the maximum kinetic energy of the photoelectrons and V_R is the retarding potential necessary to prevent their emission. By combining equations (2.13) and (2.17) we get the working equation

$$V_R = \frac{h}{e}\nu - \frac{h}{e}\nu_0 \tag{2.18}$$

which permits the evaluation of Planck's constant once the relationship between the retarding potential and the incident radiation frequency is established. The latter operation, of course, was the essence of Millikan's experiment.

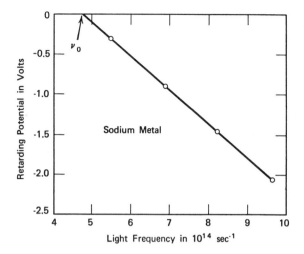

Figure 12. A plot of Millikan's data on the photo-electric effect for sodium metal. (After R. A. Millikan, A direct photoelectric determination of Planck's "h," *Phys. Rev.* 7:355-388 (1916).)

Figure 12 is a graph of certain of Millikan's data on the photoelectric effect for clean sodium metal. It is evident from the plot that Einstein's hypothesis is quite accurate. From the slope of the line and his measurement of the electronic charge, Millikan concluded that Planck's constant should have the value

$$h = 6.57 \times 10^{-27} \text{ erg-sec.}$$

which agrees very well with the value found by Max Planck in comparing his expression for the spectral energy density of blackbody radiation with experiment. The presently accepted value is

$$h = 6.6256 \pm 0.0005 \times 10^{-27} \text{ erg-sec.}$$

The validity of Einstein's hypothesis seems firmly established by the results of Millikan's experiment. Indeed, the omnipresence of the photo-electric cell and the photomultiplier tube in laboratories where physics is done attests to the very reality of light quanta. However, as was suggested before, the problem is not only one of interpreting the results of experiment, but also the incorporating of new concepts into the logical scheme of existing physical theory. Before the photoelectric effect was understood, radiation was in all of its phenomenology and in theory considered as a problem in wave

37

propagation. Now we have imparted to it some of the properties of a particle: each of the light quanta moves through the space with the speed 3×10^{10} cm/sec and bears the kinetic energy hv. This energy can be transferred to another particle, whereupon the light quantum vanishes. If a transfer of energy has occurred, however, a force must have been exerted by the light quantum upon the bombarded particle. Quite equivalently, we may state that "the momentum of the light quantum was transferred to the particle which absorbed it." This momentum is evidently expressed by the quotient of the kinetic energy and the speed; that is,

$$p = \frac{hv}{c} = \frac{h}{\lambda} \tag{2.19}$$

where λ is the wavelength of the light constituted by the quanta. When a corpuscle of radiation impinges upon an electron, then the momentum imparted to the latter particle is to be calculated from equation (2.19) in terms of the properties of the radiation and Planck's constant.

2.5. ELECTRONS AS WAVES

We should like to inquire as to the relationship between the notion of a light quantum, or *photon* (a name introduced by G. N. Lewis), and the wave theory of light. The analysis, it would seem, is best pursued by examining a situation wherein the conflict between the two ideas appears drawn most sharply: we shall attempt to explain, in terms of photon behavior, the result of an experiment already interpreted fully with the help of the wave concept.

Perhaps the most significant of such experiments is that described by Thomas Young in 1801, which completed the foundation for physical optics and provided the conclusive argument for the wave theory of light. Young, as is well known, made sunlight pass through a set of three apertures arranged as in Figure 13 and observed that a symmetrical pattern of alternating dark and bright bands was formed on a screen placed in the path of the perturbed beams. He viewed this result as a decisive piece of evidence for the proposal that light is a wave phenomenon; because light waves from the two juxtaposed slits should, upon meeting, reinforce or destroy one another much the same as do water waves under identical circumstances. Wherever the crests or troughs of two waves meet, there should be a bright region. Otherwise, varying degrees of darkness should be observed. This behavior, of course, is familiar as wave interference. The problem is to find a place for such interference within the structure of photon theory.

To begin with, we assume that photons are always radiated by a light source, whether it be the sun or an incandescent filament. The photons

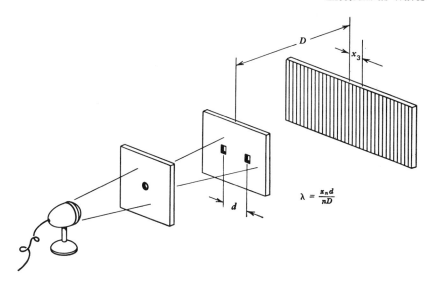

$$\lambda = \frac{x_n d}{nD}$$

Figure 13. A diagram of Young's experiment.

proceed to the set of double slits, pass through them, and, in some way, produce the observed interference pattern. By analogy with Einstein's interpretation of the photoelectric effect, we consider the brightness of a given interference fringe to be related to the number of photons per unit time that strike at the fringe position on the detecting screen. The dark bands, therefore, are the result of a small number of photon collisions, while the bright ones represent numerous impacts. Now, a quantitative measure of the rate of photon collisions per unit area can be had by placing detectors—say, photoelectric devices—in a row in front of the screen. The rate so measured, multiplied by the photon energy, should be equal in magnitude to the intensity as calculated by means of the wave theory of Young's experiment.

But suppose the source of photons is made arbitrarily weak, such that only a few photons pass through the slits at any time. We expect that the interference pattern will become very much less pronounced than before, but not that any part of it will disappear. *Even if just one photon leaves the light source per unit time, the interference fringes must still persist,* according to the wave theory with which we have agreed to concur. In order to bring the photon concept into line with our experience, we have to say, therefore, that a photon in transit through a pair of slits "knows" it must contribute to an *interference pattern* and not just to a single spot on the photoelectric sensor. This means that the interference pattern is somehow built into the photon's motion, that its motion is coupled in some way with all possible

39

trajectories from the slits to the detector. A single photon, in a sense, must be involved with each of the paths until it is actually detected and contributes to the interference pattern.

The significance of the intensity distribution calculated from the wave theory, then, is that *it expresses the relative probability of finding an interfering photon (in unit time) on the detecting screen.* It is meaningless, within the quantum concept, to ask which slit a given photon passes through or how a photon finally changes its motion from one incorporating many possible trajectories to the one corresponding to the point of detection, because the answer cannot be given without destroying the reason for asking; that is, without destroying the interference pattern. (To locate the photon, for example, we might have to cover one of the slits or intercept it on the way to the detector.) We must infer from our analysis that the correctness of the wave theory of light implies that the description of the behavior of a photon cannot be deterministic, but must be interpreted in terms of probability. It is of the utmost importance to understand here that the wave theory does not necessarily describe the average behavior of many photons. It may just as well describe a process in which only a single photon participates. The reconciliation of the wave and quantum theories of radiation has made it absolutely necessary to suppose that the interaction between light and matter is fundamentally connected with a kind of indeterminacy.

Now, if radiation can be expected at times to show the behavior of particles and at times to show the behavior of waves, perhaps matter, so well-represented as corpuscular, can also display wave phenomena. This notion was first presented by Louis de Broglie in a series of papers written in 1923 and 1924. De Broglie suggested that with the motion of an electron (or any particle of matter) there may be associated a "packet" of waves, such that the speed of the electron is equal to the group speed of the waves. The wavelength to be associated with the electron may be computed from the expression

$$\lambda = \frac{h}{p} \tag{2.20}$$

which is identical with the wavelength-momentum relation for photons. It is therefore anticipated that electrons, under the proper circumstances, will engage in reflection, refraction, interference, and the other phenomena so characteristic of light. That such behavior is not always observed is the result of the smallness of the matter wavelength, in the same sense that a small light wavelength results in the virtual disappearance of wave effects for radiation. For example, the electrons studied by J. J. Thomson moved with a speed near 10^8 cm/sec, corresponding to a wavelength of about 7 angstrom units or 1/1,000 the wavelength of visible light! Therefore, we do not see matter-wave phenomena until the behavior of particles is examined on the microscopic scale. For this very reason we believe that such phenomena may, in fact, lead to the understanding of the structure of the atom and the photons it radiates; but as yet we have no more than an interesting speculation.

2.6. ELECTRON DIFFRACTION

In July of 1925, the year following that in which de Broglie introduced the concept of the matter wave, a research assistant named Walther Elsasser published a note in which it was suggested that, if an electron by virtue of its motion may be associated with a wave, then a beam of these particles should be diffracted by a crystal. His suggestion was in reference to the well-known behavior of crystals toward high-frequency radiation, wherein they act as three-dimensional diffraction gratings. Elsasser had written his paper as an assignment from James Franck, his supervising professor, who had recently been made aware of the work of Clinton Davisson, a physicist at the Bell Telephone Laboratory, on the scattering of electrons by platinum. Davisson evidently had discovered an anomaly in the behavior of the scattered electron beam: the intensity of the beam displayed maxima at certain scattering angles. Elsasser noted in his paper that such results were to be expected from scattered waves, but not from scattered particles, which should be reflected equally at all angles from the crystal face. He further remarked that electron scattering by single crystals might prove to be the decisive phenomenon for the testing of de Broglie's idea.

Meanwhile, Davisson had discovered by accident that single crystals of nickel could be formed by slow annealing of the polycrystalline metal. He perfected the technique, then set about obtaining detailed information on the scattering of electrons. The results of this work were published two years after the appearance of Elsasser's note.

The apparatus employed by Davisson is diagrammed in Figure 14. The essential parts of this ingenious device, in the words of its creator, are:[5]

> ...the "electron gun" G, the target T and the double Faraday box collector C. The electrons constituting the primary beam are emitted thermally from the tungsten ribbon F, and are projected from the gun into a field-free enclosure containing the target and the collector; the outer walls of the gun, the target, the outer box of the collector and the

---

[5] C. J. Davisson and L. H. Germer, Diffraction of electrons by a crystal of nickel, *Phys. Rev.* **30**: 705–740 (1927).

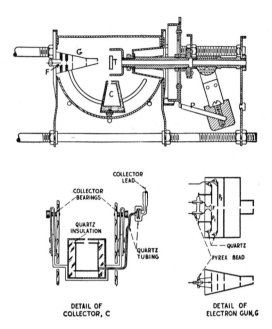

Figure 14. Davisson's diagram of his apparatus for investigating the scattering of electrons by nickel. (From C. J. Davisson and L. H. Germer, Diffraction of electrons by a crystal of nickel, *Phys. Rev.* **30**:705–740 (1927).)

box enclosing these parts are held always at the same potential. The beam of electrons meets the target at normal incidence. High speed electrons scattered within the small solid angle defined by the collector opening enter the inner box of the collector, and from thence pass through a sensitive galvanometer. Electrons leaving the target with speeds appreciably less than the speed of the incident electrons are excluded from the collector by a retarding potential between the inner and outer boxes. The angle between the axis of the incident beam and the line joining the bombarded area with the opening in the collector can be varied from 20 to 90 degrees. Also, the target can be rotated about an axis that coincides with the axis of the incident beam. It is thus possible to measure the intensity of scattering in any direction in front of the target with the exception of directions lying within 20 degrees of the incident beam.

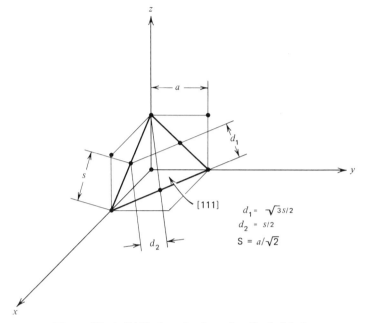

Figure 15. A [111]-plane in the unit cell of nickel.

The [111]-plane[6] of the nickel crystal was chosen for diffraction because it is the plane of greatest surface density of atoms. (The primitive cell of nickel is a face-centered cube. See Figure 15.) The atoms in this plane are arranged in a triangle that presents two kinds of (inequivalent) diffraction grating. One of them has lines parallel with the sides of the triangle, while the other has lines perpendicular to these sides. In order to avoid the complication of refraction, only the electrons scattered in directions nearly parallel with these gratings were studied in the experiment.

For a given scattering situation, the reflection angle θ is related to the electron wavelength by

$$\lambda = d \sin \theta \qquad (2.21)$$

for first-order reflection, where d is the grating constant appropriate to the bombarded plane. Equation (2.21) is the well-known plane diffraction grating formula. For values of the reflection angle near 90°, as suggested

[6] Planes of atoms in a crystal are denoted by the *Miller indices* [hkl], where h, k, and l are integers. The respective values of these integers for a given plane are determined by writing down the inverses of the x-, y-, and z-intercepts of the plane, measured in units of a primitive cell dimension. (See Figure 15.) For example, the plane whose intercepts are each at the primitive cell distance on the appropriate cartesian axis corresponds to the Miller indices [111]. The plane intercepting only the x- and y-axes at the primitive cell distance is denoted by [110].

43

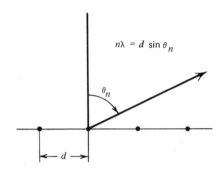

$$n\lambda = d \sin \theta_n$$

Figure 16. Essentials of diffraction by a plane grating.

in the foregoing paragraph, the expression should be adequate to describe electron diffraction. The grating constants for the two kinds of reflecting plane are given by

$$d_1 = \frac{1}{2}\left(\frac{\sqrt{3}\,a}{\sqrt{2}}\right)$$

$$d_2 = \frac{1}{2}\left(\frac{a}{\sqrt{2}}\right)$$

(2.22)

where $a = 3.517$ Å is the length of a side of the primitive cell.

From equations (2.17) and (2.20), we see that

$$\lambda = \frac{h}{(2m_e|e|V)^{\frac{1}{2}}}$$

where V is to be interpreted here as the potential difference through which the electrons are accelerated prior to diffraction. By equation (2.21),

$$(V_1)^{\frac{1}{2}} \sin \theta = 5.694$$
$$(V_2)^{\frac{1}{2}} \sin \theta = 9.862$$

for the grating constants given by equations (2.22). Now, the electron wavelength will have its maximum when the reflection angle is 90°. In this case,

$$\lambda_{max_1} = d_1 = 2.154 \text{ Å}$$
$$\lambda_{max_2} = d_2 = 1.244 \text{ Å}$$

and

$$V_{min_1} = 32.33 \text{ volts}$$

and

$$V_{min_2} = 96.99 \text{ volts},$$

respectively, for the two kinds of plane grating.

44

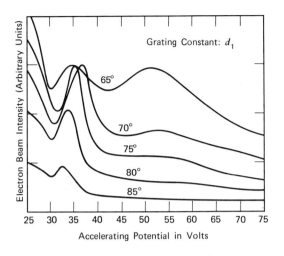

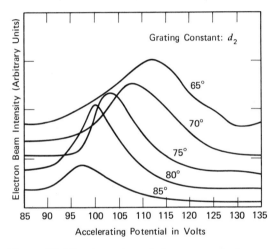

Figure 17. The intensity of the scattered electron beam as a function of the accelerating potential in Davisson and Germer's experiment. The angle of reflection is a constant parameter for each curve.

In Figure 17 are plotted some of Davisson's data on the behavior of the scattered electron beam intensity as the electron wavelength (equivalent to the accelerating potential) is decreased. Maxima are clearly evident in the intensity distributions, showing that constructive interference does occur at a certain wavelength for a given angle of reflection. But is the wavelength the

45

same as that calculated from de Broglie's relation? Table 2.1 provides the affirmative answer to this question. The observed values of the minimum accelerating potential and the product $(V^{1/2})$ sin θ are in very good agreement with the theoretical estimates. What differences exist may be readily attributed to electron refraction by the metal.

Table 2.1. Values of V and $(V)^{1/2}$ sin θ for electrons scattered by a single crystal of nickel. (Data from ref. 5, p. 725.)

θ (degrees)	V_1 (volts)	$(V_1)^{1/2}$ sin θ (volts$^{1/2}$)	V_2 (volts)	$(V_2)^{1/2}$ sin θ (volts$^{1/2}$)
85	32.5	5.68	97.5	9.83
80	34.0	5.75	100.0	9.85
75	35.0	5.72	103.5	9.83
70	36.5	5.68	108.0	9.77
65	35.0	5.37	112.5	9.62

The results of Davisson's experiment conclusively demonstrate that electrons may be associated with waves. There is a profound theoretical significance to be attributed to this fact. It is clear now that no absolute meaning can be attached to any classification of microscopic physical phenomena according to their experimental properties. Radiation and material substances *both* may be considered to be composed of fundamental particles; but these are particles whose motions are not always subject to classical mechanics. Only when the associated wavelength of the corpuscle is short compared to the scale of its immediate environment can its wave aspect be overlooked in a pragmatic sense and a deterministic description (geometrical optics or Newtonian mechanics) safely be given. In all other instances, the wave aspect must be reckoned with and the behavior of the particle can be forecast only in a statistical way. Thus we see that *determinism is a useful concept only in the limiting case of macroscopic theory.* Microscopic theory evidently must appeal to statistics and have at its foundation an intrinsically non-deterministic equation of motion. Just what form this equation of motion should have and what should be the content of this new theory is not yet apparent. It is certain, however, that a drastic revision of the formal character of mechanics will be necessary to accomplish our purpose.

PROBLEMS

1. Suppose a cathode-ray tube is placed in a magnetic field directed along the path of the electrons and that across its vertical deflection plates is established a constant potential difference. The equations of motion for the electron are then

$$\frac{d^2x}{dt^2} - \frac{|e|H}{mc}\frac{dy}{dt} = 0$$

$$\frac{d^2y}{dt^2} - \frac{|e|}{m}E + \frac{|e|H}{mc}\frac{dx}{dt} = 0$$

$$\frac{d^2z}{dt^2} = 0$$

in the notation of section 2.1.

(a) Show that, for very small values of t, the electron moves in a circle in the xy-plane and linearly in the z-plane. The total motion is thus a helix winding in the direction of the z-axis.

(b) We define the *pitch* of the helix as the distance the electron travels while rotating through 2π radians. Thus,

$$p = 2\pi\frac{v_z}{\omega}$$

where

$$\omega = |e|H/mc$$

is the angular frequency of rotation. By defining the polar coordinates r, φ in the xy-plane, where

$$r = \frac{2|e|E}{m\omega}\,\tau \sin\left(\frac{\omega\tau}{2}\right)$$

$$\varphi = \frac{\omega\tau}{2}$$

$\tau = l/v_z$ being the (short) time the beam spends between the deflection plates, show that the path of the electron beam in the xy-plane, as the magnetic field is increased, is a cochleoid.

(c) When the cochleoid passes through the x-axis, how are the pitch and length of a deflection plate related?

(d) With the help of the approximation

$$\tfrac{1}{2}m_e v_z^2 = |e|V$$

where V is the potential difference accelerating the electrons down the cathode-ray tube, show that

$$\frac{|e|}{m_e} = 2V\left(\frac{2\pi c}{Hp}\right)^2.$$

Can you think of an experiment, based upon this expression, for measuring the charge-to-mass ratio of the electron?

2. By analogy with the Franck-Hertz experiment, explain the fluorescence of the glass near the anode in an evacuated cathode-ray tube.

3. Neglecting Thomson's determination of the charge-to-mass ratio, how well is the idea that cathode rays are constituted of electromagnetic waves supported by the experimental results quoted in section 2.1?

4. What would be the result of illuminating a discharging cathode-ray tube with ultraviolet light?

5. In the case of sodium metal, the magnitude of $h\nu_0$ is 3.75×10^{-12} erg. If light of wavelength 3,000 Å impinges upon the metal, what will be the maximum speed of the resulting photoelectrons? What retarding potential would be necessary to stop photoemission?

6. The mineral tourmaline has remarkable property as regards light. When a beam of light passes into the crystal, only the radiation plane-polarized perpendicularly to the optic axis is let through. If the beam is plane-polarized at an angle θ relative to the optic axis, then, according to the wave theory of light, the intensity of the passed beam will be $\sin^2 \theta$ times the intensity of the beam completely passed by the mineral. By analogy with the arguments given in section 2.5, interpret this phenomenon in terms of the photon concept. What can be said about the perpendicular and horizontal plane-polarization of a given photon?

7. Show how the momentum of an electron can be measured by forcing a beam of these particles through small apertures arranged as in Young's experiment.

8. Suppose a beam of electrons is made to create an interference pattern as in Young's experiment. We might try to find out how the electrons interfere by letting only one of them go through the slits at any given instant and then locating them by scattering photons off their sides. Consider such a detection method from the point of view of momentum conservation and show that the electron will no longer contribute its share to an interference pattern if it is located in this way.

9. Here is a method for measuring the radius of a photon. Pass a beam of these particles through an opening whose diameter can be varied. Close down the opening until *the beam* no longer makes a spot, after passing through it, on some detecting screen. The size of the opening at this time gives an estimate of the photon radius. Analyze this procedure from the point of view of physical optics. Is the measurement meaningful?

PART II

BASIC QUANTUM THEORY

CHAPTER 3

Fundamentals of Quantum Mechanics

Physical reality is a ray in Hilbert space.

3.1. HILBERT SPACE

The need for a microscopic physical theory profoundly different in content from Newtonian mechanics has been amply illustrated by our study of atomic radiation and the non-relativistic electron. However, the precise form of the new theory is not to be deduced from this analysis, nor even from the most careful scrutiny of available empirical data. As with all fundamental notions, physical intuition must provide the basic formalism of quantum mechanics. This fact necessarily requires that the structure of the formalism rest upon a set of postulates. The postulates, in turn, rely for a great deal of their content on existing mathematical theory. It is for this reason, then, that we must first deal with the concept of a linear space.

Let us imagine a collection of mathematical quantities, to be called *vectors*, which has the following attributes:

(a) Any two members of the collection, or *set*, can be "added," the "addition" always producing a third vector that is in the set. ("The set is *closed* under the operation of 'addition'.") The properties of this "addition" are:

$$f \oplus g = g \oplus f \qquad \text{(commutation)}$$
$$(f \oplus g) \oplus h = f \oplus (g \oplus h) \qquad \text{(association)} \tag{3.1}$$

where f, g, and h are any three members of the set and $\oplus$ is the symbol for "addition"; that is, *vector addition*.

(b) Any of the members of the set can be multiplied by a scalar. The set is closed under the operation of multiplication by a scalar; the multiplication is distributive and associative:

$$\left.\begin{aligned} (a + b)f &= af \oplus bf \\ a(f \oplus g) &= af \oplus ag \end{aligned}\right\} \quad \text{(distribution)}$$

$$(ab)f = a(bf) \qquad \text{(association)} \tag{3.2}$$

where a and b are arbitrary scalars and $+$ means scalar addition.

(c) There exists a null vector N such that

$$f \oplus N = f \tag{3.3}$$

and

$$af = N \qquad \text{implies } a = 0 \text{ or } f = N.$$

(d) There exists an inverse member $(-f)$ such that

$$f \oplus (-f) = N \tag{3.4}$$

for any f in the set.

These four axioms may be used to form all the other rules of calculation, such as subtraction, multiplication with subtraction, and so on. A set of mathematical quantities for which these axioms hold is called a *linear space*.[1] Many kinds of linear space may be imagined. (Newtonian mechanics, for example, takes place in a three-dimensional linear space.) The one of importance to quantum mechanics is characterized by two properties in addition to those mentioned above. The first is that all scalars are in general complex numbers. The second is that a complex-valued *scalar product* is defined on the space. We specify this property by the definition[2]

$$(f, g) \equiv \overline{(g, f)} \tag{3.5}$$

---

[1] Physicists like to say that the vectors reside *in* the space, while mathematicians say the vectors *are* the space. This fact may help explain the great preoccupation with ether theories that plagued physics for two centuries.

[2] Notice that we do not give a "rule of composition" here, but only a general property of the scalar product. The former can be stipulated only when the structure of the linear space is known explicitly. For example, when a vector is specified to be a directed line segment, as is done in classical mechanics, we have the familiar rule of composition

$$(\mathbf{f}, \mathbf{g}) = \|\mathbf{f}\| \, \|\mathbf{g}\| \cos \theta_{fg}$$

where $\|\mathbf{f}\|$ means the length of the vector $\mathbf{f}$, and θ_{fg} is the acute angle between the directed line segments $\mathbf{f}$ and $\mathbf{g}$.

The same comment applies to equations (3.1) through (3.4), of course, since we have not yet stated *how* vectors are to be added or multiplied by a scalar.

where the bar means the complex conjugate has been formed, (f, g) being the symbol for the scalar product of the vectors f and g. Scalar multiplication is also distributive,

$$(f_1 \oplus f_2, g) = (f_1, g) + (f_2, g),$$
$$(f, g_1 \oplus g_2) = (f, g_1) + (f, g_2),$$

and antilinear in f, linear in g,

$$(af, g) = \bar{a}(f, g),$$
$$(f, ag) = \overline{(ag, f)} = a\overline{(g, f)} = a(f, g). \tag{3.6}$$

Equation (3.5) implies that (f, f) is real. (Put $g = f$ in that expression.) We further restrict this scalar product by stipulating that it be positive-definite:

$$(f, f) \geqslant 0 \tag{3.7}$$

where the equality is to hold only if $f = N$. Equation (3.7) is of no small significance, because it permits a definition of the *length* or *magnitude* of a vector. The length of a vector is expressed by

$$\|f\| \equiv \sqrt{(f, f)} \tag{3.8}$$

where the sign of the square root is chosen positive. We note in passing that if f in equation (3.8) is multiplied by the scalar $\exp(i\alpha)$, where $i = \sqrt{-1}$ and α is any real scalar, then a new vector $e^{i\alpha}f$ results. However, the length of this new vector is the same as that of f since, by equations (3.6),

$$(e^{i\alpha}f, e^{i\alpha}f) = e^{-i\alpha}(f, e^{i\alpha}f) = (e^{-i\alpha}e^{i\alpha})(f, f) = (f, f).$$

The scalar $\exp(i\alpha)$ is a number of unit absolute value and is sometimes called a *phase factor*. The set of all vectors of the form $c \exp(i\alpha)f$, where c and α are arbitrary real numbers and f is a fixed vector, is said to constitute a *ray* in the linear space.

One kind of infinite linear space possessing a finite, complex-valued scalar product is a *Hilbert space*. The theory of Hilbert space provides the mathematical foundation for quantum mechanics. It is therefore of great importance to understand, besides the defining axioms just enumerated, something of the structure of this space. Let us begin the excursion by considering the expression

$$a_1 f_1 \oplus a_2 f_2 \oplus \cdots \oplus a_m f_m = N. \tag{3.9}$$

We shall assume that the integer m can be made arbitrarily large. If equation (3.9) implies (besides the trivial possibility that all the f are null vectors) only the equations

$$a_1 = a_2 = a_3 = \cdots = a_m = 0$$

53

the vectors f_i $(i = 1, \ldots, m)$ are said to be *linearly independent*. Furthermore, if for all scalars a and b a subset of vectors contains $(af \oplus bg)$ whenever it contains f and g, this subset is called a *linear manifold*. The set of vectors f_i are said to *span* the linear manifold if for every g in the manifold there exists b_i $(i = 1, \ldots, m)$ such that

$$g = b_1 f_1 \oplus b_2 f_2 \oplus \cdots \oplus b_m f_m. \tag{3.10}$$

Finally, if the f_i $(i = 1, \ldots, m)$ are linearly independent and span a linear manifold, they form a set of *basis vectors* in the manifold. This means, simply, that the representation of g by equation (3.10) is unique. If it were not so, then we could equally write

$$g = c_1 f_1 \oplus c_2 f_2 \oplus \cdots \oplus c_m f_m.$$

But this implies

$$(b_1 - c_1)f_1 \oplus (b_2 - c_2)f_2 \oplus \cdots \oplus (b_m - c_m)f_m = N$$

which, in turn, means $b_i = c_i$ $(i = 1, \ldots, m)$ because the f_i are linearly independent. Therefore, equation (3.10) is a unique representation of the vector g in the linear manifold spanned by the f_i. This being the case, we may say that the manifold is of *dimension m*. It follows readily that any set of $(m + 1)$ vectors in the manifold must be linearly *dependent*.

A concrete example of these notions is provided by the three cartesian basis vectors $\hat{\mathbf{x}}$, $\hat{\mathbf{y}}$, and $\hat{\mathbf{z}}$, which are used in the classical mechanical description of a single particle. The equation

$$a\hat{\mathbf{x}} \oplus b\hat{\mathbf{y}} \oplus c\hat{\mathbf{z}} = \mathbf{0}$$

can only mean that each of the scalars a, b, and c is identically zero since each of the basis vectors points at right angles relative to the other two. Thus the basis vectors are linearly independent (as they must be!) and span a linear manifold of dimension three. In particular, the null vector $\mathbf{0}$ (in the usual notation) is defined by the above equation as that vector corresponding to the set of coefficients $(0, 0, 0)$.

Two vectors of the kind met in classical mechanics are perpendicular if their scalar product vanishes. As a generalization of this, we say that two vectors f and g in Hilbert space are orthogonal[3] and normalized, or *orthonormal*, if

$$(f, g) = \begin{cases} 1 \text{ when } f = g & \text{(normality)} \\ 0 \text{ when } f \neq g & \text{(orthogonality)} \end{cases}$$

Two linear manifolds are orthogonal if each member of one is orthogonal to each member of the other. In an m-dimensional linear manifold, an orthonormal set of vectors containing m members is said to be *complete*. Indeed,

---

[3] Orthogonal is from the Greek word for "right angle."

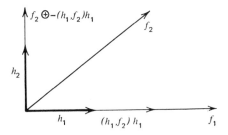

Figure 18. A geometric picture of the
Gram-Schmidt process.

the scalar multiplication of any one of the orthonormal vectors f_i ($i = 1, \ldots,$
m) with the expression

$$a_1 f_1 \oplus a_2 f_2 \oplus a_3 f_3 \oplus \cdots \oplus a_m f_m = N$$

leads immediately to $a_i = 0$ ($i = 1, \ldots, m$), showing that a set of m orthonormal vectors in an m-dimensional manifold are linearly independent. The
significance of this result is made greater by the fact that a set of orthonormal
vectors can always be constructed from an *arbitrary* set of linearly independent
vectors in the space, or in a linear manifold of the space. The procedure for
doing this, known as the *Gram-Schmidt process*, goes as follows. Call the
basis vectors f_i, the orthogonal vectors, g_i, and the orthonormal vectors h_i.
Then put (here $\sum$ means vector addition)

$$h_1 \equiv \frac{f_1}{\|f_1\|}$$

$$g_2 \equiv f_2 \oplus -(h_1, f_2)h_1$$

$$h_2 \equiv \frac{g_2}{\|g_2\|},$$

$$g_m \equiv f_m \oplus -\sum_{k=1}^{m-1} (h_k, f_m)h_k$$

$$h_m \equiv \frac{g_m}{\|g_m\|}.$$

We note in particular that

$$(h_1, h_2) = \frac{(f_1, f_2)}{\|f_1\| \|g_2\|} - \frac{(f_1, (h_1, f_2)h_1)}{\|f_1\| \|g_2\|}$$

$$= \frac{(f_1, f_2)}{\|f_1\| \|g_2\|} - \frac{(h_1, f_2)}{\|g_2\|}(h_1, h_1) = \frac{(f_1, f_2)}{\|f_1\| \|g_2\|} - \frac{(f_1, f_2)}{\|f_1\| \|g_2\|} = 0.$$

The orthonormality of the h_i ($i = 1, \ldots, m$) is thus assured. This property is denoted by a special symbol called the Kronecker delta:

$$(h_i, h_j) = \delta_{ij} \equiv \begin{cases} 1 & i = j \\ 0 & i \neq j \end{cases} \qquad (i, j = 1, \ldots, m). \tag{3.11}$$

We conclude from the analysis that *an arbitrary vector in Hilbert space may be represented by a series composed of the orthonormal basis vectors of a linear manifold containing the vector.* The series coefficients are found easily by using equation (3.11). For example, suppose

$$g = \sum_{i=1}^{m} a_i f_i. \tag{3.12}$$

If we take the scalar product of equation (3.12) with the vector f_j ($1 \leqslant j \leqslant m$), we find, by the orthonormality of the f_i,

$$(f_j, g) = a_j \qquad (j = 1, \ldots, m). \tag{3.13}$$

The length of the vector g may also be found easily, since, by equations (3.6) and (3.11),

$$(g, g) = \sum_{i=1}^{m} \sum_{j=1}^{m} \bar{a}_i a_j (f_i, f_j) = \sum_{i=1}^{m} |a_i|^2. \tag{3.14}$$

Thus far our consideration of Hilbert space has given little attention to what form the vectors might have. For an initial acquaintance with quantum mechanics we shall specify the vectors to be twice-differentiable, square-integrable, complex-valued functions of the three space coordinates and the time. (The reason for choosing them to be so will be made clear in section 3.3.) To see that an infinite set of such functions can be a Hilbert space we need only state that vector addition in this case is to be understood as ordinary addition (of complex numbers), that the null vector is the function whose value everywhere is zero, and that the "rule of composition" for the scalar product is

$$(\psi, \varphi) \equiv \int_{\text{all space}} \bar{\psi}(\mathbf{r}, t) \varphi(\mathbf{r}, t) \, d\mathbf{r}$$

where ψ and φ are the vectors. Notice that the scalar product does not include an integration over the time variable. This is because we shall be treating time as just a parameter in our discussion and not a variable with the same physical status as $\mathbf{r}$. The reason for this will become apparent in section 3.3.

A vector in Hilbert space must have a finite length. Therefore, $\psi(\mathbf{r}, t)$ is not in our brand of Hilbert space unless

$$(\psi, \psi) = \int_{\text{all space}} |\psi(\mathbf{r}, t)|^2 \, d\mathbf{r}$$

is finite and positive-definite. If a set of functions $\{\varphi_i(\mathbf{r}, t)\}$ forms a complete set of orthonormal basis vectors in some linear subspace[4] of a Hilbert space, then, according to equation (3.11),

$$(\varphi_i, \varphi_j) = \int_{\text{all space}} \bar{\varphi}_i(\mathbf{r}, t)\varphi_j(\mathbf{r}, t) \, d\mathbf{r} = \delta_{ij}.$$

Moreover, if $\psi(\mathbf{r})$ is some function in the subspace, we can always write, using equations (3.12) and (3.13),

$$\psi(\mathbf{r}) = \sum_{i=1}^{m} (\varphi_i, \psi)\varphi_i(\mathbf{r}) \qquad (3.15)$$

$$(\psi, \psi) = \sum_{i=1}^{m} |(\varphi_i, \psi)|^2. \qquad (3.16)$$

What about the uniqueness of a vector $\psi(\mathbf{r})$ in Hilbert space? Earlier it was pointed out that any vector multiplied by a phase factor is to be regarded as more or less the same vector, at least insofar as its magnitude is concerned. We wish now to generalize this statement somewhat and say that any two vectors $\psi(\mathbf{r})$ and $\varphi(\mathbf{p})$ in a pair of Hilbert spaces, where $\mathbf{r}$ and $\mathbf{p}$ denote different sets of coordinates, are "essentially the same" if the lengths of the two vectors are equal. More precisely, we shall say that $\varphi(\mathbf{p})$ is *equivalent* to $\psi(\mathbf{r})$ if the transformation of $\psi(\mathbf{r})$ into $\varphi(\mathbf{p})$ preserves the scalar product (ψ, ψ). One such transformation is of great importance for quantum mechanics because it permits two equivalent physical interpretations of Hilbert space to be made. The transformation is defined for functions of a single coordinate by (with a any real positive scalar)

$$\varphi(p) \equiv \left(\frac{a}{2\pi}\right)^{1/2} \int_{-\infty}^{\infty} \psi(x) \exp\left(-iapx\right) dx. \qquad (3.17)$$

The vector $\varphi(p)$ is called the *Fourier transform* of $\psi(x)$. The equivalence of $\psi(x)$ and its Fourier transform is assured by defining the inverse transformation

$$\psi(x) \equiv \left(\frac{a}{2\pi}\right)^{1/2} \int_{-\infty}^{\infty} \varphi(p) \exp\left(iapx\right) dp. \qquad (3.18)$$

A mathematical justification of equation (3.18), along with some remarks on Fourier transforms, is given in section A.1 of the Appendix. Now let us

---

[4] A linear manifold which is spanned by a complete set of vectors is called a *linear subspace*.

calculate (ψ, ψ), using equation (3.18), to see if it is equal to (φ, φ). We have

$$(\psi, \psi) = \int_{-\infty}^{\infty} \bar{\psi}(x)\psi(x) \, dx$$

$$= \left(\frac{a}{2\pi}\right) \int_{-\infty}^{\infty} \int_{-\infty}^{\infty} \bar{\varphi}(p') \exp(-iap'x) \, dp' \int_{-\infty}^{\infty} \varphi(p) \exp(iapx) \, dp \, dx.$$

Now we shall *assume* that all the integrals converge absolutely so that the order of integration may be altered. The integration over x is to be carried out first:

$$\int_{-\infty}^{\infty} \int_{-\infty}^{\infty} \bar{\varphi}(p') \exp(-iap'x) \, dp' \int_{-\infty}^{\infty} \varphi(p) \exp(iapx) \, dp \, dx$$

$$= \int_{-\infty}^{\infty} \bar{\varphi}(p') \int_{-\infty}^{\infty} \varphi(p) \int_{-\infty}^{\infty} \exp[ia(p - p')x] \, dx \, dp' \, dp.$$

But the third integral vanishes when $p \neq p'$ and otherwise is indeterminate. Therefore, the change in the integration sequence is strictly not permissible in this case. However, we shall blithely ignore the apparent disappearance of rigor and define a quantity $\delta(p - p')$, called a delta-"function," such that

$$\int_{-\infty}^{\infty} \exp[ia(p - p')x] \, dx \equiv \frac{2\pi}{|a|} \delta(p - p') = 0 \qquad \text{for } p \neq p' \quad (3.19)$$

$$\int_{-\infty}^{\infty} \varphi(p')\delta(p - p') \, dp' \equiv \varphi(p). \qquad (3.20)$$

In this way we find immediately

$$(\psi, \psi) = \int_{-\infty}^{\infty} \bar{\varphi}(p')\delta(p - p') \int_{-\infty}^{\infty} \varphi(p) \, dp' \, dp = \int_{-\infty}^{\infty} \bar{\varphi}(p)\varphi(p) \, dp = (\varphi, \varphi)$$

which demonstrates the equivalence of $\psi(x)$ and $\varphi(p)$. Now, equation (3.20) implies

$$\int_{-\infty}^{\infty} \delta(x) \, dx = 1. \qquad (3.21)$$

[Put $\varphi(p') \equiv 1$.] Rigorously speaking, equation (3.21) is meaningless. No matter what definition of the integral is used, the integral of a function whose only non-zero value is at a single point simply does not exist. This fact, however, should not cause us to abandon equation (3.20). Rather than do that we shall just say that $\delta(x)$ is not a function! The delta-"function" can be made rigorous if it is defined as a *symbolic function*. Briefly, a symbolic function is specified by the integral equation

$$\int_{-\infty}^{\infty} s(x)f(x) \, dx = F(f)$$

where $s(x)$ is the symbolic function, $f(x)$ is any continuous function possessing derivatives of all orders and which vanishes outside of some finite interval, and $F(f)$ is called a *distribution*. A distribution is a continuous functional on the space of $f(x)$, such that

$$F(f_1 + f_2) = F(f_1) + F(f_2)$$
$$F(af) = aF(f)$$

where a is any scalar. The theory of distributions is discussed in section A.2 of the Appendix. For the present it is sufficient to cite some of the properties of the delta-"function" that are of most significance when doing quantum mechanics.

$$(-1)^n \psi^{(n)}(x) = \int_{-\infty}^{\infty} \delta^{(n)}(p - x)\psi(p)\, dp \qquad (n = 0, 1, 2, \ldots) \qquad (3.22)$$

$$\delta^{(n)}(x) = (-1)^n \delta^{(n)}(-x) \qquad (n = 0, 1, 2, \ldots) \qquad (3.23)$$

$$\delta(\psi(x)) = \frac{\delta(x - x_0)}{|\psi^{(n)}(x_0)|} \qquad (3.24)$$

where $\delta^{(n)}(p - x)$ means $(d^n \delta/dp^n)$,

$\psi^{(n)}(x)$ means $(d^n \psi/dp^n)p = x$,

$\psi^{(n)}(x_0)$ means $(d^n \psi/dx^n)x = x_0$.

The delta-"function" is to be used as if it were an ordinary function to be multiplied by another and integrated between infinite limits. We shall never ask "what is its value," nor shall we think of it in any context save where it appears under an integral sign as just described.

3.2. OPERATORS IN HILBERT SPACE

An *operator* is a function whose domain is a subset of a vector space and whose values are also in the vector space. It establishes a relationship between a vector f of the subset and the vector Of, which expresses the value of the operator O. We say that *an operator maps a vector space into itself*. For example, the rotation of a vector about an axis in ordinary three-dimensional space is a mapping, by the "rotation operator," of that vector into another vector—the one resulting from rotation. A mapping is said to be *linear* if it takes place on a linear manifold and the operator responsible satisfies

$$O(f \oplus g) = Of \oplus Og$$
$$O(af) = aOf$$

We shall find it sufficient to consider here only linear operators. The algebra of linear operators is much like the algebra of a linear vector space. We have, for example,

$$(O_1 \boxplus O_2)f \equiv O_1 f \oplus O_2 f$$
$$(O_1 O_2)f \equiv O_1(O_2 f) \tag{3.25}$$

as the definitions of operator addition $\boxplus$ and multiplication, and so on. [We should be reminded that the operations involving two or more different operators are defined only in the common parts of their respective domains (the subsets of vectors acted upon).]

If Of is single-valued and not the null vector, then the operator O may have an *inverse*. The inverse has the symbol O^{-1} and is defined by

$$O^{-1}g \equiv f \tag{3.26}$$

where f and g are vectors such that

$$Of \equiv g \quad (g \neq N).$$

Obviously,

$$OO^{-1} = I \quad (If \equiv f), \tag{3.27}$$

where I is the *identity operator*. The *adjoint* of the operator O is given the symbol $O^\dagger$. It is an operator that has the same domain as does O and for which

$$(Of, g) = (f, O^\dagger g) = \overline{(g, Of)} \tag{3.28}$$

is true. The relationship expressed by equation (3.28) is quite symmetric; for, if f and g are interchanged and the complex conjugate of both sides of the resulting equation is taken, we find

$$(Og, f) = (g, O^\dagger f)$$

or

$$(O^\dagger f, g) = (f, Og). \tag{3.29}$$

Thus O is the adjoint of $O^\dagger$. The analogs of equations (3.25) for the adjoint operator are

$$(O_1 \boxplus O_2)^\dagger f = O_1^\dagger f \oplus O_2^\dagger f$$
$$(aO)^\dagger f = \bar{a}(O^\dagger f), \quad (O_1 O_2)^\dagger f = O_2^\dagger O_1^\dagger f. \tag{3.30}$$

If the adjoint of an operator is also its inverse, then the operator is said to be *unitary*. Evidently,

$$U^\dagger U f \equiv U U^\dagger f \equiv U U^{-1} f \equiv If = f \tag{3.31}$$

where U is any unitary operator. *Unitary operators, therefore, do not affect the scalar product*:

$$(Uf, Ug) = (f, U^\dagger U g) = (f, g). \tag{3.32}$$

If the adjoint of an operator is the operator itself, the operator is said to be *self-adjoint* or *Hermitian*. An Hermitian operator H may be defined by analogy with equation (3.28):

$$(Hf, g) = (f, Hg). \tag{3.33}$$

In general, of course, an operator will not be Hermitian. However, if O is some operator, then $OHO^{\dagger}$ *will* be Hermitian. Indeed, by equations (3.28), (3.30), and (3.33),

$$(OHO^{\dagger}f, g) = (f, (OHO^{\dagger})^{\dagger}g) = (f, OH^{\dagger}O^{\dagger}g) = (f, OHO^{\dagger}g).$$

Now, the identity operator is obviously Hermitian:

$$(If, g) = (f, g) = (f, Ig).$$

Therefore, *the operator $OO^{\dagger}$ is always Hermitian*, where O is any operator possessing an adjoint. On the other hand, the product $O_1 O_2^{\dagger}$ will not necessarily be Hermitian; for,

$$(O_1 O_2^{\dagger} f, g) = (f, (O_1 O_2^{\dagger})^{\dagger} g) = (f, O_2 O_1^{\dagger} g).$$

Therefore, $O_1 O_2$ will not be Hermitian unless the difference

$$O_1 O_2^{\dagger} - O_2 O_1^{\dagger}$$

vanishes identically. In particular, if the operators O_1 and O_2 are self-adjoint, we require that

$$O_1 O_2 = O_2 O_1.$$

The difference of products just written down is denoted by the symbol

$$[O_1, O_2] \equiv O_1 O_2 - O_2 O_1 \tag{3.34}$$

and is called the *commutator* of O_1 and O_2. When the commutator vanishes, the operators O_1 and O_2 are said to *commute*. Evidently, *every* operator commutes with the identity operator:

$$OIf = Of = IOf.$$

The importance of Hermitian operators for quantum mechanics is not to be overestimated. One of the central problems in the theory of microscopic phenomena consists of discovering the set of vectors in Hilbert space which satisfy the equation

$$Hf = \lambda f \tag{3.35}$$

where H is an Hermitian operator and λ is a scalar not equal to zero. Equation (3.35) defines what is called an *eigenvalue problem*. Now, it may be that several f (*eigenvectors*, we say) exist which satisfy equation (3.35) for a given

λ (an eigenvalue[5]), and many λ and f in general can be found. The set of all eigenvalues belonging to a given H is called the *spectrum* of H. The *invariant subspace* of H is the subspace of Hilbert space containing Hf whenever it contains f. Several properties of eigenvalues and eigenvectors may be derived by using the concepts presented in this and the preceding section:

(a) *If f is a solution of the eigenvalue problem, then af is also a solution, where a is any scalar.* Indeed, all of these solutions form a ray in Hilbert space.

(b) *Eigenvectors corresponding to different eigenvalues of H are orthogonal.* To see this we note that by equations (3.6), (3.33), and (3.35)

$$(Hf, g) - (f, Hg) = 0 = (\lambda_1 f, g) - (f, \lambda_2 g) = \bar{\lambda}_1(f, g) - \lambda_2(f, g)$$
$$= (\bar{\lambda}_1 - \lambda_2)(f, g)$$

which, if $\bar{\lambda}_1 \neq \lambda_2$, implies that (f, g) vanishes identically. We recall from section 3.1 that a consequence of the orthogonality of the eigenvectors is that they are linearly independent.

(c) *Eigenvectors corresponding to the same eigenvalue of H are not, in general, mutually orthogonal.* However, they can always be replaced by a set of orthonormal eigenvectors. By property (a) given above, any linear combination of the eigenvectors corresponding to a single eigenvalue is also a solution of the eigenvalue problem. It follows that the eigenvectors are linearly independent and that we may employ the Gram-Schmidt process to create an orthonormal set from them, if need be.

(d) *The eigenvectors of H form a complete set.* By this we mean that the eigenvectors are basis vectors for at least a subspace of Hilbert space and that *any* vector in that subspace may be expressed as a linear combination of the eigenvectors. The dimensions of the subspace in general will be infinite.

For a given eigenvalue, there will be m linearly independent eigenvectors which, therefore, span an m-dimensional subspace. We say in this case that the eigenvalue is *m-fold degenerate* or has *multiplicity m*.

(e) *The eigenvalues belonging to an Hermitian operator are real.* By equations (3.6), (3.29), (3.33), and (3.35), we have

$$(f, Hf) = (f, \lambda f) = \lambda(f, f), \qquad (Hf, f) = (\lambda f, f) = \bar{\lambda}(f, f)$$

and

$$(f, Hf) = (H^\dagger f, f) = (Hf, f)$$

where f is any eigenvector corresponding to λ.

---

[5] This word is a partial translation of the German *Eigenwert*, which means "proper value."

(f) *The eigenvalues belonging to a given H may form a **discrete spectrum** [a finite or countably infinite[6] set of numbers λ_i ($i = 1, 2, \ldots$)] or a **continuous spectrum** (the values of a function $\lambda(l)$, where $l \geqslant 1$).* The eigenvectors corresponding to a continuous spectrum satisfy the orthonormality condition[7]

$$(f(l), f(m)) = \delta(l - m) \tag{3.36}$$

where $\delta(l - m)$ is the delta-"function." Continuous eigenvalues, like their discrete counterparts, are real and may be degenerate.

(g) *Any vector in an invariant subspace of H may be expressed in terms of the eigenvectors corresponding to discrete and continuous spectra:*

$$g = \sum_{i=1}^{n} \sum_{j=1}^{m_i} a_i^{(j)} f_i^{(j)} + \int \sum_{j=1}^{m_{(l)}} b^{(j)}(l) \varphi^{(j)}(l) \, dl \tag{3.37}$$

where

$$a_i^{(j)} = (f_i^{(j)}, g)$$
$$b^{(j)}(l) = (\varphi^{(j)}(l), g)$$

$f_i^{(j)}$ is the jth eigenvector corresponding to the ith eigenvalue (whose multiplicity is m_i), and $\varphi^{(j)}(l)$ is the jth eigenvector corresponding to the (continuous) eigenvalue $\lambda(l)$ [whose multiplicity is $m(l)$]. This property of the eigenvectors is closely associated with property (d) discussed above. Actually, not all Hermitian operators possess a complete set of orthonormal eigenvectors, but those encountered in doing quantum mechanics are restricted to fulfill this requirement. Therefore, we shall not pursue the matter further here; the validity of property (g) will be accepted without question in further discussions.

3.3. THE QUANTUM-THEORETICAL INTERPRETATION OF HILBERT SPACE

The postulates from which quantum mechanics is molded are rooted deeply in the concepts concerning light and matter introduced by Einstein and de Broglie, and in the theory of Hilbert space. Our study of photons and matter waves led us to the conclusion, albeit one based upon tenuous evidence, that we cannot ask of the quantum theory the same privilege of determinism so closely associated with Newtonian mechanics. However, we expect that the

---

[6] By "countably infinite" we mean that the eigenvalues can be put into one-to-one correspondence with the set of all positive whole numbers.

[7] By definition the "eigenvectors" corresponding to continuous "eigenvalues" are not defined on Hilbert space: the scalar product is not always finite and the vectors form an uncountably infinite set. (This point is discussed further in section A.2 of the Appendix.) Our reason for introducing them here is that they are analogous to eigenvectors in Hilbert space and are computationally useful in the same sense as is the delta-"function."

behavior of a particle will be governed by an equation of motion which preserves our sense of time at least in a probabilistic way. Thus we can say that a photon will leave a light source, pass through a pair of slits, and create an interference pattern; but *not* that the photon will pass through *this* slit at *this* time before proceeding to strike a detecting screen at *this* point. In the same way we may not inquire as to which electron was diffracted by *this* groove in a plane grating so as to appear at *this* reflection angle. We see that the term "state" will have an entirely different meaning in quantum mechanics from the meaning it has in the classical theory. In the latter discipline, the state of a particle in motion is completely specified by its position and velocity at a given instant of time. It evidently makes no sense to prescribe these quantities with absolute exactness in the quantum theory; indeed, we have seen that the outcome of an experiment involving either photons or electrons, when the microscopic environment is of the appropriate scale, in no way depends on such a prescription.

We shall ask only that a quantum-mechanical "state" be sufficiently well defined that the results of experiment may be predicted with reasonable certainty. This conception of "state" is certainly more physical intrinsically than that found within the framework of the classical theory. As an illustration of this fundamental weakness of classical mechanics, let us consider the motion of a free electron wherein the particle flies toward a pair of slits placed between it and a large galvanometric sensor. According to the deterministic view, the state of the particle may be deduced at any time from a knowledge of its initial position and velocity. Therefore, we may determine whether or not and at what time it will enter one of the slits. Suppose that it does enter one of them. The classical theory is then able to specify exactly

Figure 19. The interference pattern created when electrons pass through a pair of slits.

the "spot" on the screen to which the electron will contribute when it is detected by the galvanometer. However, this beautiful prediction is of absolutely no value insofar as the real outcome of putting an electron through a pair of slits is concerned, because *an interference pattern is obtained, not just a single spot on the detecting screen.* The "state" of the electron has been *overspecified* by the classical scheme, and, as a result, it is unphysical.

Within the framework of the theory presented in the foregoing sections it is possible to make a definition of "state" that meets nearly all of the requirements which might be imposed upon a microscopic theory. The concept may be expressed as:

The State Postulate. The state of a physical system may be represented by a ray in Hilbert space. All the meaningful information concerning the system can be derived from any vector of non-vanishing length in the ray.

We call the vectors in Hilbert space that specify the behavior of a physical system *state vectors* and distinguish them from other, unphysical vectors by the *Dirac notation*

$$f \equiv |f\rangle, \qquad f_i \equiv |i\rangle, \qquad (f_i, f_i) \equiv \langle i|i\rangle,$$

where i is any positive integer. It follows immediately from equation (3.12) that any state vector can be represented in terms of the basis state vectors spanning the linear subspace upon which it is defined. For example,

$$g = \sum_{i=1}^{m} a_i f_i = \sum_{i=1}^{m} (f_i, g) f_i \equiv \sum_{i=1}^{m} |i\rangle\langle i|g\rangle = |g\rangle, \tag{3.38}$$

where $g \equiv |g\rangle$ is an arbitrary state vector and the $f_i \equiv |i\rangle$ ($i = 1, \ldots, m$) form a complete orthonormal set. The physical interpretation of equation (3.38) is that the $|i\rangle$ ($i = 1, \ldots, m$) are each possible linearly independent states of the physical system represented by $|g\rangle$. The state $|g\rangle$ is said to be a *pure state* whenever it may be expressed in the form of equation (3.38).[8] In general, any pure state of a physical system is a linear combination of all the possible linearly independent states of the system. This statement, formalized in equation (3.38), is often given the imposing name *Principle of Superposition*. We see that mathematically it is nothing more than a direct result of the linear nature of Hilbert space.

The coefficient $\langle i|g\rangle$ ($1 \leqslant i \leqslant m$) may be construed, in a certain sense, to represent the contribution of $|i\rangle$ ($1 \leqslant i \leqslant m$) to the state $|g\rangle$. However, the

---

[8] States which cannot be so expressed are called *mixed states* and may be viewed as a statistical ensemble of pure states. We shall discuss mixed states further in Chapter 6.

relation is not direct since the coefficients are in general complex scalars. A better definition is

$$\omega_i \equiv \frac{|\langle i | g \rangle|^2}{\langle g | g \rangle} \tag{3.39}$$

where ω_i ($1 \leqslant i \leqslant m$) is defined to be the fraction of the state $|g\rangle$ contributed by the state $|i\rangle$ ($1 \leqslant i \leqslant m$) or *the probability that the system represented by $|g\rangle$ is in the state $|i\rangle$*. By equation (3.14) we see that

$$\sum_{i=1}^{m} \omega_i = \sum_{i=1}^{m} \frac{|\langle i | g \rangle|^2}{\langle g | g \rangle} = 1$$

as should be for a probability function. The interpretation just given to equation (3.38) is not really postulatory, because it follows in a consistent way from the fundamental notion of "state" in quantum physics, as discussed earlier. We asked only that a "state" be well enough defined to lead to predictions in agreement with experiment. Thus we need learn no more of the state of an electron moving toward a pair of slits than is necessary to prescribe the interference pattern resulting after it passes through them. If we let the basis states $|\mathbf{r}\rangle$ represent the electron in all possible positions as it strikes the detecting screen, then

$$|g\rangle = \int \langle \mathbf{r} | g \rangle |\mathbf{r}\rangle \, d\mathbf{r}$$

represents the state of the electron in transit through the slits and $\omega_r = |\langle \mathbf{r} | g \rangle|^2 / \langle g | g \rangle$ is the probability that the electron will be in the state $|\mathbf{r}\rangle$ as it contributes to the interference pattern. We cannot be more determinate than this because *it is impossible to perform an experiment prior to the formation of the interference pattern that will indicate which of the $|\mathbf{r}\rangle$ is indeed to be the state of the electron.*

Now, it is not enough to know only the state vector representing a physical system. Besides this quantity, the concept of "property" must be given a quantum-theoretical significance. We shall consider only those properties that can be determined by experiment; such properties we shall call *observables.* The problem, then, is to prescribe the representation of observables in Hilbert space in such a way that they permit the derivation of physical information from the state vectors. The following postulate satisfies our requirements admirably.

The Observable Postulate. An observable of a physical system may be represented by an Hermitian operator mapping in Hilbert space.

According to this postulate, the properties of a system have values as obtained from the action of a properly designated Hermitian operator on the appropriate state vector in Hilbert space. The *average value* of a property associated with the Hermitian operator H is expressed by

$$\langle H \rangle \equiv \frac{(g, Hg)}{(g, g)} \equiv \frac{\langle g|H|g \rangle}{\langle g|g \rangle} \tag{3.40}$$

where $|g\rangle$ is the state of the system having the property H. To see that this definition is consistent with our notion of state, we need only note that, by equations (3.38) and (3.39), the average may also be written

$$\langle H \rangle = \sum_{i=1}^{m} \sum_{j=1}^{m} \frac{\langle g|i \rangle \langle j|g \rangle}{\langle g|g \rangle} \langle i|H|j \rangle = \sum_{i=1}^{m} \sum_{j=1}^{m} \frac{\langle g|i \rangle \langle j|g \rangle}{\langle g|g \rangle} E_j \delta_{ij}$$

$$= \sum_{j=1}^{m} \frac{|\langle j|g \rangle|^2}{\langle g|g \rangle} E_j = \sum_{j=1}^{m} \omega_j E_j$$

$$\tag{3.41}$$

where $\{|j\rangle\}$ is an orthonormal set of eigenvectors of H and, therefore, is a basis for $|g\rangle$, $\{E_j\}$ is the spectrum of H, and ω_j must be interpreted as the probability that $\langle H \rangle$ is actually E_j. This interpretation of ω_j $(j = 1, \ldots, m)$ is directly related to the role that quantity plays in the formation of pure states: if the system in the state $|g\rangle$ has the probability ω_j of being in the state $|j\rangle$, then the average value of H in $|g\rangle$ must have the same chance ω_j of being E_j, which is the average value of H in $|j\rangle$. We note also that in the special case

$$\omega_j = \delta_{jk},$$

where $1 \leqslant k \leqslant m$, we have

$$\langle H \rangle = E_k.$$

We interpret this result to mean that "$|g\rangle$ is certainly the state $|k\rangle$." $\langle H \rangle$ then becomes a determinate quantity, in the sense of classical mechanics. However, we shall find that the price of this exact knowledge is the exclusion of any information about certain other properties of the system. There are for this reason two kinds of indeterminacy in connection with the prediction of mean values for observables. One arises whenever a quantum-physical state must be represented as a linear combination of all possible basis states of the system in question. The other occurs in a less direct way, as just suggested, because of the very definition of "state" in the quantum theory.

It is possible to develop a unique criterion for differentiating states which are eigenvectors from those which are not. By virtue of the fact that $\langle H \rangle$ becomes a determinate quantity when $|g\rangle$ is an eigenvector of H, we expect that the mean square deviation of H from $\langle H \rangle$ will vanish for states which are eigenvectors. More formally, we require

$$(\Delta H)^2 \equiv \langle (H - \langle H \rangle)^2 \rangle = 0. \tag{3.42}$$

By equation (3.40), we can rewrite this as

$$(\Delta H)^2 = \frac{\langle g|(H - \langle H\rangle)^2|g\rangle}{\langle g|g\rangle} = \frac{\langle g|H^2 - 2\langle H\rangle H + \langle H\rangle^2|g\rangle}{\langle g|g\rangle}$$

$$= \frac{\langle g|H^2|g\rangle}{\langle g|g\rangle} - \langle H\rangle^2$$

so that in general

$$(\Delta H)^2 = \langle H^2\rangle - \langle H\rangle^2. \tag{3.43}$$

Since H is Hermitian,

$$\langle H^2\rangle = \frac{\langle g|H^2|g\rangle}{\langle g|g\rangle} = \frac{\langle Hg|Hg\rangle}{\langle g|g\rangle}. \tag{3.44}$$

The criterion for a state to be an eigenvector is, by equations (3.40), (3.42), (3.43), and (3.44), that the relation

$$\langle g|H|g\rangle^2 = \langle g|g\rangle\langle Hg|Hg\rangle \tag{3.45}$$

be satisfied for any $|g\rangle \neq N$.

3.4. THE EQUATION OF MOTION FOR STATE VECTORS

During our first analysis of Young's double-slit experiment in section 2.5, it was concluded that the intensity of the wave associated with the photon expressed the probability per unit time of discovering this particle at some point on the detecting screen. The state of the photon upon impact with the detector, then, was specified by the intensity distribution function, which, in turn, relies on the wave theory of light for its existence. It would appear, following de Broglie's hypothesis, that the same statement can be made concerning the behavior of an electron passing through a pair of slits toward a detector. Evidently, the wave representing this particle is sufficient to establish the nature of the interference pattern, so that a knowledge of its wave is equivalent to a knowledge of the state of the electron, insofar as electron interference is concerned. But now this concept must be brought into line with the foregoing interpretation of Hilbert space. It seems that we must identify the state vector for an electron with its associated wave. We have seen already that the relevant measurable quantity for electron interference is

$$\omega_r = \frac{|\langle g|\mathbf{r}\rangle|^2}{\langle g|g\rangle}$$

where $|\mathbf{r}\rangle$ is a basis state vector representing an electron at the position $\mathbf{r}$ in space, and $|g\rangle$ is the state of the electron during the interference process. If we further suppose that $|g\rangle$ is of unit length and specifies the state of the electron at a given time t, then the quantity

$$|\langle g|\mathbf{r}\rangle|^2 \equiv |\psi(\mathbf{r}, t)|^2,$$

a function of the three cartesian space coordinates and the time, is of most physical significance here. The function $\psi(\mathbf{r}, t)$ is in general complex-valued. For a reason to be clear shortly, we shall also consider it to be continuous and twice-differentiable. The square of the absolute value of $\psi(\mathbf{r}, t)$ plays the same role for the electron wave as does the intensity distribution for a light wave. Moreover, as with the photon concept, $\psi(\mathbf{r}, t)$ is related to a probabilistic dynamical behavior, rather than a deterministic one. The association between $\psi(\mathbf{r}, t)$ and the wave aspect of the electron thus may be expressed precisely in the following postulate, first suggested by Max Born in 1926.

The Probability Postulate. The quantity $|\psi(x, y, z, t)|^2$ is to be interpreted as the intensity distribution function for a matter wave. Rigorously,

$$|\psi(x, y, z, t)|^2 \, dx \, dy \, dz$$

is the probability that the physical system it represents occupies the volume in space between (x, y, z) and $(x + dx, y + dy, z + dz)$ at the time t.

The quantity $\psi(x, y, z, t)$ is often called a *wavefunction*. We note that the wavefunction itself does not have a direct physical meaning, but is of significance only when multiplied by its complex conjugate. It follows immediately from the probability postulate that

$$\int_{-\infty}^{\infty} \int_{-\infty}^{\infty} \int_{-\infty}^{\infty} \bar{\psi}(x, y, z, t)\psi(x, y, z, t) \, dx \, dy \, dz = 1. \tag{3.46}$$

Otherwise, $|\psi|^2$ could not be interpreted as a probability distribution function. On purely mathematical grounds, equation (3.46) is just the condition that $|g\rangle$ have unit length:

$$\langle g|g\rangle = \iint \langle g|\mathbf{r}'\rangle\langle \mathbf{r}'|\mathbf{r}\rangle\langle \mathbf{r}|g\rangle \, d\mathbf{r}' \, d\mathbf{r} = \int |\langle \mathbf{r}|g\rangle|^2 \, d\mathbf{r}$$

$$= \int |\psi(\mathbf{r}, t)|^2 \, d\mathbf{r} \tag{3.47}$$

We have seen in section 3.1 that a function like $\psi(\mathbf{r}, t)$ has the qualifications to be a vector in Hilbert space. Therefore, we may consider it as the *bona fide* representative of the state vector $|g\rangle$ which is a function of the coordinates and time. In this sense we may write $|g\rangle \Rightarrow \psi(\mathbf{r}, t)$ and regard the wavefunction as the *coordinate representative of the state vector.*

It follows that the statements made about state vectors can be just as well made about wavefunctions. For example, by equation (3.40), the mean value of an observable H is now expressed

$$\langle H \rangle = \frac{(\psi, H\psi)}{(\psi, \psi)} = \frac{\sum\limits_{i=1}^{m} |(\psi_i, \psi)|^2 (\psi_i, H\psi_i)}{\sum\limits_{i=1}^{m} |(\psi_i, \psi)|^2} \tag{3.48}$$

where the ψ_i $(i = 1, \ldots, m)$ form a complete orthonormal set of possible wavefunctions defined in coordinate-time space. In agreement with the probability postulate, $|(\psi_i, \psi)|^2/(\psi, \psi)$ is the probability that the average value of H is $(\psi_i, H\psi_i)$ $(1 \leqslant i \leqslant m)$; that is, that the particle represented by $\psi(x, y, z, t)$ is in the state $\psi_i(x, y, z, t)$ $(1 \leqslant i \leqslant m)$. If it *is* in this state, then, by equation (3.46) we must have $|(\psi_i, \psi)|^2/(\psi, \psi)$ equal to unity. Usually, of course, $|(\psi_i, \psi)|^2/(\psi, \psi)$ will have values between zero and one. Let us be careful to distinguish the two probabilities:

$|\psi(x, y, z, t)|^2 \, dx \, dy \, dz \, dt =$ probability of being between (x, y, z) and $(x + dx, y + dy, z + dz)$ at the time t;

$\dfrac{|(\psi_i, \psi)|^2}{(\psi, \psi)} =$ probability of being in the state represented by $\psi_i(x, y, z, t)$.

If the latter probability is unity, then, according to equation (3.45), the wavefunction is a solution of

$$|(\psi, H\psi)|^2 = (H\psi, H\psi) \tag{3.49}$$

where H is any observable.

Other than equation (3.49), no criterion for determining the behavior of $\psi(x, y, z, t)$ during the course of time has been set forth. We should like to have some precise statement of the time development of the state of a system in order that the quantum theory may have at least as firm a mathematical basis as does Newtonian mechanics. Thus far we have postulated that the ordinary, three-dimensional linear vector space encountered in the classical theory must be replaced by the infinite-dimensioned Hilbert space, and that the state of a physical system is no more its displacement and velocity vectors at a given time, but is instead a function of the position coordinates and the time. It remains to ask what is the equation of motion in the quantum theory which is to replace that in the Newtonian theory. As emphasized previously, we cannot deduce the form of this expression from the existing body of physical knowledge. However, we can list what restrictions the results of experiment and the already-stated postulates do impose upon the equation.

Two of these have been mentioned before; we restate them now, along with four more conditions:

(a) *The equation of motion must be intrinsically non-deterministic.* This condition is met if the equation prescribes the time evolution of the wavefunction, which is inextricably connected to probabilistic concepts.

(b) *The equation of motion must be valid for all possible superposition states.* For this reason it must be a linear and homogeneous differential equation. A differential equation is linear and homogeneous if it is of the form

$$\sum_{l=0}^{n} a_l Y^{(l)}(x) = 0$$

where $Y^{(l)}(x)$ is the lth derivative of the solution of the equation and the a_l are arbitrary functions of x. If $Y_1(x)$, $Y_2(x)$, etc. are solutions of the equation, direct substitution shows that

$$g(x) = \sum_n c_n Y_n(x)$$

is also a solution.

(c) *The equation of motion must be such as to preserve the normality of the wavefunction throughout its time development.* More precisely, we ask that

$$\frac{\partial}{\partial t} \int_{-\infty}^{\infty} \int_{-\infty}^{\infty} \int_{-\infty}^{\infty} \bar{\psi}(x, y, z, t)\psi(x, y, z, t) \, dx \, dy \, dz = 0.$$

(d) *The equation of motion must be a differential equation of the first order with respect to the time.* It must be a differential equation because it is to specify the change in the wavefunction for arbitrarily small time displacements. The equation must contain only first time-derivatives so that the wavefunction is uniquely prescribed, once its value at time zero is given.

(e) *The equation of motion must be valid for all matter waves and, in particular, must lead to results, in the limit of short wavelengths, which agree with those obtained through Newtonian mechanics.* This is a severe restriction, but one that is absolutely necessary if the quantum theory is to have a logical structure and seriously undertake the description of microscopic physical phenomena.

An equation of motion for coordinate representatives of the state vector was first written down by Erwin Schrödinger in 1926. It may be formulated as:

The Schrödinger Postulate. The Hermitian operators to be associated with the position and momentum of a single particle are

$$\mathbf{r}_{op}\Psi(\mathbf{r}, t) \equiv \mathbf{r}\Psi(\mathbf{r}, t)$$
$$\mathbf{p}_{op}\Psi(\mathbf{r}, t) \equiv -i\hbar\nabla\Psi(\mathbf{r}, t)$$

(3.50)

where "op" means "operator" and $\hbar$ is Planck's constant divided by 2π. The infinitesimal time displacements of the wavefunction are generated through the equation

$$i\hbar \frac{\partial \Psi}{\partial t} = \mathcal{H}\Psi(\mathbf{r}, t) \tag{3.51}$$

where

$$\mathcal{H} \equiv \frac{p_{op}^2}{2m} + V(\mathbf{r}_{op}, t)$$

$$= -\frac{\hbar^2 \nabla^2}{2m} + V(x, y, z, t)$$

$$\nabla^2 = \frac{\partial^2}{\partial x^2} + \frac{\partial^2}{\partial y^2} + \frac{\partial^2}{\partial z^2},$$

m being the mass of the single particle described by $\Psi(\mathbf{r}, t)$ and $V(\mathbf{r}, t)$, a potential function.

The operator $\mathcal{H}$ is called the *Hamiltonian operator* because it is the quantum-mechanical form of

$$H(\mathbf{p}, \mathbf{r}, t) = \frac{1}{2m}(p_x^2 + p_y^2 + p_z^2) + V(x, y, z, t), \tag{3.52}$$

the Hamiltonian function of classical mechanics.[9]

Schrödinger's postulate states, then, that if the Hamiltonian function for a physical system is known, the quantum-mechanical equation of motion follows directly.

We see immediately that the equation of motion (3.51) satisfies the criteria (a), (b), and (d) mentioned above. Let us develop the expression further to see if the other criteria are fulfilled as well. Suppose that the potential function is time-independent and that

$$\Psi(x, y, z, t) \equiv \psi(x, y, z)\chi(t).$$

If this expression for $\Psi(x, y, z, t)$ is introduced into equation (3.51), we get

$$-\frac{\hbar^2}{2m}\chi(t)\nabla^2\psi + \chi(t)V(x, y, z)\psi = i\hbar\psi(x, y, z)\frac{\partial \chi}{\partial t}$$

or, upon dividing by $\Psi(x, y, z, t)$,

$$-\frac{\hbar^2}{2m}\frac{\nabla^2\psi}{\psi(x, y, z)} + V(x, y, z) = \frac{i\hbar}{\chi(t)}\frac{\partial \chi}{\partial t}. \tag{3.53}$$

---

[9] For a discussion of the Hamiltonian, see J. B. Marion, *Classical Dynamics*, Academic Press, New York, 1965, Chapter 9.

Equation (3.53) contains only functions of the position coordinates on the left-hand side and only functions of the time on the right-hand side. Therefore, both sides of the equation cannot be equal for all values of the independent variables unless they are equal to a constant. Let us call the constant E. Then

$$-\frac{\hbar^2}{2m}\nabla^2\psi + V(x, y, z)\psi = E\psi(x, y, z) \tag{3.54}$$

$$i\hbar\frac{\partial\chi}{\partial t} = E\chi(t). \tag{3.55}$$

Equation (3.55) has the formal solution

$$\chi(t) = \chi_0 \exp(-iEt/\hbar), \tag{3.56}$$

where χ_0 is an arbitrary constant. Equation (3.54) is not so straightforward. It is known, by correspondence with equation (3.35), as the *Schrödinger eigenvalue problem*. Because the Hamiltonian function can represent the total energy of a particle in classical mechanics when the potential function is time-independent,[9] the eigenvalue E belonging to $\mathscr{H}$ is called the total quantum-mechanical energy or just total energy. *The central problem of the quantum theory is to determine the values of E and the eigenvectors corresponding to the operator $\mathscr{H}$.*

If $\psi(\mathbf{r})$ is a solution of

$$\mathscr{H}\psi = E\psi, \tag{3.57}$$

it follows that $\psi(\mathbf{r})$ obeys equation (3.49). To see this, consider equations (3.49) and (3.57):

$$|(\psi, \mathscr{H}\psi)|^2 = |(\psi, E\psi)|^2 = E^2(\psi, \psi)^2$$
$$= (E\psi, E\psi) = (\mathscr{H}\psi, \mathscr{H}\psi)$$

provided that $\psi(\mathbf{r})$ is of unit length and that $\mathscr{H}$ is Hermitian. The latter requirement is certainly fulfilled, since

$$(\mathscr{H}\psi, \psi) = E(\psi, \psi) = (\psi, E\psi) = (\psi, \mathscr{H}\psi).$$

If $\Psi(\mathbf{r}, t)$ is a normalized wavefunction, its normalization does not change with the passage of time, in agreement with criterion (c):

$$\frac{\partial}{\partial t}(\Psi, \Psi) = \left(\frac{\partial\Psi}{\partial t}, \Psi\right) + \left(\Psi, \frac{\partial\Psi}{\partial t}\right)$$

$$= \frac{-1}{i\hbar}(\mathscr{H}\Psi, \Psi) + \frac{1}{i\hbar}(\Psi, \mathscr{H}\Psi)$$

$$= \frac{1}{i\hbar}[-(\Psi, \mathscr{H}\Psi) + (\Psi, \mathscr{H}\Psi)] = 0$$

by equations (3.51) and (3.6) and the Hermitian nature of $\mathscr{H}$.

The Hilbert-space formalism developed in sections 3.2 and 3.3 may be applied directly to the solutions of the Schrödinger eigenvalue problem. We note for convenience the most significant portions.

(a) If $\psi(\mathbf{r})$ is a solution of the eigenvalue problem, then $\alpha\psi(\mathbf{r})$ is also a solution, where α is any scalar.

(b) Wavefunctions corresponding to different total energies are orthogonal. Thus

$$(\psi_n, \psi_{n'}) = 0 \quad (n \neq n') \qquad \text{or} \qquad (\varphi(l), \varphi(l')) = 0 \quad (l \neq l')$$

where ψ_n corresponds to E_n and $\varphi(l)$ corresponds to $E(l)$.

(c) Any wavefunction defined on an invariant subspace of Hilbert space belonging to $\mathscr{H}$ may be expressed as a linear combination (superposition!) of all the eigenfunctions of $\mathscr{H}$:

$$\Phi(x, y, z) = \sum_{i=1}^{n} \sum_{j=1}^{m_i} (\psi_i^{(j)}, \Phi)\psi_i^{(j)}(x, y, z) + \int \sum_{j=1}^{m(l)} (\varphi^{(j)}(l), \Phi)\varphi^{(j)}(l)\, dl \quad (3.58)$$

(d) The total energy E is the mean value of the Hamiltonian operator. This follows from equation (3.40):

$$\langle \mathscr{H} \rangle = \frac{(\psi, \mathscr{H}\psi)}{(\psi, \psi)} = E\frac{(\psi, \psi)}{(\psi, \psi)} = E.$$

The mean value for a superposition state is given by

$$\langle \mathscr{H} \rangle = \frac{(\Phi, \mathscr{H}\Phi)}{(\Phi, \Phi)} = \sum_{k=1}^{n} \sum_{j=1}^{m_k} \frac{|(\psi_k^{(j)}, \Phi)|^2}{(\Phi, \Phi)} (\psi_k^{(j)}, \mathscr{H}\psi_k^{(j)})$$

$$+ \int \sum_{j=1}^{m(k)} \frac{|(\varphi^{(j)}(k), \Phi)|^2}{(\Phi, \Phi)} (\varphi^{(j)}(k), \mathscr{H}\varphi^{(j)}(k))\, dk$$

$$= \sum_{k=1}^{n} \omega_k E_k + \int \omega(k)E(k)\, dk \quad (3.59)$$

where ω_k and $\omega(k)$ are the probabilities that $\langle \mathscr{H} \rangle$ is equal to E_k and $E(k)$, respectively. The physical significance of equation (3.41) is clearly apparent in equation (3.59).

(e) The general, formal solution of the Schrödinger equation (3.51) is

$$\Phi(x, y, z, t) = \sum_{k=1}^{n} \sum_{j=1}^{m_k} (\psi_k^{(j)}, \Phi)\psi_k^{(j)}(x, y, z) \exp\left(-iE_k t/\hbar\right)$$

$$+ \int \sum_{j=1}^{m(k)} (\varphi^{(j)}(k), \Phi)\varphi^{(j)}(k, x, y, z) \exp\left(-iE(k)t/\hbar\right) dk. \quad (3.60)$$

Little has been said so far about the uniqueness of the probability postulate. It is not difficult to see that any state vector equivalent to $\Psi(x, y, z, t)$ could be put into the foregoing equations and still leave us with an acceptable formalism. In particular, it would seem that the wavefunction $\Phi(p_x, p_y, p_z, t)$, where the p_q ($q = x, y, z$) are cartesian components of the momentum could just as well be used as the wavefunction in the Schrödinger equation of motion. It is worthwhile to develop the appropriate formalism resulting from this transformation of the coordinate representative of the state vector, because the *momentum wavefunction* $\Phi(\mathbf{p}, t)$ turns out to be more convenient in certain applications of quantum theory than is the coordinate wavefunction $\Psi(\mathbf{r}, t)$.

We shall define the momentum representative of the state vector to be the Fourier transform of the coordinate representative of the state vector:

$$\Phi(\mathbf{p}, t) \equiv (2\pi\hbar)^{-3/2} \int \Psi(\mathbf{r}, t) \exp(-i\mathbf{p}\cdot\mathbf{r}/\hbar)\, d\mathbf{r} \qquad (3.61)$$

The Hermitian operators in the momentum representation are just the Fourier transforms of those in the coordinate representation. For example, the momentum operator for a single particle becomes

$$\frac{-i\hbar}{(2\pi\hbar)^{3/2}} \int \boldsymbol{\nabla}\Psi \exp(-i\mathbf{p}\cdot\mathbf{r}/\hbar)\, d\mathbf{r}$$

$$= \frac{-i\hbar}{(2\pi\hbar)^{3/2}} \int [\boldsymbol{\nabla}(\Psi(\mathbf{r}, t) \exp(-i\mathbf{p}\cdot\mathbf{r}/\hbar)) - \Psi(\mathbf{r}, t)\boldsymbol{\nabla}\exp(-i\mathbf{p}\cdot\mathbf{r}/\hbar)]\, d\mathbf{r}$$

$$= \frac{-i\hbar}{(2\pi\hbar)^{3/2}} (\Psi(\mathbf{r}, t) \exp(-i\mathbf{p}\cdot\mathbf{r}/\hbar))\Big|_{x,y,z=-\infty}^{x,y,z=+\infty}$$

$$+ \frac{\mathbf{p}}{(2\pi\hbar)^{3/2}} \int \Psi(\mathbf{r}, t) \exp(-i\mathbf{p}\cdot\mathbf{r}/\hbar)\, d\mathbf{r} = \mathbf{p}\Phi(\mathbf{p}, t) \qquad (3.62)$$

where we have imposed the physical condition that $\Psi(\mathbf{r}, t)$ vanish when $|x| = |y| = |z| = \infty$. We see that the operator to be associated with the momentum in the momentum representation multiplies the wavefunction in that representation by just the momentum itself. This result is analogous and quite symmetric to the postulate that the position operator in the coordinate representation simply multiplies the coordinate wavefunction by the position vector. The operator to be associated with the potential energy in the momentum representation may also be found without very much difficulty:

$$(2\pi\hbar)^{-3/2} \int V(\mathbf{r})\Psi(r, t) \exp(-i\mathbf{p}\cdot\mathbf{r}/\hbar)\, d\mathbf{r} = \int \Phi(\mathbf{p}', t)V(\mathbf{p} - \mathbf{p}')\, d\mathbf{p}',$$

where $V(\mathbf{p} - \mathbf{p}')$ is the Fourier transform of $V(\mathbf{r} - \mathbf{r}')$ and we have employed the convolution theorem for Fourier transforms (see section A.1 of the

Appendix) to obtain the last result. With these expressions, the Schrödinger equation may now be written

$$\frac{p^2}{2m} \Phi(\mathbf{p}, t) + \int \Phi(\mathbf{p}', t) V(\mathbf{p} - \mathbf{p}')\, d\mathbf{p}' = i\hbar \frac{\partial \Phi}{\partial t}. \tag{3.63}$$

This equation of motion may be reduced in the same way as was equation (3.51) by putting $\Phi(\mathbf{p}, t) \equiv \varphi(\mathbf{p})\chi(t)$. We find, then,

$$\frac{p^2}{2m} \varphi(\mathbf{p}) + \int \varphi(\mathbf{p}') V(\mathbf{p} - \mathbf{p}')\, d\mathbf{p}' = E\varphi(\mathbf{p}) \tag{3.64}$$

$$i\hbar \frac{\partial \chi}{\partial t} = E\chi(t), \tag{3.65}$$

the second equation being the same as obtained earlier. Equation (3.64) is the Schrödinger eigenvalue problem in the momentum representation. It is an integral equation for the momentum wavefunction, whereas the corresponding equation in the coordinate representation is a differential equation.

3.5. AN EXAMPLE: THE FREE ELECTRON

It is certainly time to put all this formalism to work! We shall choose as our first problem the motion of an electron in infinite and bounded free spaces. Let us consider the infinite free space first. The Schrödinger equation, in the coordinate representation, follows directly from equation (3.51) as

$$-\frac{\hbar^2}{2m_e} \nabla^2 \Psi = i\hbar \frac{\partial \Psi}{\partial t}$$

where m_e is now the mass of the (non-relativistic) electron. This equation may be separated into time-dependent and coordinate-dependent parts in the usual way, with the result

$$-\frac{\hbar^2}{2m_e} \nabla^2 \psi = E\psi(x, y, z) \tag{3.66}$$

$$\chi(t) = \chi_0 \exp\left(-iEt/\hbar\right) \tag{3.67}$$

E being the total energy and χ_0, some constant. We are interested chiefly in the solution of equation (3.66) since this wavefunction should lead to a calculation of the total energy. The equation may be solved by following a general procedure well known in the theory of linear differential equations.[10] Suppose we have the linear, homogeneous differential equation

$$\sum_{l=0}^{n} a_{n-l} \frac{d^l y}{dx^l} = 0$$

[10] See E. D. Rainville, *Elementary Differential Equations*, The Macmillan Co., New York, 1964, pp. 113–120.

where the a_{n-l} ($l = 0, \ldots, n$) are constants. This expression may be written as

$$\sum_{l=0}^{n} a_{n-l} D^l y(x) = 0 \qquad (3.68)$$

where, obviously, $D^l \equiv d^l/dx^l$. Using equation (3.68) we can form the *auxiliary equation*

$$\sum_{l=0}^{n} a_{n-l} m^l = 0 \qquad (3.69)$$

where m is a variable whose values are determined by equation (3.69). According to theory,[10] the n roots of equation (3.69), if they are distinct, are related to $y(x)$ by

$$y(x) = \sum_{k=1}^{n} c_k \exp(m_k x) \qquad (3.70)$$

where the c_k ($k = 1, \ldots, n$) are arbitrary constants and m_k ($1 \leqslant k \leqslant n$) is the kth root of equation (3.69).

Now, to get back to equation (3.66). We assume the solution can be separated into three components:

$$\psi(x, y, z) \equiv X(x) Y(y) Z(z).$$

If this definition is introduced into the differential equation, we get

$$\frac{1}{X(x)} \frac{d^2 X}{dx^2} + \frac{1}{Y(y)} \frac{d^2 Y}{dy^2} + \frac{1}{Z(z)} \frac{d^2 Z}{dz^2} = -\frac{2m_e E}{\hbar^2}. \qquad (3.71)$$

As with the initial decomposition of the free-electron Schrödinger equation, equation (3.71) will not be valid for all values of x, y, and z unless the three terms on the left-hand side are equal to a constant. We shall prescribe these constants in the most convenient way:

$$\frac{1}{X(x)} \frac{d^2 X}{dx^2} = -\frac{2m_e E_x}{\hbar^2}; \qquad \frac{1}{Y(y)} \frac{d^2 Y}{dy^2} = -\frac{2m_e E_y}{\hbar^2};$$

$$\frac{1}{Z(z)} \frac{d^2 Z}{dz^2} = -\frac{2m_e E_z}{\hbar^2} \qquad (3.72)$$

where

$$E_x + E_y + E_z \equiv E.$$

It appears that the solution of any one of equations (3.72) will provide the format for the general solution of equation (3.66). Consider the equation for $X(x)$. We have

$$\frac{d^2 X}{dx^2} + \frac{2m_e E_x}{\hbar^2} X(x) = 0 \qquad (3.73)$$

which is, interestingly, a linear, homogeneous differential equation. The auxiliary equation is

$$m^2 + \frac{2m_eE_x}{\hbar^2} = 0$$

and has two solutions

$$m_1 = \frac{i}{\hbar}(2m_eE_x)^{1/2}$$

$$m_2 = \frac{-i}{\hbar}(2m_eE_x)^{1/2}.$$

The solution of equation (3.73) is then

$$X(x) = c_1 \exp\left[(i/\hbar)(2m_eE_x)^{1/2}x\right] + c_2 \exp\left[(-i/\hbar)(2m_eE_x)^{1/2}x\right] \quad (3.74)$$

where c_1 and c_2 are arbitrary. Now, the two functions on the right-hand side of equation (3.74) are orthogonal, since

$$\int_{-\infty}^{\infty} \exp\left[(i/\hbar)(2m_eE_x)^{1/2}x\right] \exp\left[(i/\hbar)(2m_eE_x)^{1/2}x\right] dx = 0$$

and, therefore, they are linearly independent. For this reason we can write

$$X(x) = N \exp\left[(\pm i/\hbar)(2m_eE_x)^{1/2}x\right] \quad (3.75)$$

as the general solution of equation (3.73), where N is a constant whose value derives from the normalization of $X(x)$. The normalization integral is

$$\int_{-\infty}^{\infty} |X(x)|^2 \, dx \equiv 1$$

or, if the positive-exponent solution is used,

$$N^2 \int_{-\infty}^{\infty} \exp\left[(-i/\hbar)(2m_eE_x)^{1/2}x\right] \exp\left[(i/\hbar)(2m_eE_x)^{1/2}x\right] dx = N^2 \int_{-\infty}^{\infty} dx.$$

It seems as if it is impossible to normalize the $X(x)$. Not quite. Recall that for some $E_x' \neq E_x$ we have

$$N^2 \int_{-\infty}^{\infty} \exp\left[(-i/\hbar)(2m_eE_x')^{1/2}x\right] \exp\left[(i/\hbar)(2m_eE_x)^{1/2}x\right] dx$$

$$= N^2 \int_{-\infty}^{\infty} \exp\left[(i/\hbar)[(2mE_x)^{1/2} - (2m_eE_x')^{1/2}]x\right] dx$$

$$= N^2 2\pi\hbar\delta[(2m_eE_x)^{1/2} - (2m_eE_x')^{1/2}],$$

78

by equation (3.19). The normalized solution of equation (3.73), is, therefore,

$$X_{E_x}(x) = (2\pi\hbar)^{-\frac{1}{2}} \exp\left[(\pm i/\hbar)(2m_e E_x)^{\frac{1}{2}}x\right] \tag{3.76}$$

subject to the orthonormality condition

$$\int_{-\infty}^{\infty} X_{E_x'}(x)X_{E_x}(x)\,dx = \delta[(2m_e E_x)^{\frac{1}{2}} - (2m_e E_x')^{\frac{1}{2}}]. \tag{3.77}$$

According to equation (3.36), this is just the condition for the E_x to form a continuous spectrum. Therefore, we may conclude that *the energy spectrum of an electron in infinite free space is continuous.* We notice that, by equation (3.76), the eigenvalues for the free electron are doubly degenerate.

Are there any quantities besides the total energy which are constants of the motion for the free electron? Because the potential energy is zero, the total and kinetic energies must be equal, so that the momentum should also be a constant of the motion. To prove this deduction we calculate

$$\langle p_x \rangle = \left(X_{E_x'}, -i\hbar\frac{dX_{E_x}}{dx}\right) = -i\hbar\left(X_{E_x'}, \frac{dX_{E_x}}{dx}\right)$$

$$= \frac{-i\hbar}{(2\pi\hbar)}\frac{i}{\hbar}(2m_e E_x)^{\frac{1}{2}}2\pi\hbar\delta[(2m_e E_x')^{\frac{1}{2}} - (2m_e E_x)^{\frac{1}{2}}]$$

$$= (2m_e E_x)^{\frac{1}{2}}\delta[(2m_e E_x')^{\frac{1}{2}} - (2m_e E_x)^{\frac{1}{2}}]$$

as expected. Thus we can rewrite equation (3.76) as

$$X_{E_x}(x) = (2\pi\hbar)^{-\frac{1}{2}} \exp\left[(\pm i/\hbar)p_x x)\right]$$

and the solutions of equation (3.66) become

$$\psi(\mathbf{r}, \mathbf{p}) = (2\pi\hbar)^{-\frac{3}{2}} \exp(\pm i\mathbf{p}\cdot\mathbf{r}) \tag{3.78}$$

where it is understood that $\mathbf{p}$ can take on any magnitude or direction.[11]

Now let us place our free electron into a cubical enclosure of length L. Speaking in terms of potential energy, this act is equivalent to imposing the condition

$$V(q) = \begin{cases} \infty & q < 0 \\ 0 & 0 \leqslant q \leqslant L \quad (q = x, y, z) \\ \infty & q > L \end{cases} \tag{3.79}$$

---

[11] It should be understood that $\psi(\mathbf{r}, \mathbf{p})$ does not really represent a physical state because it is not a vector in Hilbert space. The reason it is not, of course, is because it does not possess a finite length—the lack of which is intimately connected with the unphysical unboundedness of the electron's enclosure. When the electron is enclosed in a finite volume, we shall find solutions of the Schrödinger eigenvalue problem which are truly state vectors. Therefore, we shall consider the significance of equation (3.78) to be more philosophical and computational than physical.

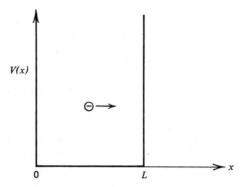

Figure 20. The potential energy function for an electron confined to a cubical enclosure, as seen in one dimension.

where the left front corner is presumed to be at the origin. (See Figure 20.) Nine eigenvalue equations can be constructed using equation (3.79). We shall consider for convenience sake only the motion in the x-direction once again. For the region inside the enclosure we have

$$\frac{\hbar^2}{2m_e} \frac{d^2\psi}{dx^2} + E_x\psi(x) = 0 \qquad (0 \leqslant x \leqslant L).$$

The solutions follow from equation (3.75) as

$$\psi(x) = N \exp\left[(\pm i/\hbar)(2m_eE_x)^{\frac{1}{2}}x\right]. \tag{3.80}$$

Because these solutions are linearly independent, any linear combination of them is also a solution. In particular we choose

$$\sin\left[\frac{(2m_eE_x)^{\frac{1}{2}}x}{\hbar}\right] \equiv \frac{1}{2i}\left\{\exp\left[(i/\hbar)(2m_eE_x)^{\frac{1}{2}}x\right] - \exp\left[(-i/\hbar)(2m_eE_x)^{\frac{1}{2}}x\right]\right\}$$

so that

$$\varphi(x) = C \sin\left[\frac{(2m_eE_x)^{\frac{1}{2}}x}{\hbar}\right] \tag{3.81}$$

is one new solution. (By taking the *sum* of the exponentials we could get another solution, namely, the cosine function. The set of the two new solutions then would be equivalent in all respects to the pair of solutions first obtained.) The normalization constant must satisfy

$$C^2 \int_0^L \sin^2\left[\frac{(2m_eE_x)^{\frac{1}{2}}x}{\hbar}\right] dx = 1.$$

Thus we have

$$C^2 \int_0^L \sin^2\left[\frac{(2m_eE_x)^{1/2}x}{\hbar}\right] dx = \frac{C^2\hbar}{(2m_eE_x)^{1/2}} \int_0^{(2m_eE_x)^{1/2}L/\hbar} \sin^2 z \, dz$$

$$= \frac{C^2\hbar}{(2m_eE_x)^{1/2}} \left\{\frac{z}{2} - \tfrac{1}{4}\sin 2z\right\}\Bigg|_0^{(2m_eE_x)^{1/2}L/\hbar}$$

$$= \frac{C^2L}{2} - \frac{C^2}{4}\frac{\hbar}{(2m_eE_x)^{1/2}}\sin\left[\frac{(8m_eE_x)^{1/2}L}{\hbar}\right] \equiv 1.$$

from which the value of C can be calculated without difficulty. We have not yet specified what the wavefunction must do on the boundary of the enclosure. Because the potential energy becomes infinite there, we shall impose the condition

$$\varphi(x) \equiv 0 \qquad (|x| \geqslant L).$$

This restriction leads immediately to

$$\left[\frac{(2m_eE_x)^{1/2}}{\hbar}\right]L \equiv n_x\pi \qquad (n_x = 1, 2, \ldots)$$

or

$$E_{n_x} = \frac{n_x^2}{8m_e}\left(\frac{2\pi\hbar}{L}\right)^2 \qquad (n_x = 1, 2, \ldots). \tag{3.82}$$

We see that *an electron in an enclosure of finite dimensions possesses a discrete energy spectrum.* As L gets very large, however, consecutive energy eigenvalues become closer and closer in magnitude until, at $L \to +\infty$, they form a continuum, in agreement with our earlier analysis.

The orthonormal solution of our problem is finally

$$\varphi_{n_x}(x) = \left(\frac{2}{L}\right)^{1/2}\sin\left(\frac{\pi n_x x}{L}\right) \qquad (n_x = 1, 2, \ldots). \tag{3.83}$$

The first few eigenstates are shown in Figure 21. We note, that, unlike the situation as described by Newtonian mechanics, the electron is not to be found everywhere with probability one. Indeed, there are portions of the enclosure where the electron has zero probability of existing. These points are called the *nodes* of the wavefunction. In this case, there are $(n_x - 1)$ nodes for the n_xth wavefunction.

The total energy of an electron in a cubical enclosure is found by generalizing equation (3.82):

$$E_n = \frac{n^2}{8m_e}\left(\frac{2\pi\hbar}{L}\right)^2 \qquad (n = 3, 4, \ldots), \tag{3.84}$$

where

$$n^2 = n_x^2 + n_y^2 + n_z^2 \qquad (n_x, n_y, n_z = 1, 2, \ldots).$$

The total energy of the electron is degenerate for all n greater than three.

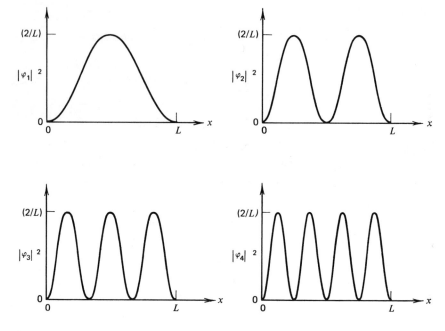

Figure 21. The probability distributions for the first four eigenstates of an electron confined to a cubical enclosure, as seen in one dimension.

To conclude our first examination of the free electron, let us consider what happens when one of these particles travels in the x-direction and is incident upon a "potential barrier" of "height" V_0 (Figure 22). The potential energy is expressed by

$$V(x) = \begin{cases} 0 & x < 0 \\ V_0 & 0 \leqslant x \leqslant d \\ 0 & x > d \end{cases} \qquad (3.85)$$

where V_0 is a positive constant. The three Schrödinger eigenvalue equations that can be constructed by using (3.85) are:

$$\frac{d^2}{dx^2}\psi_{\mathrm{I}} + \frac{2m_e E}{\hbar^2}\psi_{\mathrm{I}}(x) = 0 \qquad (x < 0),$$

$$\frac{d^2}{dx^2}\psi_{\mathrm{II}} + \frac{2m_e}{\hbar^2}(E - V_0)\psi_{\mathrm{II}}(x) = 0 \qquad (0 \leqslant x \leqslant d),$$

$$\frac{d^2}{dx^2}\psi_{\mathrm{III}} + \frac{2m_e E}{\hbar^2}\psi_{\mathrm{III}}(x) = 0 \qquad (x > d).$$

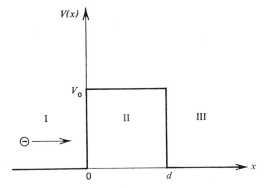

Figure 22. A finite potential barrier impinged upon by an electron.

The interesting case is when $E < V_0$. The solutions of these equations can be found readily with the method outlined earlier. They are

$$\psi_{\mathrm{I}}(x) = c_1 \exp(ikx) + c_2 \exp(-ikx) \qquad (3.86)$$

$$\psi_{\mathrm{II}}(x) = c_3 \exp(k_1 x) + c_4 \exp(-k_1 x) \qquad (3.87)$$

$$\psi_{\mathrm{III}}(x) = c_5 \exp(ikx) \qquad (3.88)$$

where

$$k \equiv \frac{(2m_e E)^{\frac{1}{2}}}{\hbar}$$

and

$$k_1 \equiv \frac{(2m_e(V_0 - E))^{\frac{1}{2}}}{\hbar}.$$

We have put $c_6 = 0$ in equation (3.88) because $\exp(-ikx)$ corresponds to the electron moving in the negative x-direction after passing through the barrier—a physical situation we are excluding here.[12] We shall also require that the wavefunction be continuous and possess a continuous first derivative on the boundaries of the potential barrier. Therefore, we have the conditions

$$c_1 + c_2 = c_3 + c_4$$

$$c_3 \exp(k_1 d) + c_4 \exp(-k_1 d) = c_5 \exp(ikd)$$

$$c_1 ik - c_2 ik = c_3 k_1 - c_4 k_1$$

$$c_3 k_1 \exp(k_1 d) - c_4 k_1 \exp(-k_1 d) = c_5 ik \exp(ikd)$$

[12] Exp $(-ikx)$ is an eigenstate of the momentum operator corresponding to momentum $-\hbar k$.

from which we may find four of the c's in terms of the remaining one. If we define, reasonably, the transmitted fraction of the electron as

$$T \equiv \frac{|\psi_{\mathrm{III}}|^2}{|\psi_{\mathrm{I}}(\mathrm{trans})|^2}$$

where

$$\psi_{\mathrm{I}}(\mathrm{trans}) \equiv c_1 \exp(ikx),$$

and the reflected fraction of the electron as

$$R \equiv \frac{|\psi_{\mathrm{I}}(\mathrm{ref})|^2}{|\psi_{\mathrm{I}}(\mathrm{trans})|^2}$$

where

$$\psi_{\mathrm{I}}(\mathrm{ref}) \equiv c_2 \exp(ikx),$$

we find without difficulty for large values of $k_1 d$,

$$T_{k_1 d \to \infty} \frac{16k^2}{4k^2 + [k_1 - (k^2/k_1)]^2} \exp(-2k_1 d).$$

We see that there is a non-zero transmission fraction which decays rapidly to zero as the thickness of the potential barrier or the descrepancy between E and V_0 increases. The surprising thing is that an electron with a total energy less than the barrier height can get through at all. Such behavior is clearly prohibited by the laws of Newtonian mechanics and thus must be a distinctly quantum-mechanical phenomenon. This effect is sometimes called "tunneling" and may be compared in some ways to the passage of a light wave through an absorber. In Figure 23 a plot of the real portion of the wavefunction is shown; it may be seen that $\psi_{\mathrm{II}}(x)$ is far from being zero.

It is not very hard to generalize our analysis of the behavior of an electron incident upon a potential barrier to make it work for the situation wherein the particle is confronted by an infinite chain of these barriers. In that case, equation (3.85) becomes

$$V(x) = \begin{cases} 0 & [d + n(a + d)] < x < [(n + 1)(a + d)] \\ V_0 & n(a + d) \leqslant x \leqslant [d + n(a + d)] \end{cases}$$

where $n = 0, \pm 1, \pm 2, \ldots$, and a is the separation between consecutive barrier "walls". (See Figure 24.) This potential function differs essentially from those considered previously in that it is *periodic*,

$$V(x + a + d) = V(x), \tag{3.89}$$

with period $(a + d)$. The Schrödinger eigenvalue problem is now

$$\frac{d^2\psi_{\mathrm{I}}}{dx^2} + k^2\psi_{\mathrm{I}}(x) = 0$$

$$\frac{d^2\psi_{\mathrm{II}}}{dx^2} - k_1{}^2\psi_{\mathrm{II}}(x) = 0$$

$$(3.90)$$

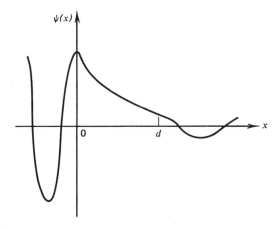

Figure 23. A sketch of the real part of the wave-function for an electron striking a finite potential barrier while possessing a total energy less than the barrier height.

for the regions between and within the potential barriers, respectively. However, because of the periodic nature of $V(x)$, the solutions are not quite the same as those given in equations (3.86) and (3.87). Instead, we have (for $E < V_0$)

$$u_K^{\text{I}}(x) = c_1 \exp [i(k - K)x] + c_2 \exp [-i(k + K)x] \qquad (3.91)$$

$$u_K^{\text{II}}(x) = c_3 \exp [(k_1 - iK)x] + c_4 \exp [-(k_1 + iK)x] \qquad (3.92)$$

where K is a constant. These equations result from equations (3.86), (3.87), and (3.89) in the following way. Since $V(x)$ is periodic, $\psi(x + a + d)$ is a

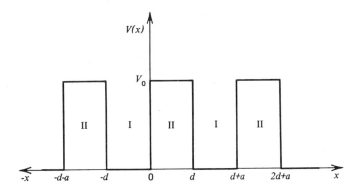

Figure 24. An infinite chain of one-dimensional potential barriers.

85

solution of the eigenvalue problem (3.90) if $\psi(x)$ is a solution. On the other hand, the two linearly independent solutions of equations (3.90), say, $\psi_1(x)$ and $\psi_2(x)$, are sufficient to express *any* other solution, according to the basic properties of Hilbert space. Thus, we can write

$$\psi_1(x + a + d) = \alpha_1\psi_1(x) + \alpha_2\psi_2(x)$$

or

$$\psi_2(x + a + d) = \beta_1\psi_1(x) + \beta_2\psi_2(x)$$

or even

$$\psi(x + a + d) = A\psi_1(x + a + d) + B\psi_2(x + a + d).$$

If we substitute the first two of these equations into the third one, we get

$$\psi(x + a + d) = (A\alpha_1 + B\beta_1)\psi_1(x) + (A\alpha_2 + B\beta_2)\psi_2(x)$$
$$\equiv \delta A\psi_1(x) + \delta B\psi_2(x) = \delta\psi(x)$$

where δ is a constant subject to

$$A\alpha_1 + B\beta_1 = \delta A$$
$$A\alpha_2 + B\beta_2 = \delta B.$$

Now let us define

$$\delta \equiv \exp [iK(a + d)]$$
$$u_K(x) \equiv \exp (-iKx)\psi(x).$$

The latter function, $u_K(x)$, has a very interesting property:

$$u_K(x + a + d) = \exp [-iK(x + a + d)]\psi(x + a + d)$$
$$= \exp [-iK(x + a + d)] \exp [iK(a + d)]\psi(x)$$
$$= \exp (-iKx) \exp (iKx)u_K(x)$$
$$= u_K(x).$$

$u_K(x)$ is periodic with the periodicity of $V(x)$! Evidently, then, the solution of equations (3.90) can always be written as the product of an exponential function and a function with the periodicity of $V(x)$, which is to say,[13]

$$\psi(x) = \exp (iKx)u_K(x). \tag{3.93}$$

It follows, from equations (3.86), (3.87), and (3.93), that equations (3.91) and (3.92) are proper solutions of equations (3.90).

---

[13] Equation (3.93) is sometimes called *Bloch's Theorem*. The solution, $\psi(x)$ is accordingly named a *Bloch function*.

If we proceed, as earlier, to determine the c's in equations (3.91) and (3.92), we find

$$c_1 + c_2 = c_3 + c_4$$

$$c_1 \exp\left[i(k - K)d\right] + c_2 \exp\left[-i(k + K)d\right]$$
$$= c_3 \exp\left[(k_1 - iK)d\right] + c_4 \exp\left[-(k_1 + iK)d\right]$$

$$c_1 i(k - K) - c_2 i(k + K) = c_3(k_1 - iK) - c_4(k_1 + iK)$$

$$c_1 i(k - K) \exp\left[i(k - K)d\right] - c_2 i(k + K) \exp\left[-i(k + K)d\right]$$
$$= c_3(k_1 - iK) \exp\left[(k - iK)d\right] - c_4(k_1 + iK) \exp\left[-(k + iK)d\right].$$

It is known from the theory of simultaneous equations that all the c's in the foregoing expressions will be identically zero unless the determinant of their coefficients vanishes.[14] This requirement, it turns out, after lots of algebraic manipulation, implies

$$\frac{k_1{}^2 - k^2}{2kk_1} \sinh(k_1 d) \sin(kd) + \cosh(k_1 d) \cos(kd) = \cos(2Kd). \quad (3.94)$$

We see that K is, in general, a *complex* number, since the left-hand side of equation (3.94) can be either larger than $+1$ or smaller than -1. On the other hand, equation (3.93) tells us that K must not be other than a *real* number if $\psi(x)$ is to remain in Hilbert space (that is, remain finite as $|x| \to \infty$ and so be square-integrable). It appears that those values of k and k_1 which lead to imaginary K are physically inadmissible, according to the state postulate. Since k and k_1 depend on the total energy, E, we conclude that certain values of the latter quantity are *forbidden* to the electron. The energy spectrum for this particle, when among an infinite chain of potential barriers, then, must consist of regions where *all* values of the energy are allowed (since k and k_1 depend continuously upon K) and regions, corresponding to imaginary K, where *no* value is allowed. The former are called *energy bands*, the latter, *energy gaps*. The concept of energy bands, arising from a fairly simple one-dimensional model, is of great significance in the theory of solids, wherein the chain of potential barriers represents ion cores interacting with the electron.

3.6. THE CORRESPONDENCE PRINCIPLE

The relationship between the behavior of a free electron in an enclosure of infinite extent and one in a cubical container is most clearly spelled out if the

---

[14] See, for example, H. Margenau and G. M. Murphy, *The Mathematics of Physics and Chemistry*, D. Van Nostrand Co., Princeton, 1956, Vol. I, pp. 313*f*.

total energy of the latter particle is expressed in terms of its momentum. In particular, we have

$$\langle p_x{}^2 \rangle = \left(\varphi_{n_x}, -\hbar^2 \frac{d^2}{dx^2} \varphi_{n_x} \right)$$

$$= -\frac{2\hbar^2}{L} \int_0^L \sin\left(\frac{n_x \pi x}{L}\right) \frac{d^2}{dx^2} \sin\left[\left(\frac{n_x \pi}{L}\right) x\right] dx$$

$$= \frac{2\hbar^2}{L^2} \left(\frac{n_x \pi}{L}\right)^2 \int_0^L \sin^2\left[\left(\frac{n_x \pi}{L}\right) x\right] dx$$

$$= \frac{2\hbar^2}{L} \left(\frac{n_x \pi}{L}\right)\left(\frac{n_x \pi}{2}\right) = n_x{}^2 \frac{(2\pi\hbar)^2}{4L^2} \tag{3.95}$$

so that equation (3.82) can be written

$$\langle E_{n_x} \rangle = \frac{\langle p_x{}^2 \rangle}{2m_e}. \tag{3.96}$$

For the particle in an infinite enclosure we found

$$E_x = \frac{p_x{}^2}{2m_e} \tag{3.97}$$

where p_x can take on a continuum of values. Equations (3.96) and (3.97) have a striking formal similarity. However, they differ intrinsically in that the first is consonant with the quantum-theoretical concept of state while the second conforms well with the expectations in classical physics. As a direct result, equation (3.96) is definitely *not* satisfied for all imaginable values of $\langle p_x{}^2 \rangle$, but only those for which

$$\langle p_x{}^2 \rangle = \frac{n_x{}^2}{4} \left(\frac{2\pi\hbar}{L}\right)^2 \qquad (n_x = 1, 2, \ldots) \tag{3.95}$$

holds true. In other words, the quantum-mechanical expression portrays a dynamical condition on mean values which is greatly similar to but much stronger than the relationship dictated by the classical-mechanical expression. However, we notice that as n_x takes on very large values the difference between consecutive values of $\langle p_x{}^2 \rangle$ becomes very small relative to $\langle p_x{}^2 \rangle$ itself and, in the limit of infinite n_x, approaches zero. Therefore, we may say that equation (3.96) *approaches* equation (3.97) in a particular sense. More specifically, equation (3.96) becomes very nearly the same as equation (3.97) in the limit of large "quantum numbers," n_x. This behavior, as a matter of fact, is required by the last of the criteria set forth in section 3.3 concerning any quantum-mechanical equation of motion. All we have done here is to verify that the criterion is met in the case of the free electron. But this example permits us to make a more precise statement concerning the relationship between quantum mechanics and classical mechanics. This statement, called the *Correspondence Principle* by Niels Bohr (who first wrote it down), may be expressed as:

The Correspondence Postulate. The mathematical formulation of the quantum theory must retain all of the physical content of classical mechanics in the sense that

(a) the dynamical relations of the classical theory will always have formally similar, quantum-theoretical analogs, and

(b) the two theories will always lead to identical results in the limit of very large quantum numbers.

In making this statement we have stipulated that there can never be a true inconsistency between Newtonian and quantum concepts, for the two will always be in agreement whenever it is physically proper to employ the former notions in a description of natural phenomena. As regards microscopic behavior, the classical theory can provide formal relations (such as Newton's laws) which can be expected to hold in an average sense, the averages being calculable from the postulates of quantum mechanics.[15]

There is one other important physical consequence of the correspondence postulate. By its very statement we are led to believe that the quantum theory is inextricably connected with the appearance of whole numbers in the description of empirical results. We shall find this to be quite a correct prediction when we return to the discussion of atomic line spectra. The appearance of integers in the quantum theory is the direct consequence of the form of the Schrödinger equation, which factors immediately to produce an eigenvalue problem. That this is so is no accident; Schrödinger was familiar with equation (1.1), and actively sought a differential equation that would lead to expressions containing integers.

3.7. THE UNCERTAINTY RELATIONS

In several discussions on the subject of Young's experiment (double-slit interference) involving either photons or electrons, the restrictions placed on the quantum-theoretical concept of "state" have been pointed out. Thus far we have established that state vectors are to provide no more information about the system they represent than is necessary to predict the results of experiments that might be performed on the system. As regards Young's experiment, the pertinent information concerns the number distribution of

[15] Examples of this aspect of the correspondence postulate will be given in Chapter 6.

photons or electrons on a detecting screen placed in front of the slits. We ask that the quantum theory permit a calculation of this distribution, or, as it turns out, probability function, but not that it should tell us anything more. It is to be clearly understood that *this is not an inherent weakness of the theory*; the empirical behavior of photons and electrons *requires* the definition of state we have adopted. That such is the case was carefully indicated in our first encounter with photon interference. At that time it was shown that a precise knowledge of the trajectory of an interfering photon, or of the process of self-interference, was precluded by a precise knowledge of the structure of the interference pattern. The most dramatic evidence for this assertion is the fact that, if one of the slits is covered in an effort to determine the trajectory of the photon, the interference pattern itself just disappears. The whole experiment, then, has no meaning. As the interference pattern leads to a calculation of the wavelength of the photon (Figure 13) and thus, through equation (2.20) to the value of its momentum, the foregoing statements can be epitomized by saying that *it is impossible to measure the position and momentum of a photon engaged in double-slit interference, at one and the same instant.*

But have we proved our point? Is it not so that even *single* slits produce interference phenomena? Can we not put a photon or electron through a single aperture, thereby determining its position with an arbitrarily high precision, and then accurately calculate its momentum from a simultaneous measurement of the spacing of the interference bands? This possibility is certainly worth examining in detail, so let us create the hypothetical experiment and make a close analysis of its content. Suppose an electron were to pass through a single slit of width d, as shown in Figure 25. By this very passage we have determined the position of the electron at a certain instant to within the distance $\Delta y \sim d$, where Δy is a measure of the imprecision in our estimate of the y-coordinate. Now, from experiment we know the electron will be diffracted by the slit. The expression governing the diffraction phenomenon is equation (2.21), introduced in connection with Davisson's study of electron diffraction by nickel. Thus we have

$$\lambda = d \sin \theta \qquad (3.98)$$

where λ is the de Broglie wavelength of the electron and θ is the angle, relative to the line of incidence, through which the electron is diffracted. The electron, then, is to be found moving in a direction whose inclination relative to that of incidence is between zero and θ at most. The y-component of its momentum must, therefore, be uncertain by at least

$$\Delta p_y \sim \frac{h}{\lambda} \sin \theta. \qquad (3.99)$$

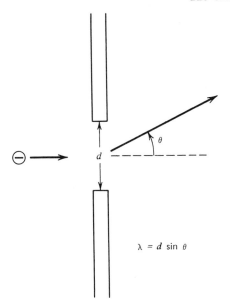

$$\lambda = d \sin \theta$$

Figure 25. Single-slit diffraction of an
electron.

If equations (3.98) and (3.99) are combined, taking account of the definition
of Δy, we find

$$\Delta y \, \Delta p_y \sim h. \tag{3.100}$$

Our attempt has failed. We have discovered that even with the single slit, the
fact that the electron exhibits wave properties leads unalterably to a fundamental imprecision in any attempt to measure position and momentum
simultaneously, for if the slit is vanishingly small, then by equation (3.100)
the uncertainty in momentum will be arbitrarily large. It is of significance
that a measure of the fundamental uncertainty appears to be Planck's constant. Let us formulate the problem more carefully and see if this is indeed
the case.

Consider the always-positive (or zero) quantity[16]

$$\|(x_{\mathrm{op}} - \langle x \rangle)\psi + i\lambda(p_{\mathrm{op}} - \langle p \rangle)\psi\|^2$$

where λ is any scalar and $\psi(x)$ is a normalized wavefunction which vanishes

---

[16] The argument to be given is largely to be found in A. Messiah, *Quantum Mechanics*,
John Wiley and Sons, New York, 1961, Vol. I, p. 133*f*.

identically outside some finite interval. If we expand the scalar product in this quantity we have

$$\|(x_{op} - \langle x \rangle)\psi\|^2 - i\lambda[((p_{op} - \langle p \rangle)\psi, (x_{op} - \langle x \rangle)\psi)$$
$$- ((x_{op} - \langle x \rangle)\psi, (p_{op} - \langle p \rangle)\psi)] - \lambda^2\|(p_{op} - \langle p \rangle)\|^2$$
$$= (\Delta x)^2 - i\lambda[(p_{op}\psi, x_{op}\psi) - (x_{op}\psi, p_{op}\psi)] - \lambda^2(\Delta p_x)^2$$

since

$$((p_{op} - \langle p \rangle)\psi, (x_{op} - \langle x \rangle)\psi) = (p_{op}\psi, x_{op}\psi) - \langle x \rangle(p_{op}\psi, \psi)$$
$$-\langle p \rangle(\psi, x_{op}\psi) + \langle p \rangle\langle x \rangle(\psi, \psi) = (p_{op}\psi, x_{op}\psi) - \langle x \rangle\langle p \rangle$$

and

$$((p_{op} - \langle p \rangle)\psi, (x_{op} - \langle x \rangle)\psi) = \overline{((x_{op} - \langle x \rangle)\psi, (p_{op} - \langle p \rangle)\psi)}.$$

Now, the coefficients of λ above may be integrated by parts to give

$$i\hbar \int \left(\frac{d\bar{\psi}}{dx} x\psi(x) + x\bar{\psi}\frac{d\psi}{dx}\right) dx = i\hbar \int \frac{d}{dx}(\bar{\psi}x\psi) \, dx - i\hbar \int \bar{\psi}\psi \, dx$$
$$= -i\hbar(\psi, \psi) = -i\hbar$$

because $(\bar{\psi}x\psi)$ is identically zero on the boundary. Thus

$$(\Delta x)^2 - \lambda\hbar - \lambda^2(\Delta p_x)^2 \geqslant 0$$

is our inequality. Now let us take the first derivative of the last expression:

$$\frac{d}{d\lambda}[(\Delta x)^2 - \lambda\hbar + \lambda^2(\Delta p_x)^2] = -\hbar + 2\lambda(\Delta p_x)^2.$$

when $\lambda = \hbar/2(\Delta p_x)^2$ the expression has its minimum value. (The second derivative of the expression is always positive.) Thus we require our inequality to hold even when

$$\lambda = \frac{\hbar}{2(\Delta p_x)^2}.$$

This implies

$$(\Delta x)^2 \geqslant -\frac{\hbar^2}{4(\Delta p_x)^2} + \frac{\hbar^2}{2(\Delta p_x)^2} = \frac{\hbar^2}{4(\Delta p_x)^2}$$

or

$$(\Delta x)^2(\Delta p_x)^2 \geqslant \left(\frac{\hbar}{2}\right)^2$$

or

$$(\Delta x)(\Delta p_x) \geqslant \frac{\hbar}{2}. \tag{3.101}$$

Equation (3.101) expresses in an exact way the fundamental uncertainty suggested by our analysis of electron diffraction. It follows easily by the same method that we can write

$$(\Delta x)(\Delta p_x) \geqslant \frac{\hbar}{2}$$

$$(\Delta y)(\Delta p_y) \geqslant \frac{\hbar}{2} \qquad\qquad (3.102)$$

$$(\Delta z)(\Delta p_z) \geqslant \frac{\hbar}{2}$$

which are known as the *Heisenberg uncertainty relations*. They express the restriction imposed by natural behavior upon the accuracy with which a vector in Hilbert space may specify the state of a single particle. Within the context of the quantum theory, the precise determination of state properties has meaning only to within the extent prescribed by equations (3.102). *The failure of the classical theory, then, lies with its implicit hypothesis that the magnitude of Planck's constant is near enough to zero to be neglected even for microscopic phenomena,* so that position and momentum remain well-defined entities at all levels of dimension.

PROBLEMS

1. One example of a linear space is that composed of directed line segments —the ordinary, real vectors in three-dimensional space. Verify and interpret geometrically equations (3.1) through (3.8) for these vectors.

2. Show that the vectors

$$\mathbf{A} = \hat{\mathbf{x}} \oplus 2\hat{\mathbf{y}} \oplus 2\hat{\mathbf{z}} \qquad \text{and} \qquad \mathbf{B} = 2\hat{\mathbf{x}} \oplus -3\hat{\mathbf{y}} \oplus 2\hat{\mathbf{z}}$$

are orthogonal. Are these vectors normalized?

3. Using the Gram-Schmidt process, create an orthonormal set from the vectors

$$\mathbf{A} = \hat{\mathbf{x}} \oplus 2\hat{\mathbf{y}} \oplus -2\hat{\mathbf{z}} \qquad \mathbf{B} = 2\hat{\mathbf{x}} \oplus -2\hat{\mathbf{y}} \oplus \hat{\mathbf{z}} \qquad \mathbf{C} = -2\hat{\mathbf{x}} \oplus \hat{\mathbf{y}} \oplus 2\hat{\mathbf{z}}$$

4. Show that the functions

$$f_0 = L^{-\frac{1}{2}} \qquad f_1(x) = \left(\frac{2}{L}\right)^{\frac{1}{2}} \cos\frac{n\pi x}{L} \qquad f_2(x) = \left(\frac{2}{L}\right)^{\frac{1}{2}} \sin\frac{n\pi x}{L},$$

where n takes on positive integral values, form an orthonormal set in the domain $0 \leqslant x \leqslant L$.

5. Demonstrate the orthonormality of the functions

$$f_m(x) = L^{-\frac{1}{2}} \sin \frac{m\pi x}{L} \qquad (m = 1, 2, \ldots)$$

in the domain $-L \leqslant x \leqslant L$. Show that these functions are linearly independent and that, for any odd function $u(x)$ defined in the interval $(-L \leqslant x \leqslant L)$,

$$u(x) = \sum_{m=1}^{\infty} a_m f_m(x).$$

The expression above is called a *Fourier sine series*. What are the Fourier sine coefficients a_m in terms of the $f_m(x)$ and $u(x)$? What happens if $u(x)$ is an even function? Do the $f_m(x)$ form a *complete* orthonormal set?

6. Calculate the Fourier sine coefficients for the function

$$f(x) = x \qquad (-L \leqslant x \leqslant L)$$

and show that

$$x = 2L \sum_{m=1}^{\infty} \frac{(-1)^{m+1}}{(m\pi)} \sin \frac{m\pi x}{L}.$$

7. Create an orthonormal set from the functions

$$f_0 = 1 \qquad f_1(x) = x \qquad f_2(x) = x^2 \qquad (-1 \leqslant x \leqslant 1)$$

Call the orthonormal functions $P_0(x)$, $P_1(x)$, and $P_2(x)$, respectively. These new functions are known as the *Legendre polynomials* and are of importance in the quantum theory of the hydrogen atom.

8. Show that for any pair of functions $\{f(x), g(x)\}$ for which the integrals

$$\int_{-\infty}^{\infty} |f(x)|^2 \, dx \qquad \int_{-\infty}^{\infty} |g(x)|^2 \, dx$$

are finite, the *Schwarz inequality* holds:

$$|(f, g)| \leqslant \|f\| \, \|g\|.$$

Also, verify the Schwarz inequality for ordinary real vectors in three-dimensions. (*Hint*: Compute (h, h), where $h = f + ag$ with $a = -(g, f)/(g, g)$.)

9. Consider the function

$$f(x) = \exp\left[-(x/x_0)^2\right] \qquad (-\infty < x < \infty)$$

Show that its Fourier transform is (for $a = 1$)

$$\varphi(p) = \frac{x_0}{\sqrt{2}} \exp\left[-(px_0/2)^2\right] \qquad (-\infty < p < \infty)$$

(*Hint*: $[(x/x_0) + (ipx_0/2)]^2 = (x/x_0)^2 + ipx - (p^2 x_0^2/4)$.)

10. What is the Fourier transform of the delta-"function"?

11. Show that, if H is an Hermitian operator and f and g are vectors in Hilbert space such that (f, Hf) and (g, Hg) are positive-definite,

$$|(f, Hg)|^2 \leqslant (f, Hf)(g, Hg).$$

This result is often called the *generalized Schwarz inequality*.

12. Prove the following commutator identities. (Here $+$ is operator addition.)

(a) $\qquad\qquad [A, B + C] = [A, B] + [A, C]$

(b) $\qquad\qquad [A, BC] = [A, B]C + B[A, C]$

13. With the appropriate operator associations, show that

$$[q_{op}, q'_{op}] = [p_{qop}, p_{q'op}] = 0 \qquad [q'_{op}, p_{qop}] = ih I \delta_{q'q},$$

where q_{op} and p_{qop} are any pair of the operators associated with the q-components of position and momentum for a single particle.

14. Consider the angular momentum operators, defined by

$$\mathscr{L}_r \equiv (q_{sop} p_{top} - q_{top} p_{sop}) \qquad (r = x, y, z; s = x, y, z; t = x, y, z)$$

where q_{sop} is the sth component of the position operator, etc., and *rst are to be taken in cyclic order* $(x, y, z; y, z, x; z, x, y)$. Show that

$$[\mathscr{L}_r, \mathscr{L}_s] = ih\mathscr{L}_t \qquad (r, s, t \text{ in cyclic order}).$$

15. Suppose the basis states of a system are represented by

$$\psi_n(x) = \left(\frac{2}{L}\right)^{1/2} \sin\frac{n\pi x}{L} \qquad (0 \leqslant x \leqslant L)$$

What is the mean value of x^2 in the superposition state

$$\psi(x) = \sum_{n=1}^{\infty} a_n \psi_n(x)?$$

16. A one-dimensional electron moves in a "cubical enclosure," subject to

$$V(x) = \begin{cases} V_0 & x < 0 \\ 0 & 0 \leqslant x \leqslant L \\ V_0 & x > L \end{cases}$$

Set up the Schrödinger eigenvalue problem for each value of the potential function and find the eigenvectors for the case $E < V_0$.

95

17. A free, one-dimensional electron in an enclosure of finite extent is represented by the wavefunction

$$\psi(x) = (2L)^{-\frac{1}{2}} \exp\left(ip_0 x/\hbar\right) \qquad (-L \leqslant x \leqslant L).$$

What is the corresponding momentum wavefunction?

18. The square of the absolute value of a momentum wavefunction may be interpreted as the probability that the momentum of a system lies in the range **p** to $(\mathbf{p} \oplus d\mathbf{p})$, where **p** is the argument of the wavefunction. For example, in the foregoing problem we find

$$\varphi(p) = \left(\frac{\hbar}{\pi L}\right)^{\frac{1}{2}} \frac{\sin\left[(p_0 - p)L/\hbar\right]}{(p_0 - p)}$$

to be the momentum wavefunction. The probability density is accordingly $|\varphi(p)|^2$. Draw a graph of this function and interpret it physically:

(a) Can a particle confined to a finite enclosure be associated with a monochromatic de Broglie wave?

(b) What is the most likely range of momenta for the particle in this case?

19. Interpret the free-electron wavefunctions dealt with in Problem 17 in terms of the uncertainty relations by calculating (or estimating) Δx and Δp with the coordinate and momentum probability distributions.

20. Suppose the eigenvalue problem for an electron were

$$\frac{\hbar^2}{2m_e} \frac{d^2\psi}{dx^2} - e^2\delta(x)\psi(x) = E\psi(x) \qquad (-\infty < x < \infty)$$

where e is the electronic charge. Find the value of E for $E < 0$ and the form of the corresponding wavefunction which is continuous for all values of x. (*Hint*: Solve for the wavefunction when $x \neq 0$, remembering that it must be finite throughout its domain so as to be square-integrable. Then integrate the eigenvalue problem between $-\epsilon < x < \epsilon$ and take the limit as $|\epsilon| \to 0$ to find $\psi(0)$. The exact value of $\psi(0)$ is deduced by actually differentiating $\psi(x)$ for $x \neq 0$, evaluating the derivative at $|x| = \epsilon$, and taking the limit. Use the value of $\psi(0)$ so found to calculate the normalization constant for $\psi(x)$ and to compute E.)

CHAPTER 4

The Hydrogen Atom

As the twig's bent, the tree's inclined.

4.1. THE LINE SPECTRUM OF HYDROGEN

Hydrogen gas, when illuminated in a discharge tube, produces a dull red glow that can be resolved into four bright lines in the visible region. (See Figure 26.) This is by far the spectrum of simplest structure, which fact corresponds well with the observation that hydrogen is the lightest chemical element known. It therefore seems natural that the line spectrum of hydrogen was the first to be described quantitatively and that the hydrogen atom was the first to have its internal structure examined in the context of the quantum theory.

After 1860, when Kirchhoff and Bunsen pointed out that each chemical element may be associated uniquely with its line spectrum and that, to a large degree, the line spectrum of a compound could be uniquely decomposed into the spectra of the constituents, a good deal of work was directed toward finding sequential patterns in spectra, the expectation being that such patterns would lead to an understanding of atomic structure. The endeavor had proved

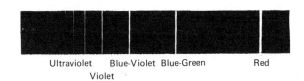

Ultraviolet   Blue-Violet Blue-Green   Red
        Violet

Figure 26. The line spectrum of hydrogen gas in the visible and ultraviolet regions. (From *Introduction to Atomic and Nuclear Physics*, by H. E. White, Copyright © 1964, by Litton Educational Publishing, Inc., by permission of Van Nostrand Reinhold Company.)

to be qualitatively successful by the early 1880's. It had been discovered, for example, that the best way to study line spectra is to arrange them according to wavenumber (the inverse of wavelength). By following this procedure, European scientists had been able to show that the intervals between the three members of any triplet line in the zinc spectrum were independent of wavenumber; that the lines in the spectra of the alkali metals were of decreasing separation as their wavenumbers increased, and that lines were regularly arranged in many spectra.

These discoveries were strong evidence that some numerical expression could be found to describe spectral series. The behavior of the triplet components in the zinc spectrum suggests the subtraction of wavenumbers, while the ordering in alkali metal spectra may be related to whole number indices. These suggestions culminated in 1885 with the publication of a paper by a Swiss mathematician named Johann Balmer.[1] Balmer considered the line spectrum of hydrogen not so much in the light of earlier work on more complex spectra, but as a problem in number theory. The question he sought to answer, was, simply, can an expression involving integers be derived to reproduce the wavelengths of the four lines in the visible spectrum of hydrogen? The result of his study was affirmative; the expression he found is

$$\lambda_n = \lambda_0 \frac{n^2}{n^2 - 4} \qquad (n = 3, 4, 5, 6) \qquad (4.1)$$

where λ_n is the wavelength of the $(n - 2)$th line and $\lambda_0 = 3645.6$ Å. In Table 4.1 a comparison between the observed and calculated wavelengths in the visible region for hydrogen is shown. Balmer's expression is seen to be in very good agreement with experiment. Now, equation (4.1) may be rewritten as a wavenumber expression:

$$\frac{1}{\lambda_{2n}} = R\left(\frac{1}{4} - \frac{1}{n^2}\right) \qquad (n = 3, 4, 5, 6) \qquad (4.2)$$

where $R = 1.0968 \times 10^5$ cm^{-1}. Equation (4.2) is a special case of equation (1.1). The form of the expression is that given it by Rydberg; R is for this reason called the *Rydberg constant*.

The relation between Balmer's expression and the structure of the hydrogen atom becomes the next item for consideration. According to our analysis in Chapter 1, the atom is to be conceived as a microscopic planetary system, with the single electron playing the role of satellite for the nucleus. Experiment

---

[1] Some interesting remarks on Balmer's work are given by L. Banet, Evolution of the Balmer series, *Am. J. Phys.* **34**: 496–503 (1966). Balmer's paper itself appears in translation in W. R. Hindmarsh, *Atomic Spectra*, Pergamon Press, New York, 1967, pp. 101–107.

Table 4.1. A comparison of Balmer's expression with experiment. The wavelength data were taken in air.

Color	n	$\lambda_{calcd.}$ Å	$\lambda_{obsd.}$ Å
Red	3	6562.1	6562.793
Blue-Green	4	4860.8	4861.327
Blue	5	4340.0	4340.466
Violet	6	4101.3	4101.738

has shown that the radiation for this atom must consist of photons, and that these photons must be uniquely related to the energy state of the electron. Moreover, the energy state is reflected in the assignment of fixed dimensions and orientations of the electronic orbits. The first theoretical synthesis of these empirical requirements was accomplished by Niels Bohr in 1913. He based his quantum mechanics on two postulates.

(a) The energy states permitted an electron are prescribed by certain quantum conditions. Specifically, it is required that the angular momentum of the electron be an integral multiple of $\hbar$.
(b) The radiative process consists of a change in the energy state of the electron and the production or absorption of photons.

The quantity of electronic energy involved can never be less than the energy of the concomitant photon; that is,

$$|\Delta E| = h\nu \tag{4.3}$$

where ΔE is the change in electronic energy and ν is the frequency of the photon.

With these postulates and the assumption of circular orbits for the electron in the hydrogen atom, Bohr was able to deduce an expression for the electron-nucleus separation:

$$r_n = \frac{n^2\hbar^2}{\mu_e e^2} \qquad (n = 1, 2, 3, \ldots) \tag{4.4}$$

where μ_e is the reduced mass of the electron and $\hbar$ and e have their usual meanings. The quantity

$$r_1 \equiv a_0 \equiv \frac{\hbar^2}{\mu_e e^2} = 5.29167 \pm 0.00007 \times 10^{-9}\,\text{cm}$$

is accordingly termed the *Bohr radius* for the hydrogen atom. An elementary, classical-mechanical calculation of the total energy of the electron,[2] with the help of equation (4.4), then gives

$$E_n = -\frac{e^2}{2a_0}\left(\frac{1}{n^2}\right) \qquad (n = 1, 2, 3, \ldots). \tag{4.5}$$

The total energies E_n ($n = 1, 2, 3, \ldots$) *are the only ones permitted the electron.* By combining this expression with equation (4.3), we can deduce a relation between the integer n and the wavenumber λ^{-1} of an emitted photon:

$$\frac{1}{\lambda_{n_0 n}} = \frac{e^2}{2a_0 hc}\left(\frac{1}{n_0^2} - \frac{1}{n^2}\right) \qquad (n > n_0,\, n_0 = 1, 2, 3, \ldots) \tag{4.6}$$

where n_0 is associated with the final energy state and n is associated with the initial state. By comparison with equation (4.2), we conclude that the Rydberg constant must be

$$R_H = \frac{e^2}{2a_0 hc} = \frac{(4.80298)^2 \times 10^6}{(2)(5.29167)(6.6256)(2.997925)}$$
$$= 1.09678 \times 10^5 \,\text{cm}^{-1}$$

in complete agreement with experiment.

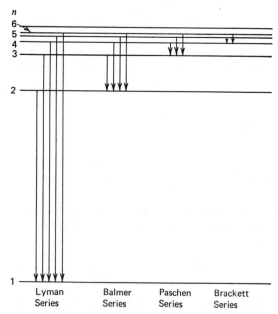

Figure 27. Some of the spectral series for hydrogen.

<hr />

[2] Detailed accounts of this derivation can be found in almost all texts on elementary quantum theory. Bohr's original work is reprinted in W. R. Hindmarsh, *Atomic Spectra*, Pergamon Press, New York, 1967, pp. 117–144.

The Bohr theory fully explains the line spectrum of hydrogen in the visible region if we put $n_0 = 2$. We can say, then, that *the visible radiation from hydrogen is the result of electronic transitions to the energy state associated with the quantum number "two" from energy states associated with the quantum numbers "three," "four," "five,"* and *"six."* Other values of n_0 and n should give rise to other portions of the spectrum, presumably in the infrared and ultraviolet regions. This is indeed the case. Other series of spectral lines for hydrogen are the Lyman series ($n_0 = 1$), the Paschen series ($n_0 = 3$), and the Brackett series ($n_0 = 4$). The first lies entirely in the ultraviolet, the second in the infrared, and the third in the far infrared region.

Despite this successful treatment of the hydrogen spectrum, Bohr's quantum theory encountered serious difficulties not many years after its inception. For example, to get agreement with experiment, half-integral values of the quantum numbers had to be introduced arbitrarily; in certain cases, such as the line spectrum of helium, no agreement could be had by any ad hoc means. Because of this failure, a new theory had to be devised. The new theory, in part, was presented in Chapter 3. It is our task now to take up the study of the hydrogen atom in this fresh context.

4.2. THE SCHRÖDINGER EQUATION FOR THE HYDROGEN ATOM

In keeping with our earliest conception of atomic structure, we imagine the hydrogen atom to be composed of a single electron whirling about a central nucleus—the proton. The Hamiltonian function for this system, according to equation (3.52), must be

$$H(\mathbf{p}_e, \mathbf{p}_n, \mathbf{r}_e, \mathbf{r}_n) = \frac{p_e{}^2}{2m_e} + \frac{p_n{}^2}{2m_n} - \frac{e^2}{|\mathbf{r}_e - \mathbf{r}_n|} \tag{4.7}$$

where $\mathbf{p}_e$ is the momentum of the electron, $\mathbf{p}_n$ is that of the nucleus (proton), m refers to mass, and $e^2/|\mathbf{r}_e - \mathbf{r}_n|$ is the familiar coulomb potential energy. It is certainly an assumption on our part that the electron-proton potential function is specified by the coulomb expression, but we shall adopt this tack nonetheless and hope the result is in agreement with experiment. By introducing center-of-mass and relative coordinates into equation (4.7), we find

$$H(\mathbf{P}, \mathbf{p}, \mathbf{r}) = \frac{P^2}{2M} + \frac{p^2}{2\mu_e} - \frac{e^2}{r} \tag{4.8}$$

where, as usual,

$$\mathbf{P} = \mathbf{p}_e + \mathbf{p}_n$$
$$\mathbf{p} = (\mathbf{v}_e - \mathbf{v}_n)\mu_e$$

$$\mathbf{r} = \mathbf{r}_e - \mathbf{r}_n$$
$$M = m_e + m_n$$
$$\mu_e = \frac{m_e m_n}{(m_e + m_n)}.$$

The first term on the right-hand side of equation (4.8) refers to the motion of the atom as a whole ($\mathbf{P}$ is the total momentum!), while the second and third terms represent the motion of the electron relative to that of the proton. It is the relative motion that interests us now.

The "relative" Hamiltonian function

$$H(\mathbf{p}, \mathbf{r}) = \frac{p^2}{2\mu_e} - \frac{e^2}{r} \tag{4.9}$$

may be used to construct the Schrödinger eigenvalue equation once the operators appropriate to $\mathbf{p}$ and $\mathbf{r}$ have been found. With $\mathbf{r}$, according to the Schrödinger postulate, we associate the operator $\mathbf{r}_{op}$. With p^2 we must associate $-\hbar^2 \nabla^2$, where

$$\nabla^2 = \frac{\partial^2}{\partial x^2} + \frac{\partial^2}{\partial y^2} + \frac{\partial^2}{\partial z^2}.$$

This form of ∇^2, the Laplacian operator, is not particularly suitable in the present case, however, because the hydrogen atom exhibits spherical rather than rectangular symmetry. Thus we should write down ∇^2 in terms of the

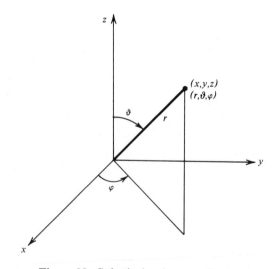

Figure 28. Spherical polar coordinates.

spherical polar coordinates (r, ϑ, φ) and not the cartesian coordinates (x, y, z). (See Figure 28.) Now, in general,[3]

$$\nabla_q^2 = (Q_1 Q_2 Q_3)^{-1} \left\{ \frac{\partial}{\partial q_1} \left[\frac{Q_2 Q_3}{Q_1} \frac{\partial}{\partial q_1} \right] + \frac{\partial}{\partial q_2} \left[\frac{Q_1 Q_3}{Q_2} \frac{\partial}{\partial q_2} \right] + \frac{\partial}{\partial q_3} \left[\frac{Q_1 Q_2}{Q_3} \frac{\partial}{\partial q_3} \right] \right\}$$
(4.10)

where the q_i $(i = 1, 2, 3)$ are any three orthogonal coordinates and

$$Q_i = \left[\left(\frac{\partial x}{\partial q_i} \right)^2 + \left(\frac{\partial y}{\partial q_i} \right)^2 + \left(\frac{\partial z}{\partial q_i} \right)^2 \right]^{\frac{1}{2}} \quad (i = 1, 2, 3).$$

In spherical polar space we have

$$x = r \sin \vartheta \cos \varphi, \quad (0 \leqslant r \leqslant \infty)$$
$$y = r \sin \vartheta \sin \varphi, \quad (0 \leqslant \vartheta \leqslant \pi)$$
$$z = r \cos \vartheta, \quad (0 \leqslant \varphi \leqslant 2\pi)$$

so that

$$Q_1 = Q_r = 1$$
$$Q_2 = Q_\vartheta = r$$
$$Q_3 = Q_\varphi = r \sin \vartheta$$

and

$$\nabla_r^2 = \frac{1}{r^2} \frac{\partial}{\partial r} \left(r^2 \frac{\partial}{\partial r} \right) + \frac{1}{r^2 \sin \vartheta} \frac{\partial}{\partial \vartheta} \left(\sin \vartheta \frac{\partial}{\partial \vartheta} \right) + \frac{1}{r^2 \sin^2 \vartheta} \frac{\partial^2}{\partial \varphi^2} \quad (4.11)$$

Using equations (4.9) and (4.11) and the Schrödinger postulate, we can now write down

$$-\frac{\hbar^2}{2\mu_e} \left\{ \frac{1}{r^2} \frac{\partial}{\partial r} \left(r^2 \frac{\partial \psi}{\partial r} \right) + \frac{1}{r^2 \sin \vartheta} \frac{\partial}{\partial \vartheta} \left(\sin \vartheta \frac{\partial \psi}{\partial \vartheta} \right) \right.$$
$$\left. + \frac{1}{r^2 \sin^2 \vartheta} \frac{\partial^2 \psi}{\partial \varphi^2} \right\} - \frac{e^2}{r} \psi(\mathbf{r}) = E\psi(\mathbf{r}). \quad (4.12)$$

Equation (4.12) is the Schrödinger eigenvalue problem for the hydrogen atom, as obtained from the classical-mechanical Hamiltonian function.

Now let us try to solve the equation. We begin by *assuming* that the equation is separable; that is, we assume that $\psi(\mathbf{r})$ has the form

$$\psi(\mathbf{r}) \equiv R(r)\theta(\vartheta)\Phi(\varphi).$$

[3] H. Margenau and G. M. Murphy, *The Mathematics of Physics and Chemistry*, D. Van Nostrand Co., Princeton, 1956, Vol. I, p. 175.

The Hydrogen Atom

Upon introducing this definition into the expression we get

$$-\frac{\hbar^2}{2\mu_e}\left\{\frac{1}{R(r)r^2}\frac{d}{dr}\left(r^2\frac{dR}{dr}\right) + \frac{1}{r^2\sin\vartheta\theta(\vartheta)}\frac{d}{d\vartheta}\left(\sin\vartheta\frac{d\theta}{d\vartheta}\right)\right.$$

$$\left. + \frac{1}{r^2\sin^2\vartheta\Phi(\varphi)}\frac{d^2\Phi}{d\varphi^2}\right\} - \left(E + \frac{e^2}{r}\right) = 0.$$

If this equation is multiplied through by $r^2\sin^2\vartheta$, it may be rewritten as

$$\left\{\frac{\sin^2\vartheta}{R(r)}\frac{d}{dr}\left(r^2\frac{dR}{dr}\right) + \frac{\sin\vartheta}{\theta(\vartheta)}\frac{d}{d\vartheta}\left(\sin\vartheta\frac{d\theta}{d\vartheta}\right)\right\} + \frac{2\mu_e}{\hbar^2}\left[\left(E + \frac{e^2}{r}\right)r^2\sin^2\vartheta\right]$$

$$= -\frac{1}{\Phi(\varphi)}\frac{d^2\Phi}{d\varphi^2}. \quad (4.13)$$

Because the two sides of equation (4.13) are functions of different variables (the one side, of r and ϑ; the other, of φ alone), the equality can hold only if

$$-\frac{1}{\Phi(\varphi)}\frac{d^2\Phi}{d\varphi^2} \equiv m^2 \quad (4.14)$$

where m^2 is a constant. Equation (4.13), then, factors into equation (4.14) and

$$-\frac{\hbar^2}{2\mu_e}\left\{\frac{1}{r^2R(r)}\frac{d}{dr}\left(r^2\frac{dR}{dr}\right) + \frac{1}{r^2\sin\vartheta\theta(\vartheta)} - \frac{m^2}{r^2\sin^2\vartheta}\right\} - \left(E + \frac{e^2}{r}\right) = 0.$$

$$(4.15)$$

It is not difficult to see that equation (4.15) may also be factored; for we can write

$$\left\{\frac{1}{R(r)}\frac{d}{dr}\left(r^2\frac{dR}{dr}\right)\right\} + \frac{2\mu_e}{\hbar^2}r^2\left(E + \frac{e^2}{r}\right) \quad \cdot \text{Missing}$$

$$= \frac{-1}{\sin\vartheta\theta(\vartheta)}\frac{d}{d\vartheta}\left(\sin\vartheta\frac{d\theta}{d\vartheta}\right) + \frac{m^2}{\sin^2\vartheta} \equiv \beta$$

which implies

$$\left[\frac{1}{r^2}\frac{d}{dr}\left(r^2\frac{dR}{dr}\right)\right] + \frac{2\mu_e}{\hbar^2}\left[\left(E + \frac{e^2}{r}\right) - \frac{\beta}{r^2}\right]R(r) = 0 \quad (4.16)$$

$$\frac{1}{\sin\vartheta}\frac{d}{d\vartheta}\left(\sin\vartheta\frac{d\theta}{d\vartheta}\right) - \left[\frac{m^2}{\sin^2\vartheta} - \beta\right]\theta(\vartheta) = 0 \quad (4.17)$$

where β is a constant. The problem of the hydrogen atom is thus reduced to solving equations (4.14), (4.16), and (4.17), and calculating the eigenvalues m, β, and E.

4.3. THE SOLUTIONS OF THE ANGULAR EQUATIONS OF MOTION

We might well inquire before going on as to what is the physical significance of the eigenvalues m and β. Because they are not associated with a differential equation containing a term in the coordinate r, we know that they are not energy eigenvalues. Indeed, they must be angular momentum eigenvalues, since the operators to which they belong were derived from the kinetic energy part of the Hamiltonian function. Let us prove this supposition.

The angular momentum in Newtonian mechanics is defined by

$$\mathbf{L} = \mathbf{r} \times \mathbf{p}.$$

According to the Schrödinger postulate, the quantum-theoretical operator associated with $\mathbf{L}$ must then be

$$\mathscr{L} \equiv \mathbf{r} \times (-i\hbar\nabla)$$

where

$$\nabla = \hat{\mathbf{x}}\frac{\partial}{\partial x} \oplus \hat{\mathbf{y}}\frac{\partial}{\partial y} \oplus \hat{\mathbf{z}}\frac{\partial}{\partial z}.$$

It is not difficult to see that

$$\mathscr{L}_x = i\hbar\left(z\frac{\partial}{\partial y} - y\frac{\partial}{\partial z}\right)$$

$$\mathscr{L}_y = i\hbar\left(x\frac{\partial}{\partial z} - z\frac{\partial}{\partial x}\right)$$

$$\mathscr{L}_z = i\hbar\left(y\frac{\partial}{\partial x} - x\frac{\partial}{\partial y}\right).$$

With the help of the equations relating x, y, and z to the spherical polar coordinates, we can also write

$$\mathscr{L}_x = i\hbar\left(\sin\varphi\frac{\partial}{\partial\vartheta} + \cot\vartheta\cos\varphi\frac{\partial}{\partial\varphi}\right)$$

$$\mathscr{L}_y = i\hbar\left(\cot\vartheta\sin\varphi\frac{\partial}{\partial\varphi} - \cos\varphi\frac{\partial}{\partial\vartheta}\right) \tag{4.18}$$

$$\mathscr{L}_z = -i\hbar\frac{\partial}{\partial\varphi}$$

so that

$$\mathscr{L}^2 = -\hbar^2\left\{\frac{1}{\sin\vartheta}\frac{\partial}{\partial\vartheta}\left(\sin\vartheta\frac{\partial}{\partial\vartheta}\right) + \frac{1}{\sin^2\vartheta}\frac{\partial^2}{\partial\varphi^2}\right\}. \tag{4.19}$$

The Hydrogen Atom

Equation (4.19) is a representation, in spherical polar space, of the operator associated with the square of the total angular momentum. By comparison of this expression and equations (4.18) with equations (4.14) and (4.17), we see that the angular Schrödinger equations for the hydrogen atom may be rewritten as

$$\mathcal{L}_z^2 \Phi_m(\varphi) = (m\hbar)^2 \Phi_m(\varphi) \tag{4.20}$$

and

$$\mathcal{L}^2 \theta_\beta(\vartheta) = \beta \hbar^2 \theta_\beta(\vartheta). \tag{4.21}$$

The physical meaning of the two eigenvalues is clear now. The eigenvalue $m\hbar$ is the average value of the angular momentum about the z-axis for the electron in hydrogen, while $\beta\hbar^2$ is the average value of the square of the total angular momentum. That these two quantities appear in the theory is to be expected on the basis of the correspondence postulate, since it follows from Newtonian mechanics that the angular momentum of a particle subject to a central force is a constant of the motion. *Equation (4.21) may be thought of as the quantum-theoretical statement of the law of conservation of angular momentum for central-force interactions.* It is clear that the conservation law, as in the classical scheme, follows directly from the spherical symmetry of the physical system in question.

Equation (4.20), that is,

$$\hbar^2 \frac{d^2 \Phi_m}{d\varphi^2} + (m\hbar)^2 \Phi_m(\varphi) = 0 \tag{4.20}$$

is a linear, homogeneous differential equation. Its solution follows, by the methods employed in Chapter 3, as

$$\Phi_m(\varphi) = N \exp(im\varphi) \qquad (m = 0, \pm 1, \pm 2),$$

where N is a normalization constant. We have imposed upon m integer values because $\Phi_m(\varphi)$ must be a single-valued function of φ in the domain $(0, 2\pi)$ in order to be an acceptable wavefunction.[4] Otherwise, there could be two values of $\Phi_m(0)$—a difficult thing to interpret physically. The value of N is calculated in the usual way:

$$N^2 \int_0^{2\pi} |\Phi_m(\varphi)|^2 \, d\varphi \equiv 1 = N^2 \int_0^{2\pi} d\varphi = 2\pi N^2.$$

---

[4] It may be well argued that this requirement is too stringent. If it is simply asked that the Φ_m be orthogonal, then it follows that the difference between any two m-values must be an integer. That the m-values must themselves be integers can be shown from symmetry arguments. For details on this problem, see M. Whippman, Orbital angular momentum in quantum mechanics, *Am. J. Phys.* **34**: 656–659 (1966).

Thus we have

$$\Phi_m(\varphi) = (2\pi)^{-\frac{1}{2}} \exp(im\varphi) \qquad (m = 0, \pm 1, \pm 2, \ldots) \qquad (4.22)$$

as the orthonormal solution of equation (4.20).

The other angular equation is not so easy to solve. To begin the task, let us put $z = \cos\vartheta$ in equation (4.21). We then have

$$\frac{d}{dz}\left\{(1 - z^2)\frac{dP(z)}{dz}\right\} + \left\{\beta - \frac{m^2}{(1 - z^2)}\right\}P(z) = 0, \qquad (4.23)$$

where $P(z) \equiv \theta(\vartheta)$ and the domain of $P(z)$ is $-1 \leqslant z \leqslant 1$. Notice that one term of the equation becomes infinite when z takes on either of its extremum values. For this reason the points $z = \pm 1$ are called the *singular points* of the differential equation (4.23). We should like to study the behavior of the function $P(z)$ at these points before proceeding further. If the transformation $\epsilon \equiv (1 - z)$ is inserted into the equation, we find [putting $S(\epsilon) \equiv P(z)$]:

$$\frac{d}{d\epsilon}\left(\epsilon(2 - \epsilon)\frac{dS}{d\epsilon}\right) + \left\{\beta - \frac{m^2}{\epsilon(2 - \epsilon)}\right\}S(\epsilon) = 0. \qquad (4.24)$$

If the solution is fairly regular, it should be expressible as a power series in ϵ:

$$S(\epsilon) = \sum_{n=0}^{\infty} a_n \epsilon^{n+c} \qquad (4.25)$$

where a_n and c are constants. Substituting equation (4.25) into equation (4.24), we get

$$\epsilon(2 - \epsilon)\sum_{n=0}^{\infty}(n + c)(n + c - 1)a_n\epsilon^{n+c-2}$$

$$+ 2(1 - \epsilon)\sum_{n=0}^{\infty}(n + c)a_n\epsilon^{n+c-1}$$

$$+ \left\{\beta - \frac{m^2}{\epsilon(2 - \epsilon)}\right\}\sum_{n=0}^{\infty}a_n\epsilon^{n+c} = 0$$

or

$$\sum_{n=0}^{\infty}[(n + c)(n + c - 1) + 2(n + c) - \beta]a_n\epsilon^{n+c+2}$$

$$+ \sum_{n=0}^{\infty}[2\beta - 6(n + c) - 4(n + c)(n + c - 1)]a_n\epsilon^{n+c+1}$$

$$+ \sum_{n=0}^{\infty}[4(n + c)(n + c - 1) + 4(n + c) - m^2]a_n\epsilon^{n+c} = 0$$

upon collecting all terms having common powers of ϵ. If we put $n \to n - 2$ in the first sum and $n \to n - 1$ in the second, everything in the equation becomes a coefficient of ϵ^{n+c}:

$$\sum_{n=2}^{\infty} [(n + c - 2)(n + c - 3) + 2(n + c - 2) - \beta]a_{n-2}\epsilon^{n+c}$$

$$+ \sum_{n=1}^{\infty} [2\beta - 6(n + c - 1) - 4(n + c - 1)$$

$$- 4(n + c - 1)(n + c - 2)]a_{n-1}\epsilon^{n+c}$$

$$+ \sum_{n=0}^{\infty} [4(n + c)(n + c - 1) + 4(n + c) - m^2]a_n\epsilon^{n+c} = 0.$$

This equation must be satisfied identically, which means that the sum of all coefficients of ϵ^{n+c} for a given n value must vanish. For $n = 0$ we require

$$(4c(c - 1) + 4c - m^2)a_0 = 0$$

which, in turn, implies that $c = |m|/2$ for a_0 arbitrary. This very same result is obtained if we put $\eta \equiv (1 + z)$ in equation (4.23) and repeat the foregoing analysis.

It seems that a regular solution of equation (4.23) may be found if we put

$$P(z) \equiv \epsilon^{|m|/2}\eta^{|m|/2}F(z) = (1 - z^2)^{|m|/2}F(z).$$

The differential equation for $F(z)$ is then

$$(1 - z^2)\frac{d^2F}{dz^2} - 2(|m| + 1)z\frac{dF}{dz} + [\beta - |m|(|m| + 1)]F(z) = 0 \quad (4.26)$$

where we assume

$$F(z) = \sum_{n=0}^{\infty} c_n z^n. \quad (4.27)$$

The substitution of equation (4.27) into equation (4.26) leads to

$$\sum_{n=-2}^{\infty} c_{n+2}(n + 1)(n + 2)z^n$$

$$- \sum_{n=0}^{\infty} [n(n - 1) + 2n(|m| + 1) - \beta + |m|(|m| + 1)]c_n z^n = 0.$$

The coefficients of z^n must in general satisfy

$$(n + 1)(n + 2)c_{n+2} - [n(n - 1) + 2n(|m| + 1) - \beta + |m|(|m| + 1)]c_n = 0$$

or

$$c_{n+2} = \frac{(n + |m|)(n + |m| + 1) - \beta}{(n + 1)(n + 2)} c_n$$

Now, we notice that

$$\gamma_n = \frac{c_{n+2}}{c_n} \sim \left(1 + \frac{|m|}{n}\right)\left(1 + \frac{|m| + 1}{n}\right) \sim 1 + 2\frac{(|m| + 1)}{n}$$

as n gets very large. This ratio is the same as that for the corresponding coefficients in the series expansion of

$$(1 - z^2)^{-|m|},$$

a function which obviously diverges when $z = \pm 1$. This means that the series (4.27) will not converge for $z = \pm 1$ and, therefore, that it must be terminated at some value of $n \equiv n_{max} < +\infty$. We have no choice but to impose the conditions

1) $c_n = 0$, $n > n_{max}$

$$\boxed{c_n = 0,} \qquad n \geqslant n_{max} \qquad 2)\ \gamma_n = 0,\ n \geqslant n_{max}$$

$$\beta \equiv (n_{max} + |m|)(n_{max} + |m| + 1) \equiv l(l + 1), \qquad (4.28)$$

where $l \equiv (n_{max} + |m|) = |m|, |m| + 1, \ldots = 0, 1, \ldots$, is called the *angular quantum number*. From equation (4.21) we conclude that

$$\langle \mathscr{L}^2 \rangle^{\frac{1}{2}} = (l(l + 1))^{\frac{1}{2}}\hbar \qquad (l = 0, 1, 2, 3, \ldots). \qquad (4.29)$$

The root-mean-square total angular momentum of the electron in an eigenstate of the hydrogen atom can take on only the values prescribed by equation (4.29). The z-axis component of the total angular momentum, because of the condition (4.28), is likewise limited to certain multiples of $\hbar$:

$$\langle \mathscr{L}_z \rangle = m\hbar \qquad (|m| = 0, 1, \ldots, l - 1, l). \qquad (4.30)$$

Nothing definite can be said about the x- and y-components of the total angular momentum because the functions $\Phi_m(\varphi)$ and $\theta_\beta(\vartheta)$ are not eigenfunctions corresponding to $\mathscr{L}_x$ and $\mathscr{L}_y$. Insofar as $\langle \mathscr{L}_x \rangle$ and $\langle \mathscr{L}_y \rangle$ are concerned, the normal state of the electron in the hydrogen atom is always a superposition state, while for $\langle \mathscr{L}^2 \rangle$ and $\langle \mathscr{L}_z \rangle$ this is not so.

Our analysis thus far is not sufficient to provide an explicit calculation of $F(z)$. A good bit of manipulation,[5] too tedious to reproduce here (but mathematically important), shows that

$$F(z) = \frac{d^{|m|}}{dz^{|m|}} P_l(z)$$

[5] H. Margenau and G. M. Murphy, *The Mathematics of Physics and Chemistry*, D. Van Nostrand Co., Princeton, 1956, Vol. I, pp. 61–69, 98–109.

where

$$P_l(z) = (2^l l!)^{-1} \frac{d^l}{dz^l} (z^2 - 1)^l.$$

The polynomial $P_l(z)$ is called a *Legendre polynomial*. The complete but unnormalized solution of equation (4.23) is now

$$P_l^m(z) = (1 - z^2)^{|m|/2} \frac{d^{|m|}}{dz^{|m|}} P_l(z).$$

The normalization constant is calculated in the usual way:[5]

$$\int_{-1}^{1} P_l^m(z) P_{l'}^m(z) \, dz = \frac{(l + |m|)!}{(l - |m|)!} \frac{2}{2l + 1} \delta_{ll'} \equiv N^{-2} \delta_{ll'}.$$

The orthonormal solution of equation (4.16) is finally

$$\theta_l^m(z) = \left\{ \frac{2l + 1}{2} \frac{(l - |m|)!}{(l + |m|)!} \right\}^{1/2} P_l^m(z). \tag{4.31}$$

4.4. THE SOLUTION OF THE RADIAL EQUATION OF MOTION

The radial equation of motion may now be written

$$\frac{d^2 R}{dr^2} + \frac{2}{r} \frac{dR}{dr} + \left\{ \frac{2\mu_e}{\hbar^2} \left(E + \frac{e^2}{r} \right) - \frac{l(l + 1)}{r^2} \right\} R(r) = 0. \tag{4.32}$$

If we put

$$k^2 = -\frac{2\mu_e E}{\hbar^2} \qquad \alpha = \frac{\mu_e e^2}{\hbar^2 k} \qquad \rho = 2kr$$

we can rewrite equation (4.32) as

$$\frac{d^2 S}{d\rho^2} + \frac{2}{\rho} \frac{dS}{d\rho} + \left\{ \frac{\alpha}{\rho} - \frac{1}{4} - \frac{l(l + 1)}{\rho^2} \right\} S(\rho) = 0 \tag{4.33}$$

where $S(\rho) \equiv R(r)$. Let us examine equation (4.33) for large values of ρ. In this case the expression reduces to

$$\frac{d^2 S}{d\rho^2} - \tfrac{1}{4} S(\rho) = 0 \tag{4.34}$$

which is a linear, homogeneous differential equation, as studied in Chapter 3. The solutions, by the usual method, are

$$S(\rho) = \exp(\pm \rho/2).$$

We choose to continue with $S(\rho) \sim \exp(-\rho/2)$ since the radial wavefunction must remain finite for large values of r. In the same way as for the angular equation, we put

$$S(\rho) = \exp(-\rho/2)F(\rho)$$

where

$$F(\rho) = \sum_{m=0}^{\infty} a_m \rho^{m+s},$$

the a_m $(m = 1, 2, \ldots)$ and s being constants. The equation for the a_m $(m = 1, 2, \ldots)$ is then

$$\sum_{m=0}^{\infty} [(m+s)(m+s-1) + 2(m+s) - l(l+1)]a_m \rho^{m+s-2}$$

$$+ \sum_{m=1}^{\infty} [\alpha - (m+s) - 1]a_{m-1} \rho^{m+s-2} = 0.$$

The first summation, for a_0 arbitrary and non-vanishing, implies

$$s^2 + s = l(l+1).$$

The solutions of this equation are

$$s = \begin{cases} l \\ -(l+1). \end{cases}$$

The second solution must be rejected for $l \geqslant 1$ because

$$S(\rho) \underset{\rho \to 0}{\sim} \rho^{-l-1}$$

and so is unnormalizable; for $l = 0$ the solution is not acceptable since

$$S(\rho) \underset{\rho \to 0}{\sim} \rho^{-1}$$

is singular enough to obliterate the first two terms in equation (4.33) and thus make the equation unsatisfiable for small ρ. Therefore, we keep only the positive-exponent solution.

Now let us put

$$S(\rho) \equiv \exp(-\rho/2)\rho^l L(\rho)$$

into equation (4.33) to get a differential equation for $L(\rho)$. We find

$$\rho \frac{d^2 L}{d\rho^2} + \{2(l+1) - \rho\}\frac{dL}{d\rho} + (\alpha - l - 1)L(\rho) = 0. \tag{4.35}$$

The Hydrogen Atom

By putting

$$L(\rho) = \sum_{m=0}^{\infty} a_m \rho^m$$

the recursion formula

$$a_{m+1} = \frac{(\alpha - l - 1 - m)}{[2(m+1)(l+1) + m(m+1)]} a_m$$

is found. For large values of m,

$$\left| \frac{a_{m+1}}{a_m} \right| \sim \frac{1}{m+1}.$$

But this is just the way the coefficients in the power series expansion for $\exp(\rho)$ behave:

$$\exp(\rho) = \sum_{m=0}^{\infty} \frac{1}{m!} \rho^m.$$

Because the multiplier of $L(\rho)$ is $[\exp(-\rho/2)]\rho^l$ we must have

$$\exp(-\rho/2)\rho^l L(\rho) \xrightarrow[\rho \to \infty]{} \exp(\rho/2) \to \infty.$$

This is certainly unsatisfactory. It appears we must once again terminate the series:

$$\alpha - l - 1 - m_{\max} \equiv 0 \equiv \alpha - n$$

where the principal quantum number n can take on the values $l + 1, l + 2, \ldots$ By definition of α,

$$\frac{\mu_e e^2}{\hbar^2} \left(\frac{-2\mu_e E}{\hbar^2} \right)^{-\frac{1}{2}} = n$$

or

$$E_n = -\frac{\mu_e e^4}{2\hbar^2 n^2} \qquad (n = 1, 2, \ldots), \qquad (4.36)$$

which is identical with equation (4.5), derived from the Bohr theory.

A little more work shows that[6]

$$L(\rho) \equiv L_{n+l}^{2l+1}(\rho) = \frac{d^{(2l+1)}}{d\rho^{(2l+1)}} \left[\exp(\rho) \frac{d^{(n+l)}}{d\rho^{(n+l)}} \left(\rho^{n+l} \exp(-\rho) \right) \right]$$

where $L_{n+l}^{2l+1}(\rho)$ is known as the *associated Laguerre polynomial* of degree $(n - l - 1)$ and order $(n + l)$. The normalized solution of equation (4.32) is thus

$$R_{nl}(r) = N \exp(-\rho/2)\rho^l L_{n+l}^{2l+1}(\rho).$$

[6] H. Margenau and G. M. Murphy, *The Mathematics of Physics and Chemistry*, D. Van Nostrand Co., Princeton, 1956, Vol. I, pp. 77f; 126–130.

The normalization constant is found in the usual way:[6]

$$N^2 \int_0^\infty R_{n'l}(r)R_{nl}(r)r^2\,dr = \frac{N^2}{8}\left(\frac{n\hbar^2}{\mu_e e^2}\right)^3 \int_0^\infty \exp(-\rho)\rho^{2l}L_{n'+l}^{2l+1}(\rho)\rho^2 L_{n+l}^{2l+1}(\rho)\,d\rho$$

$$= \frac{N^2}{8}\left(\frac{n\hbar^2}{\mu_e e^2}\right)^3 \frac{[(n+l)!]^3}{(n-l-1)!}(2n)\delta_{nn'}$$

so that

$$R_{nl}(r) = \left\{\left(\frac{2}{na_0}\right)^3 \frac{(n-l-1)!}{2n[(n+l)!]^3}\right\}^{\frac{1}{2}} \exp(-\rho/2)\rho^l L_{n+l}^{2l+1}(\rho) \qquad (4.37)$$

is the orthonormal solution of equation (4.32), a_0 being the Bohr radius of the hydrogen atom.

Our analysis of the Schrödinger eigenvalue problem for the hydrogen atom has involved two equations of the general form

$$p(x)\frac{d^2\psi}{dx^2} + \frac{dp}{dx}\frac{d\psi}{dx} - q(x)\psi + \lambda\psi(x)w(x) = 0. \qquad (4.38)$$

These are equations (4.23) and (4.33). To see this we need only put

$$p(z) = 1 - z^2, \qquad q(z) = \frac{m^2}{1-z^2}, \qquad \lambda = \beta = l(l+1), \qquad w(z) = 1$$

in equation (4.23), and put, after multiplication through by ρ,

$$p(\rho) = \rho^2, \qquad q(\rho) = \frac{\rho^2 + 4l\rho + (2l+1)^2 - 1}{4}$$

$$\lambda = \alpha + l = n + l \qquad w(\rho) = \rho$$

in equation (4.33). Differential equations of the form of equation (4.38) are called *equations of the Sturm-Liouville type*. Sturm-Liouville theory forms an important part of the study of partial differential equations. Its significance for quantum physics lies with the fact that some very general statements can be made about the solutions of equation (4.38), statements which insure that whenever the Schrödinger eigenvalue problem can be reduced to a Sturm-Liouville problem, the wavefunction found will meet the criteria set forth in section 3.4. Specifically, it can be shown that:

(a) The eigenfunctions form an orthonormal set.
(b) Every continuous function which has piecewise continuous first and second derivatives, and which satisfies the boundary conditions of the eigenvalue problem (that is, any coordinate representative of the state vector), can be expanded in a series of the eigenfunctions.

The proof of these statements, along with some remarks on the Sturm-Liouville problem, are given in section A.3 of the Appendix.

4.5. HYDROGENIC EIGENFUNCTIONS AND SPECTRA

The wavefunction for the electron in hydrogen is, according to equations (4.12), (4.22), (4.31), and (4.37),

$$\psi_{nlm}(\mathbf{r}) = \left\{ \left(\frac{2}{na_0}\right)^3 \frac{(n-l-1)!}{2n[(n+l)!]^3} \frac{(2l+1)(l-|m|)!}{4\pi(l+|m|)!} \right\}^{\frac{1}{2}}$$
$$\times \exp(-\rho/2)\rho^l L_{n+l}^{2l+1}(\rho) P_l^m(\cos\vartheta) \exp(im\varphi) \qquad (4.39)$$

where $\rho = (2/na_0)r$. If we were to neglect non-coulombic electron-nucleus interactions, electron-electron interactions, and relativistic effects, equation (4.39) might represent a good approximation to the state of an electron in *any* atom. All we need do is put $a_0 \rightarrow a_0/Z$, where Z is the atomic number. Of course, it is not realistic to neglect all these interactions; $\psi_{nlm}(\mathbf{r})$ is really not a very good wavefunction, except for one-electron atoms. This fact should not deter us from making a careful study of the hydrogenic wavefunction, however, since the wavefunction will in all likelihood be the starting point for a more quantitative formulation of the theory of many-electron atoms. So let us begin the analysis.

First, it must be understood that there are $(2l+1)$ angular wavefunctions $\theta_l^m(\vartheta)\Phi_m(\varphi)$ for every one of the eigenvalues $l(l+1)$, simply because there are that many values of m for each l: $m = -l, -l+1, \ldots, -1, 0, 1, 2, \ldots, l-1, l$. The angular wavefunctions, then, span a $(2l+1)$-dimensional subspace. Any superposition state of the electron, corresponding to a fixed total angular momentum and to any z-component of the angular momentum between $-l\hbar$ and $l\hbar$, can be represented by

$$\Psi_l(\vartheta, \varphi) = \sum_{m=-l}^{l} a_m \left\{ \frac{(2l+1)(l-|m|)!}{4\pi(l+|m|)!} \right\}^{\frac{1}{2}} P_l^m(\cos\vartheta) \exp(im\varphi) \quad (4.40)$$

where, as usual,

$$a_m = \left\{ \frac{(2l+1)(l-|m|)!}{4\pi(l+|m|)!} \right\}^{\frac{1}{2}} \int_0^{2\pi} \int_0^{\pi} P_l^m(\cos\vartheta)$$
$$\times \exp(-im\varphi)\Psi_l(\vartheta, \varphi) \sin\vartheta \, d\vartheta \, d\varphi$$

and

$$\|\Psi_l\| = \sqrt{\sum_{m=-l}^{l} |a_m|^2}.$$

The mean value of the angular momentum about the z-axis follows from equation (3.59):

$$\langle \mathscr{L}_z \rangle_l = \sum_{m=-l}^{l} \frac{|a_m|^2 m}{\|\Psi_l\|^2} \hbar \qquad (4.41)$$

where we have invoked the orthonormality of the angular wavefunction.

The function

$$(-1)^{\frac{1}{2}(m+|m|)}\left\{\frac{(2l+1)(l-|m|)!}{4(l+|m|)!}\right\}^{\frac{1}{2}}P_l^m(\cos\vartheta)\exp(im\varphi) \equiv \Upsilon_l^m(\vartheta,\varphi)$$

is known in classical physics as a *spherical harmonic*. The spherical harmonics, aside from a nonessential (for quantum physics) factor $(-1)^{\frac{1}{2}(m+|m|)}$ are the angular wavefunctions for the hydrogen atom. The first few spherical harmonics are:

$$\Upsilon_0^0(\vartheta,\varphi)=(4\pi)^{-\frac{1}{2}}$$

$$\Upsilon_1^0(\vartheta,\varphi)=\left(\frac{3}{4\pi}\right)^{\frac{1}{2}}\cos\vartheta$$

$$\Upsilon_1^{\pm1}(\vartheta,\varphi)=\mp\left(\frac{3}{8\pi}\right)^{\frac{1}{2}}\sin\vartheta\exp(\pm im\varphi)$$

$$\Upsilon_2^0(\vartheta,\varphi)=\left(\frac{5}{16\pi}\right)^{\frac{1}{2}}(3\cos^2\vartheta-1)$$

$$\Upsilon_2^{\pm1}(\vartheta,\varphi)=\mp\left(\frac{15}{8\pi}\right)^{\frac{1}{2}}\cos\vartheta\sin\vartheta\exp(\pm i\varphi)$$

$$\Upsilon_2^{\pm2}(\vartheta,\varphi)=\left(\frac{15}{32\pi}\right)^{\frac{1}{2}}\sin^2\vartheta\exp(\pm 2i\varphi)$$

$$\Upsilon_3^0(\vartheta,\varphi)=\left(\frac{7}{16\pi}\right)^{\frac{1}{2}}(5\cos^3\vartheta-3\cos\vartheta)$$

$$\Upsilon_3^{\pm1}(\vartheta,\varphi)=\mp\left(\frac{21}{64\pi}\right)^{\frac{1}{2}}\sin\vartheta(5\cos^2\vartheta-1)\exp(\pm i\varphi)$$

$$\Upsilon_3^{\pm2}(\vartheta,\varphi)=\left(\frac{105}{32\pi}\right)^{\frac{1}{2}}\sin^2\vartheta\cos\vartheta\exp(\pm 2i\varphi)$$

$$\Upsilon_3^{\pm3}(\vartheta,\varphi)=\mp\left(\frac{35}{64\pi}\right)^{\frac{1}{2}}\sin^3\vartheta\exp(\pm 3i\varphi)$$

The probability of finding the hydrogenic electron oriented about the nucleus such that its angular coordinates are between ϑ and $\vartheta+d\vartheta$ and φ and $\varphi+d\varphi$ is given by

$$P_{ml}(\vartheta,\varphi)\sin\vartheta\,d\vartheta\,d\varphi \equiv |\Upsilon_l^m(\vartheta,\varphi)|^2\sin\vartheta\,d\vartheta\,d\varphi. \tag{4.42}$$

$P_{ml}(\vartheta,\varphi)$ may be called the angular probability function. In Figure 29 the angular probability function is plotted in the domain of ϑ for several values of m and l. We notice that the electron moves in planes more nearly perpendicular with the z-axis as the z-component of the angular momentum

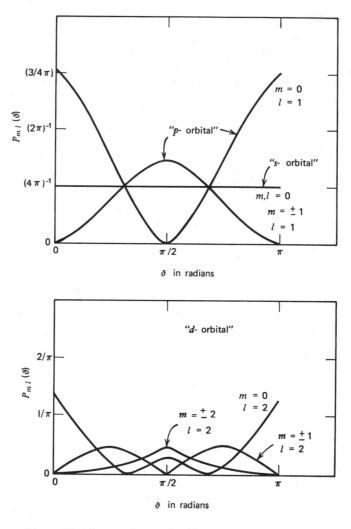

Figure 29. The angular probability function for an electron in the three lowest angular momentum eigenstates in the hydrogen atom.

increases, for a given value of l. This is the expected result, since the total angular momentum is a constant of the motion.

The radial *hydrogenic* wavefunction

$$R_{nl}(r) = \left\{ \left(\frac{2Z}{na_0}\right)^3 \frac{(n-l-1)}{2n[(n+l)!]^3} \right\}^{\frac{1}{2}} \exp\left(-2Zr/na_0\right)\left(\frac{2Zr}{na_0}\right)^l L_{n+l}^{2l+1}\left(\frac{2Zr}{na_0}\right)$$

116

corresponds to the energy eigenvalue

$$E_n = -\frac{\mu_e Z^2 e^4}{2\hbar^2 n^2} \quad (n = 1, 2, 3, \ldots) \tag{4.36a}$$

The multiplicity of E_n is that of the principal quantum number n and that of the angular quantum number l. For each l, the multiplicity is $(2l + 1)$; for each n, l can take on values from zero to $(n - 1)$. Therefore, the multiplicity of E_n is given by

$$\text{multiplicity} = \sum_{l=0}^{n-1} (2l + 1) = n(n - 1) + n = n^2$$

where we have summed the arithmetical series in the usual way. The hydrogenic wavefunctions $\psi_{nlm}(\mathbf{r})$ span, therefore, an n^2-dimensional subspace. Any superposition state of the electron corresponding to a fixed total energy, to the z-component of the angular momentum between $(1 - n)\hbar$ and $(n - 1)\hbar$, and to the total angular momentum between zero and $\sqrt{n(n - 1)}\,\hbar$ can be represented as

$$\Psi_n(r, \vartheta, \varphi) = \sum_{l=0}^{n-1} \sum_{m=-l}^{l} c_l^m R_{nl}(r) Y_l^m(\vartheta, \varphi) \tag{4.43}$$

where

$$c_l^m = \int_0^{2\pi} \int_0^{\pi} \int_0^{\infty} R_{nl}(r) \overline{Y}_l^m(\vartheta, \varphi) \Psi_n(r, \vartheta, \varphi) r^2 \, dr \sin \vartheta \, d\vartheta \, d\varphi.$$

The mean value of the total angular momentum is then

$$\langle \mathscr{L}^2 \rangle_n^{\frac{1}{2}} = \frac{\left\{ \sum_{l=0}^{n-1} \sum_{m=-l}^{l} |c_l^m|^2 l(l + 1) \right\}^{\frac{1}{2}} \hbar.}{\|\Psi_n\|} \tag{4.44}$$

To get some idea of the structure of a hydrogenic atom, we might compute the mean value of the atomic radius. We have[7]

$$\langle r \rangle_{nl} = \int_0^{\infty} \int_0^{\pi} \int_0^{2\pi} r |\psi_{nlm}(\mathbf{r})|^2 \, d\varphi \sin \vartheta \, d\vartheta r^2 \, dr = \int_0^{\infty} r |R_{nl}(r)|^2 r^2 \, dr$$

$$= \frac{a_0 n}{2Z} \frac{(n - l - 1)}{2n[(n + l)!]^3} \int_0^{\infty} \exp(-\rho) \, \rho^{2l+3} |L_{n+l}^{2l+1}(\rho)|^2 \, d\rho$$

$$= \frac{n^2 a_0}{Z} \left\{ 1 + \frac{1}{2} \left[1 - \frac{l(l + 1)}{n^2} \right] \right\}. \tag{4.45}$$

[7] The integral is carried out in detail in H. Margenau and G. M. Murphy, *Mathematics of Physics and Chemistry*, D. Van Nostrand Co., Princeton, 1956, Vol. I, p. 130.

The Hydrogen Atom

In the case of the hydrogen atom in the lowest energy state we find

$$\langle r \rangle_{10} = \tfrac{3}{2} a_0$$

which is somewhat larger than the radius predicted by the Bohr theory. The root-mean-square deviation of $\langle r \rangle_{nl}$ is found to be[8]

$$\Delta r_{nl} = [\langle r^2 \rangle_{nl} - \langle r \rangle_{nl}{}^2]^{1/2} = \frac{a_0 n^2}{2Z} \left\{ 1 - \frac{1}{n^2} (l + 1)(l - 2) \right\}^{1/2}$$

or

$$\frac{\Delta r_{nl}}{\langle r_{nl} \rangle} = \frac{\{1 - (1/n^2)(l + 1)(l - 2)\}^{1/2}}{\{3 - [l(l + 1)]/n^2\}}.$$

The relative fluctuation in the mean value of the atomic radius decreases from $(\sqrt{3})^{-1}$ in the lowest energy state ($n = 1, l = 0$) to zero as $l \to (n - 1)$ and $n \to \infty$. This behavior is in good accord with the correspondence postulate, which requires that coordinates become well-defined, in the Newtonian sense, in the limit of large quantum numbers.

The first few hydrogenic radial wavefunctions are as follows.

$$R_{10}(r) = \left(\frac{Z}{a_0}\right)^{3/2} 2 \exp(-\rho/2) \qquad \rho = \frac{2r}{n a_0}$$

$$R_{20}(r) = \frac{(Z/a_0)^{3/2}}{2\sqrt{2}} (2 - \rho) \exp(-\rho/2)$$

$$R_{21}(r) = \frac{(Z/a_0)^{3/2}}{2\sqrt{6}} \rho \exp(-\rho/2)$$

$$R_{30}(r) = \frac{(Z/a_0)^{3/2}}{9\sqrt{3}} (6 - 6\rho + \rho^2) \exp(-\rho/2)$$

$$R_{31}(r) = \frac{(Z/a_0)^{3/2}}{9\sqrt{6}} (4 - \rho) \exp(-\rho/2)$$

$$R_{32}(r) = \frac{(Z/a_0)^{3/2}}{9\sqrt{30}} \rho^2 \exp(-\rho/2)$$

The quantity ρ is defined as $(2Z/na_0)r$, as given earlier. The quantum numbers n and l, used to denote the radial wavefunctions, are sometimes given other designations in connection with the organization of spectra. By convention, the set of quantum numbers $l = 0, 1, 2, 3, \ldots$, are given the letter symbols $s, p, d, f, g, \ldots$, and so on. The principal quantum numbers are denoted either by their numerical magnitudes or by the letter designations K, L, M, N, etc.,

[8] The rms deviation is not zero because $\psi_{nlm}(\mathbf{r})$ is not an eigenfunction of $\mathbf{r}_{op}$!

representing $n = 1, 2, 3, 4, \ldots$. Thus $R_{10}(r)$ refers to the $1s$ "orbital" of the "K shell," $R_{21}(r)$ refers to the $2p$ orbital of the L shell, and $R_{32}(r)$ refers to the $3d$ orbital of the M shell. Alternatively, $\psi_{100}(\mathbf{r})$ designates a "$1s$ state," $\psi_{21m}(\mathbf{r})$ designates a $2p$ state, and so on.

The probability distribution function for the electron relative to the nucleus is

$$P_{nl}(r) \equiv |R_{nl}(r)|^2. \tag{4.46}$$

$P_{nl}(r)$ may be called the radial probability function. More important is the probability that the electron moves at a distance between r and $r + dr$ from the nucleus,[9]

$$g_{nl}(r) \, dr \equiv 4\pi r^2 |R_{nl}(r)|^2 \, dr. \tag{4.47}$$

$g_{nl}(r)$ is called the *radial distribution function*. In Figure 30 $g_{nl}(r)$ is plotted for r in units of a_0—the so-called *atomic units*. (It should be noted that because the reduced mass appears in the Bohr radius, the atomic unit will vary slightly in magnitude from atom to atom.) It can be seen from the figure that the most probable value for the radius of a hydrogen atom in the $1s$ state is the Bohr radius. Because of the asymmetry of the radial distribution function, the mean value of the radius is somewhat larger than 1 atomic unit.

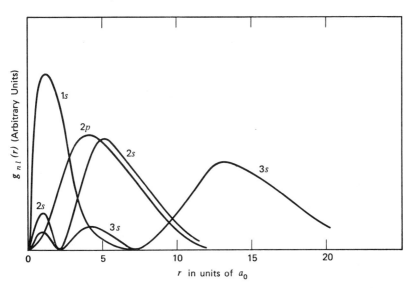

Figure 30. The radial distribution function for an electron in the hydrogen atom.

[9] Strictly speaking $g_{nl}(r)$ is the probability that the electron is in the volume element $r^2 \, dr \sin \vartheta \, d\vartheta \, d\varphi$, integrated over angles.

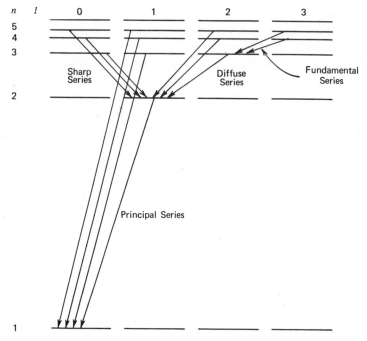

Figure 31. Some of the spectral series for a hydrogenic atom.

In the $2s$ and $3s$ states, we find that the electron resides farther and farther away from the nucleus, although there is certainly a non-zero probability of finding the particle at a distance more expected for lower energies. As the total angular momentum increases for an electron in a given shell, we find that while there is less possibility for different radii, the radial distribution function is quite broadly peaked at the likely distances.

The total energy of the electron in a given shell may be expressed by

$$E_n = -\left(\frac{Z}{n}\right)^2 \tag{4.48}$$

in *Rydberg units*, which are units of $(\mu_e e^4/2\hbar^2)$. Certain of these energy levels are indicated in Figure 31 along with the appropriate spectral series.

We should be very naive if we assumed that an electron may undergo a transition from one energy level to another indiscriminately. It turns out that the probability for a transition between energy levels will be zero unless the integral

$$\int_{\mathbf{r}} \psi_{n'l'm'}(\mathbf{r}) \mathcal{M} \psi_{nlm}(\mathbf{r}) \, d\mathbf{r} \qquad (n' \neq n)$$

does not vanish, where $\mathcal{M}$ is the operator associated with the dipole moment of the atom. We shall be concerned with this problem in more detail later on, but for now it suffices to point out that, for hydrogenic atoms, the integral vanishes for all but two instances. These two instances constitute *selection rules* for the transitions. The selection rules are:

(a) The angular quantum numbers l for the two states involved must differ by ± 1.

(b) The quantum numbers m for the two states must be of the same value or must differ by ± 1.

From these rules we deduce that transitions will occur only between s-states and p-states, p-states and d-states, d-states and f-states, etc. An emission spectral series, then, may be named for the initial state's angular quantum number. The *principal series* (p-series), mentioned in Chapter 1, corresponds to transitions into the $1s$ state from p-states. The *sharp series* (s-series) corresponds to the transitions into the $2p$ state from s-states. The *diffuse series* corresponds to transitions into the $2p$ state from d-states. Finally, the *fundamental series* (f-series) corresponds to transitions from f-states to the $3d$ state.

It must be remembered that the spectrum of a one-electron atom will not be exactly as suggested here. This is because we have neglected non-coulombic electron-nucleus interactions, relativistic effects, and especially some properties of atomic electrons which are uniquely quantum physical. When the theory is corrected for these effects it is found that the total energy depends upon the magnitude of l as well as n, and upon a new quantum number as yet undesignated. Obviously, the corrected energy levels are for the most part very close in magnitude to their original values. Indeed, the effects are least pronounced for "effectively hydrogenic" atoms. However, this fact does not alter the very great theoretical importance of the corrections which motivates our study in any case.

4.6. THE NORMAL ZEEMAN EFFECT OF HYDROGENIC ATOMS

Suppose that a gas discharge tube filled with mercury vapor at low pressure is connected to a high-voltage power supply, and that the vapor is made to radiate. The very brightly illuminated tube is then placed before the slit of a spectrometer so that the radiation in the visible region may be resolved into its components. The result is not disappointing: the spectrum is composed of numerous bright lines, many of which have structure. Our interest just now rests with the dim yellow line of wavelength 5790.7 Å. This line is of a simple structure, being a singlet, and, it turns out, is the product of an electronic transition between a d-state and a p-state. The question we have in mind is the

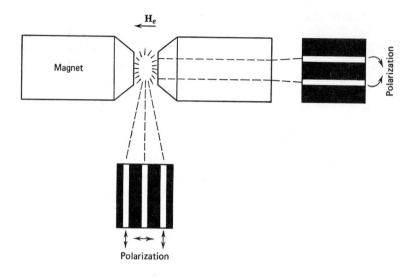

Figure 32. The normal Zeeman effect for a hydrogenic atom.

following. What happens to the line when the mercury vapor is subjected to a fairly strong magnetic field?

The best way to satisfy our curiosity is to place the discharge tube between the pole faces of an electromagnet and switch on the field. If the field is relatively homogeneous and of the order of 20,000 gauss, we find, upon looking through the spectrometer telescope once more, that the yellow singlet has disappeared and that a triplet has taken its place. But that isn't all! If we position the spectrometer so that its telescope points in a direction parallel to that of the magnet pole pieces (that is, parallel to the magnetic field direction), not three but *two* spectral lines are seen where one stood before. Moreover, if the whole experiment is repeated with the help of a polarizing device, it is observed that the triplet lines correspond to linearly polarized light, while the doublet lines are circularly polarized. (See Figure 32.) In particular, the middle line of the triplet is composed of light waves whose plane of vibration is parallel to the direction of the magnetic field; the remaining two lines are polarized in a direction perpendicular to that of the field. The doublet lines are circularly polarized in the counterclockwise (left line) and clockwise (right line) directions, respectively.

This very interesting phenomenon was first observed by Pieter Zeeman in 1896. It is for this reason known as the *normal Zeeman effect*. The modifier "normal" is used to classify those instances when a singlet splits into a triplet and doublet. It is observed in general that a spectral line may split into more than three lines, which phenomenon is called the *anomalous Zeeman effect*.

Just now, however, we shall not consider the anomalous effect. It is enough to explain the "normal" behavior!

Consider an electron moving in one of the orbits permitted a hydrogenic atom. According to what was said in section 1.3, we may consider the electronic orbit to be equivalent to a current loop which, in turn, may be associated with a magnetic dipole moment

$$\mu = \frac{I}{c} \mathbf{S} \tag{4.49}$$

where I is the current, c is the speed of light *in vacuo*, and $\mathbf{S}$ is the (vector) area of the orbit. The vector area is defined by

$$\mathbf{S} \equiv \frac{1}{2} \oint (\mathbf{r} \times d\mathbf{s}) \tag{4.50}$$

where $d\mathbf{s}$ is an increment of length along the orbit, $\mathbf{r}$ is a radius vector, and $\frac{1}{2}(\mathbf{r} \times d\mathbf{s})$ is an element of (triangular) area swept out by the radius vector. The integral in equation (4.50) is carried out over the entire orbital perimeter. Now, the current density vector $\mathbf{j}$ is defined in terms of the current and the orbital velocity by

$$\mathbf{j} \, dV = I \, d\mathbf{s} = -\frac{|e|}{\mu_e} \rho \mathbf{v} \, dV \tag{4.51}$$

where dV is an element of volume, e is the electronic charge, and ρ is the reduced mass density. With the help of equations (4.50) and (4.51), equation (4.49) may be written

$$\mu = \frac{-|e|}{2\mu_e c} \int (\mathbf{r} \times \rho \mathbf{v}) \, dV = \left(-\frac{|e|}{2\mu_e c} \right) \mathbf{L} \tag{4.52}$$

where $\mathbf{L}$ is the total angular momentum of the electron. We see that the magnetic dipole moment vector points in a direction exactly opposite that of the angular momentum vector.

The potential energy of a dipole of moment μ in a uniform magnetic field is[10]

$$U_H = -\mu \cdot \mathbf{H}_e = \frac{|e|}{2\mu_e c} (\mathbf{L} \cdot \mathbf{H}_e) \tag{4.53}$$

where $\mathbf{H}_e$ is the external field. It follows from equations (4.9) and (4.53) that the Hamiltonian function for the hydrogenic electron in a uniform magnetic field is

$$H(p, r) = \frac{p^2}{2\mu_e} - \frac{Ze^2}{r} + \frac{|e|}{2\mu_e c} (\mathbf{L} \cdot \mathbf{H}_e).$$

---

[10] See W. K. H. Panofsky and M. Phillips, *Classical Electricity and Magnetism*, Addison-Wesley Publ. Co., Reading, Mass., 1962, p. 19.

Now, if we choose the direction of $\mathbf{H}_e$ to be that of the z-axis, the scalar product $(\mathbf{L} \cdot \mathbf{H}_e)$ may be rewritten as $L_z H_e$, where L_z is the z-component of the total angular momentum. With this modification and the Schrödinger postulate, we can express the eigenvalue problem as

$$-\frac{\hbar^2}{2\mu_e}\left\{\frac{1}{r^2}\frac{\partial}{\partial r}\left(r^2\frac{\partial\psi}{\partial r}\right) + \frac{1}{r^2\sin\vartheta}\frac{\partial}{\partial\vartheta}\left(\sin\vartheta\,\frac{\partial\psi}{\partial\vartheta}\right) + \frac{1}{r^2\sin^2\vartheta}\frac{\partial^2\psi}{\partial\varphi^2}\right\}$$

$$-\frac{Ze^2}{r}\psi(\mathbf{r}) - \frac{i\hbar|e|H_e}{2\mu_e c}\frac{\partial\psi}{\partial\varphi} = E\psi(\mathbf{r}). \quad (4.54)$$

What if we were to choose $\psi(\mathbf{r})$ to be $\psi_{nlm}(\mathbf{r})$, given by equation (4.39)? $\psi_{nlm}(\mathbf{r})$ is certainly an eigenfunction of the first four terms of the Hamiltonian operator in equation (4.54). But, then, the fifth term is no problem either, because

$$\frac{\partial\psi_{nlm}}{\partial\varphi} = R_{nl}(r)\theta_l{}^m(\vartheta)\frac{\partial\Phi_m}{\partial\varphi} = im\psi_{nlm}(\mathbf{r}).$$

We conclude that the total energy E of the hydrogenic electron in a uniform magnetic field is

$$E_{nm} = -\frac{Z^2 e^4 \mu_e}{2\hbar^2 n^2} + m\mu_B H_e \qquad (m = 0, \pm 1, \ldots) \qquad (4.55)$$

where

$$\mu_B \equiv \frac{|e|\hbar}{2\mu_e c} = 9.2732 \pm 0.0006 \times 10^{-21} \text{ erg/gauss}$$

is called the *Bohr magneton*. We see that the hydrogenic energy is no longer independent of the quantum numbers l and m, but in fact is split into $(2l + 1)$ equally spread levels. For example, we expect a d-state ($l = 2$) to have five associated energy levels. (See Figure 33.) Bearing in mind the selection rules mentioned in section 4.5, we must have

$$h\nu = \Delta E_{nm} = \begin{cases} \Delta E_n - \mu_B H_e \\ \Delta E_n \\ \Delta E_n + \mu_B H_e \end{cases} \qquad (4.56)$$

since $\Delta m = -1, 0, +1$, where ΔE_n is the energy difference between states having l-values different by ± 1. It follows that a singlet will split into three lines upon the application of a magnetic field. The first line is of frequency $(\Delta E - \mu_B H_e)/h$, corresponding to $\Delta m = -1$, and appears circularly polarized in a plane perpendicular to that of the magnetic field. The second line is of the same frequency as the original singlet and is linearly polarized in a plane parallel to the direction of the field. This component of the triplet is *not* observed when the spectrum is looked at along a direction parallel to the field because the latter is no longer perpendicular to the direction of light propagation. The third line is of frequency $(\Delta E + \mu_B H_e)/h$, corresponding to

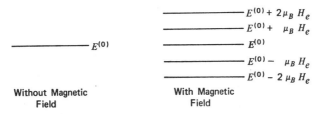

Figure 33. Normal Zeeman splitting of a d-state energy level.

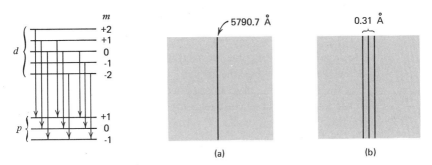

Figure 34. The normal Zeeman effect for the 5790.7-Å mercury line: (a) without an applied magnetic field, (b) with an applied field of 20 kilogauss.

$\Delta m = +1$, and appears circularly polarized in the same way as is the first line.

As regards the splitting of the 5790.7-Å line in the mercury spectrum, we see that a d-state-to-p-state transition involves eight energy levels when the magnetic field is present. (See Figure 34.) According to the selection rules, there are nine permitted transitions between these levels. The transitions $d(m = 2) \rightarrow p(m = 1)$, $d(m = 1) \rightarrow p(m = 0)$, $d(m = 0) \rightarrow p(m = -1)$ all correspond to the first line, and so on, such that there are but three observed spectral lines. Our theory of the normal Zeeman effect, then, is adequate to describe this phenomenon for the yellow mercury line. The fact that quantum number m plays such an important part in the theory has resulted in the term *magnetic quantum number* for this quantity. We shall find that the magnetic quantum number will always be of significance whenever the atom under consideration does not possess spherical symmetry or, as in the present case, it has its spherical symmetry disturbed by an external field.

PROBLEMS

1. The quantum condition mentioned in conjunction with the first of the postulates leading to Bohr's quantum mechanics may be expressed by the integral equation

$$\oint p_s \, dq_s = n_s h \qquad (s = 1, 2, 3)$$

where p_s is the sth component of the momentum of the electron, q_s is the sth component of the position, n_s is an integer, and the integral is taken around one complete orbit. For a particle acted upon by a central force, the motion always takes place in a plane, so that just two integrals are sufficient for deducing the magnitude of the total energy.

(a) In the case of the hydrogen atom, show that

$$p_s \equiv p_\theta \equiv L = n_\theta h \equiv kh \qquad (k = 0, 1, 2, \ldots)$$

where L is the total angular momentum of the electron and $n_\theta \equiv k$ is called the azimuthal quantum number. (The angle θ is that swept out by the electron in traversing its orbit, and is not to be confused with the angles ϑ and φ as defined earlier.)

(b) Show that

$$p_r = \pm \left\{ \frac{a}{r^2} + \frac{b}{r} + c \right\}^{\frac{1}{2}}$$

where $a = -k^2\hbar^2$, $b = 2\mu_e e^2$, and $c = 2\mu_e E$.

(c) The radius of the electronic orbit varies between the values of r which make p_r vanish identically. Show that these values are

$$r_+ = -\frac{b + \{b^2 - 4ac\}^{\frac{1}{2}}}{2c}$$

$$r_- = -\frac{b - \{b^2 - 4ac\}^{\frac{1}{2}}}{2c}$$

(d) By evaluating the integral

$$\oint p_r \, dr = 2 \int_{r_-}^{r_+} p_r \, dr = 2 \int_{r_-}^{r_+} \left\{ \frac{a}{r^2} + \frac{b}{r} + c \right\}^{\frac{1}{2}} dr$$

show that

$$\oint p_r \, dr = 2\pi \left[\frac{b}{2(-c)^{\frac{1}{2}}} - (-a)^{\frac{1}{2}} \right] \equiv n_r h$$

and that

$$E = -\frac{\mu_e e^4}{2\hbar^2} \frac{1}{(n_r + k)^2}.$$

Compare this expression for the total energy of the hydrogenic electron with that given in equation (4.36).

2. According to Newtonian mechanics, the frequency of the electromagnetic wave radiated by an orbiting atomic electron is

$$\nu = \frac{1}{\pi e^2}\left(\frac{2|E|^3}{\mu_e}\right)^{\frac{1}{2}}.$$

When the principal quantum number is quite large, the quantum-theoretical expression for the frequency of radiation emitted by a hydrogenic atom may be written

$$\nu \sim \frac{(dE_n/dn)}{h}$$

where E_n is $-Rhc/n^2$. Using the correspondence postulate and these two expressions for ν, deduce what must be the form of the Rydberg constant, R.

3. Construct and solve the Schrödinger eigenvalue problem for the center-of-mass motion of the hydrogen atom.

4. Show that the Legendre polynomials are the coefficients of powers of x in the Taylor series representing the function

$$f(x, z) = (x^2 - 2xz + 1)^{-\frac{1}{2}} \qquad (|z| \leq 1)$$

5. Show by direct calculation for specific values of l that

$$P_l(z) = \frac{1}{\pi}\int_0^\pi [z + \sqrt{z^2 - 1}\,\cos\varphi]^l\, d\varphi \qquad (|z| \leq 1)$$

6. The *associated Legendre functions* satisfy equation (4.23) with $\beta = l(l + 1)$. It is not too difficult to show that these functions also satisfy the recurrence relation

$$zP_l{}^m(z) = \frac{1}{2l + 1}\{(l + |m|)P_{l-1}^m(z) + (l - |m| + 1)P_{l+1}^m(z)\}.$$

With this information calculate the mean value of $\cos\vartheta$ for any angular momentum state of the hydrogenic electron.

7. The *addition theorem* for spherical harmonics is expressed by

$$P_l(\cos\gamma) = \sum_{m=-l}^{l} \frac{(l - |m|)!}{(l + |m|)!}\, P_l{}^m(\cos\vartheta_1)P_l{}^m(\cos\vartheta_2)\exp\left[im(\varphi_1 - \varphi_2)\right]$$

where γ is the angle between radius vectors to the points (ϑ_1, φ_1) and (ϑ_2, φ_2). By putting $\gamma = 0$ and noting that $P_l{}^0(1) = 1$, show that

$$\sum_{m=-l}^{l} \overline{Y_l{}^m}(\vartheta, \varphi)Y_l{}^m(\vartheta, \varphi) = \frac{2l + 1}{4\pi}.$$

This is a form of *Unsöld's theorem*: that the sum of squares of angular wavefunctions corresponding to a given value of l is a constant.

8. Hydrogenic atoms are spherically symmetric systems. However, it would appear that the z-axis represents a preferred direction, since the angular wavefunctions are eigenfunctions of $\mathscr{L}_z$ but not $\mathscr{L}_x$ or $\mathscr{L}_y$. Show that this is not the case, that it is possible to have states of the hydrogenic electron which are represented by eigenfunctions of $\mathscr{L}_x$ or $\mathscr{L}_y$ corresponding to the eigenvalue $m\hbar$ ($m = 0, \pm 1, \pm 2, \ldots, \pm l$). (*Hint*: Use equation (4.40) in your argument.)

9. Suppose a hydrogenic electron is in the state represented by the spherical harmonic $\Upsilon_l^m(\vartheta, \varphi)$. This means that $\langle \mathscr{L}^2 \rangle$ and $\langle \mathscr{L}_z \rangle$ are known precisely, but that $\langle \mathscr{L}_x \rangle$ and $\langle \mathscr{L}_y \rangle$ are only partially known.

(a) Show that

$$(\mathscr{L}_x^2 + \mathscr{L}_y^2)\Upsilon_l^m(\vartheta, \varphi) = (l^2 + l - m^2)\hbar^2\Upsilon_l^m(\vartheta, \varphi).$$

(b) Invoking the definition of the mean-square deviation, show that

$$(l^2 + l - m^2)\hbar^2 \geqslant \{(\Delta\mathscr{L}_x)^2 + (\Delta\mathscr{L}_y)^2\} \geqslant |m\hbar^2|,$$

given the "uncertainty relation" $\Delta\mathscr{L}_x \Delta\mathscr{L}_y \geqslant |(m/2)\hbar^2|$.

10. One way of picturing the lack of information about $\langle \mathscr{L}_x \rangle$ and $\langle \mathscr{L}_y \rangle$ when the hydrogenic electron is in the state $\Upsilon_l^m(\vartheta, \varphi)$ is to draw the total angular momentum as a vector, in ordinary space, of length $(l(l + 1))^{1/2}\hbar$. The z-component of this vector is, of course, $m\hbar$.

(a) Show that the angle between the vector and the z-axis is

$$\alpha = \cos^{-1}\left[\frac{m}{\sqrt{l(l + 1)}}\right].$$

(b) Show that the projection of the vector onto the xy-plane is of length $(l^2 + l - m^2)^{1/2}\hbar$.

(c) Draw a diagram incorporating the foregoing information. Interpret the lack of knowledge concerning the x- and y-components of the total angular momentum vector in terms of its surface of revolution.

(d) According to the result of the second part of problem 9, what value of m corresponds to the least uncertainty in our knowledge of $\langle(\mathscr{L}_x^2 + \mathscr{L}_y^2)\rangle$? What is the value of the angle α in this case?

11. Newtonian mechanics would have it that the total angular momentum vector shall be fixed in space throughout the motion of a hydrogenic electron. We know that such a rigid stipulation is impossible in the quantum theory, because of the uncertainty in our knowledge of the components of the total angular momentum vector. The best we can do is put $\langle \mathscr{L}_z \rangle = \pm l\hbar$, according to one of the results of problem 10. The angle between the total angular momentum vector and the z-axis is then at its minimum value. Show that this situation is in harmony with the classical theory by employing the correspondence postulate.

12. The Schrödinger eigenvalue problem for the relative motion of two particles constrained to remain a distance r_e apart is

$$-\frac{\hbar^2}{2I}\left\{\frac{1}{\sin\vartheta}\frac{\partial}{\partial\vartheta}\left(\sin\vartheta\frac{\partial\psi}{\partial\vartheta}\right) + \frac{1}{\sin^2\vartheta}\frac{\partial^2\psi}{\partial\varphi^2}\right\} = E\psi(\vartheta, \varphi)$$

where $I = \mu r_e^2$ is called the moment of inertia. Solve the differential equation and find an expression for the total energy. Draw an energy-level diagram. Can you think of a physical situation to which your results might apply?

13. Some very obvious examples of hydrogenic atoms are H, He$^+$, Li^{+2}, and Be^{+3}. The groundstate ($n = 1$) total energies of the electrons in these systems are -13.598, -54.39, -122.4, and -217.6 eV, respectively. Compare these experimental data with the quantum-theoretical predictions of the groundstate energies.

14. It can be shown that, for a hydrogenic electron, $\langle r^{-1}\rangle = Z/a_0 n^2$. Calculate the mean value of the potential energy of the electron. What must be the mean value of the kinetic energy? Show that your result is in agreement with the quantum-theoretical *virial theorem*

$$\langle T\rangle = \frac{1-k}{2}\langle V\rangle$$

where $\langle T\rangle$ is the mean value of the kinetic energy and the exponent of r in the expression for the force on the electron, $F(r)$, is $-k$.

15. What is the root-mean-square speed of the electron in the groundstate hydrogen atom?

16. Show that the associated Laguerre polynomials $L_n^k(\rho)$ are the coefficients of $(x^n/n!)$ in the Taylor expansion of

$$f(\rho, x) = \frac{(-x)^k}{(1-x)^{k+1}}\exp\left[(-\rho x)/1 - x\right].$$

(The best way to do this is by calculating two or three of the $L_n^k(\rho)$ explicitly.)

17. Let us see how good an approximation it is to consider the helium atom a hydrogenic atom. Assuming no other interactions but those between the nucleus and the two electrons, the Schrödinger eigenvalue problem is

$$-\frac{\hbar^2}{2\mu_e}\{\nabla_{\mathbf{r}_1}^2 + \nabla_{\mathbf{r}_2}^2\}\Psi(\mathbf{r}_1, \mathbf{r}_2) - \left\{\frac{2e^2}{r_1} + \frac{2e^2}{r_2}\right\}\Psi(\mathbf{r}_1, \mathbf{r}_2) = E\Psi(\mathbf{r}_1, \mathbf{r}_2)$$

where r_1 refers to the (spherical polar) coordinates of the first electron and r_2 denotes those of the second electron.

(a) What is the form of the radial wavefunction in the groundstate?

(b) The groundstate energy of the electrons in helium is -78.6 eV. Compare this value to the total energy calculated by means of the present approximation. What is the most likely cause of the discrepancy in values?

(c) The radius of the helium atom is thought to be 0.30 Å. How does this figure compare with the prediction for a hydrogenic atom of nuclear charge two?

18. The most probable radii of Li^{2+}, Be^{3+}, and B^{4+} are 0.20, 0.143, and 0.112 Å, respectively. Compare these data with the expected values for the corresponding hydrogenic atoms.

19. Draw a diagram illustrating normal Zeeman splitting in s-, p-, d-, and f-states of hydrogenic atoms.

20. Draw a diagram illustrating the electronic transitions permitted in the normal Zeeman effect for a line in the sharp series.

21. According to equations (4.55), the component of the magnetic moment vector along the field direction is for a hydrogenic atom an integral multiple of the Bohr magneton.

(a) How many different values, in a given total angular momentum state, may this vector component have?

(b) Why is the "spacial quantization" resulting from this orientation restriction on the magnetic moment vector *not* adequate to explain the result of the Stern-Gerlach experiment?

CHAPTER 5

Quantum Mechanical Approximation Methods

Half a loaf is better than none at all.

5.1. TIME-INDEPENDENT PERTURBATION THEORY

In light of the great success we have had thus far in solving the Schrödinger eigenvalue problems for free and hydrogenic electrons, it might be supposed that microscopic physical theory has thereby been reduced to the task of carefully picking Hamiltonian functions. If only this were true! Unfortunately, it is the rule rather than the exception that, even if the correct Hamiltonian operator has been constructed, there is absolutely no guarantee that the resulting eigenvalue problem can be solved exactly. For example, in the case of the helium atom—the simplest next to hydrogen—the Schrödinger eigenvalue problem for the two electrons is (to order 1/nuclear mass)

$$\left\{ -\frac{\hbar^2}{2\mu_e}(\nabla_{\mathbf{r}_1}^{\ 2} + \nabla_{\mathbf{r}_2}^{\ 2}) - 2\left(\frac{e^2}{r_1} + \frac{e^2}{r_2}\right) + \frac{e^2}{r_{12}} \right\}\psi(\mathbf{r}_1, \mathbf{r}_2) = E\psi(\mathbf{r}_1, \mathbf{r}_2)$$

referred to the center of the nucleus, where $\mathbf{r}_{12}$ is the displacement of one electron from the other. Because of the term in r_{12} this differential equation is not separable in spherical polar space (or in any other coordinate system), as was the equation for hydrogen. Indeed, this difficulty is characteristic of the Schrödinger eigenvalue problem for *every* many-electron atom. Evidently, for all but the simplest physical situations (namely, the one-body problem and its equivalents), special approximation techniques will be necessary in order to do quantum mechanics. Let us see how this may be done.

Suppose the eigenvalue problem

$$\mathcal{H}^{(0)}|N\rangle^{(0)} = E_N^{(0)}|N\rangle^{(0)} \qquad (N = 1, 2, 3, \ldots) \tag{5.1}$$

Quantum Mechanical Approximation Methods

has been completely solved, where $E_N^{(0)}$ is an energy eigenvalue of multiplicity β and $|N\rangle^{(0)}$ is an orthonormal state vector belonging to $\mathcal{H}^{(0)}$:

$$|N\rangle^{(0)} = \sum_{k=1}^{\beta} a_k^N |N_k\rangle^{(0)}. \tag{5.2}$$

(The β state vectors $|N_k\rangle^{(0)}$ belonging to $\mathcal{H}^{(0)}$ are also assumed to be orthonormal.) We should like to ask the question: What is the physical consequence of perturbing the system described by equation (5.1), in such a way that the value of the total energy $E_N^{(0)}$ is altered slightly? By hypothesis the perturbation is a small one, so we may sensibly write

$$E_N = E_N^{(0)} + gE^{(1)} \tag{5.3}$$

where g is a small number, called the *coupling constant*, and $E^{(1)}$ is the augmentation in the total energy "per coupling constant." (It is evident that we can put g equal to one at the end of our analysis to remove the ambiguity in $E^{(1)}$.) Consistently with equation (5.3) we put

$$|N\rangle = |N\rangle^{(0)} \oplus g|N\rangle^{(1)}$$
$$\mathcal{H} = \mathcal{H}^{(0)} + g\mathcal{H}' \tag{5.4}$$

and introduce equations (5.3) and (5.4) into the Schrödinger eigenvalue problem for the perturbed system

$$\mathcal{H}|N\rangle = E_N|N\rangle \qquad (N = 1, 2, 3, \ldots). \tag{5.5}$$

The result is, including only terms of the first order in the coupling constant,

$$\mathcal{H}^{(0)}|N\rangle^{(0)} \oplus g\mathcal{H}^{(0)}|N\rangle^{(1)} \oplus g\mathcal{H}'|N\rangle^{(0)}$$
$$= E_N^{(0)}|N\rangle^{(0)} \oplus gE_N^{(0)}|N\rangle^{(1)} \oplus gE^{(1)}|N\rangle^{(0)}.$$

Upon equating the coefficients of equal powers of g we find the zeroth- and first-order approximations, respectively:

$$\mathcal{H}^{(0)}|N\rangle^{(0)} = E_N^{(0)}|N\rangle^{(0)}$$
$$\mathcal{H}^{(0)}|N\rangle^{(1)} \oplus \mathcal{H}'|N\rangle^{(0)} = E_N^{(0)}|N\rangle^{(1)} \oplus E^{(1)}|N\rangle^{(0)}.$$

The zeroth-order expression is solved, by hypothesis. The first-order equation may be solved by expanding $|N\rangle^{(1)}$ in a complete set of the $|N\rangle^{(0)}$:

$$|N\rangle^{(1)} = \sum_{n=1}^{\infty} a_n |n\rangle^{(0)}. \tag{5.6}$$

Thus we have

$$\sum_{n=1}^{\infty} a_n E_N^{(0)} |n\rangle^{(0)} \oplus \mathcal{H}'|N\rangle^{(0)} = E_N^{(0)} \sum_{n=1}^{\infty} a_n |n\rangle^{(0)} \oplus E^{(1)}|N\rangle^{(0)} \tag{5.7}$$

where it is to be understood the $|N\rangle^{(0)}$ is one of the $|n\rangle^{(0)}$. Now let us take the scalar product of equation (5.7) with the state vector $|N_{k'}\rangle^{(0)}$ $(1 \leqslant k' \leqslant \beta)$. We find, using equation (5.2),

$$a_N E_N^{(0)} a_{k'}{}^N + \sum_{k=1}^{\beta} a_k{}^N {}^{(0)}\langle N_{k'}|\mathscr{H}'|N_k\rangle^{(0)} = E_N^{(0)} a_N a_{k'}{}^N + E^{(1)} a_{k'}{}^N$$

which reduces easily to

$$\sum_{k=1}^{\beta} a_k{}^N({}^{(0)}\langle N_{k'}|\mathscr{H}'|N_k\rangle^{(0)} - E^{(1)}\delta_{k'k}) = 0 \qquad (k' = 1,\ldots,\beta). \quad (5.8)$$

Equations (5.8) represent a set of homogeneous, linear equations whose solutions—the $a_k{}^N$—are not identically zero only if the condition

$$\begin{vmatrix} H'_{N_1 N_1} - E^{(1)} & \cdots & H'_{N_1 N_\beta} \\ \vdots & & \vdots \\ H'_{N_\beta N_1} & \cdots & H'_{N_\beta N_\beta} - E^{(1)} \end{vmatrix} = 0 \qquad (5.9)$$

is satisfied,[1] where

$$H'_{N_k N_k} \equiv {}^{(0)}\langle N_{k'}|\mathscr{H}'|N_k\rangle^{(0)}.$$

The determinantal equation (5.9) in general will possess as many as β distinct roots, corresponding to β possible values for $E^{(1)}$. When the roots are all or in part distinct, we say that *the perturbation $\mathscr{H}'$ removes the degeneracy in $E_N^{(0)}$*. When the $E^{(1)}$ are found, usually by numerical analysis,[2] they may be put back into equation (5.7), which is then solved to find the $a_k{}^N$ $(k = 1,\ldots,\beta)$. These coefficients are needed to establish the correct form of the zeroth-order state vector, as may be seen from equation (5.2). The coefficients a_n ($n = 1, 2,\ldots$), which are needed for calculating $|N\rangle^{(1)}$, may be found by taking the scalar product of $|N'\rangle^{(0)}$ $(N' \neq N)$ with equation (5.7). We have

$$a_{N'} E_N^{(0)} + {}^{(0)}\langle N'|\mathscr{H}'|N\rangle^{(0)} = E_N^{(0)} a_{N'}, \qquad (N' \neq N)$$

or, upon solving for $a_{N'}$

$$a_{N'} = \frac{H'_{N'N}}{E_N^{(0)} - E_{N'}^{(0)}}, \qquad (N \neq N', N' = 1, 2, 3, \ldots) \quad (5.10)$$

where

$$H'_{N'N} \equiv {}^{(0)}\langle N'|\mathscr{H}'|N\rangle^{(0)}.$$

---

[1] See H. Margenau and G. M. Murphy, *The Mathematics of Physics and Chemistry*, D. Van Nostrand Co., Princeton, 1956, Vol. I, p. 314.
[2] *Ibid.*, pp. 500–504.

Quantum Mechanical Approximation Methods

The perturbed state vector follows as

$$|N\rangle = (1 + ga_N)|N\rangle^{(0)} \oplus g\sum_{n=1}^{\infty}{}' \frac{H'_{nN}}{E_N^{(0)} - E_n^{(0)}} |n\rangle^{(0)}$$

where the prime means not to include $n = N$ in the sum. The coefficient a_N ($N \neq n$) is determined by normalizing $|N\rangle$. This operation gives

$$\langle N|N\rangle = {}^{(0)}\langle N|N\rangle^{(0)} + g({}^{(1)}\langle N|N\rangle^{(0)} + {}^{(0)}\langle N|N\rangle^{(1)}) \equiv 1$$
$$= 1 + g[\bar{a}_N + a_N]$$

to terms of first order, upon noting the first of equations (5.4) and realizing, by equation (5.6), that

$$a_N = {}^{(0)}\langle N|N\rangle^{(1)}.$$

In order to maintain normalized $|N\rangle$ to first order, we must put $\bar{a}_N = -a_N$, which means a_N is a pure imaginary number. If we call ga_N the imaginary number $i\alpha_N$, where α_N is real and positive, the expression for $|N\rangle$ becomes

$$|N\rangle = (1 + i\alpha_N)|N\rangle^{(0)} \oplus g\sum_{n=1}^{\infty}{}' a_n|n\rangle^{(0)}$$

$$= \exp(i\alpha_N)|N\rangle^{(0)} \oplus g\sum_{n=1}^{\infty}{}' a_n|n\rangle^{(0)}$$

since

$$\exp(i\alpha_N) = 1 + i\alpha_N$$

to the first order of approximation. Now, the choice of α_N is quite arbitrary because the state vector $|N\rangle^{(0)}$ is uniquely defined only to within a phase factor. We can therefore put $\alpha_N = 0$ with no loss in generality. The perturbed state vector is then

$$|N\rangle = |N\rangle^{(0)} \oplus g\sum_{n=1}^{\infty}{}' \frac{H'_{nN}}{E_N^{(0)} - E_N^{(0)}} |n\rangle^{(0)}. \tag{5.11}$$

When the multiplicity of $E_N^{(0)}$ is just one, equation (5.8) reduces to

$$E^{(1)} = H'_{NN} \tag{5.12}$$

where we have dropped the now unnecessary subscript k, so that

$$E_N = E_N^{(0)} + gH'_{NN} \tag{5.13}$$

is the total energy in the perturbed state. We see that, in this special case, *the first correction to the unperturbed energy is the average value of the perturbation Hamiltonian operator.* The correction to the unperturbed state vector is still expressed by equation (5.11), of course.

134

In order to calculate the second-order correction to $E_N^{(0)}$ and $|N\rangle^{(0)}$, we must write, by analogy with equations (5.3) and (5.4),

$$E_N = E_N^{(0)} + gE^{(1)} + g^2 E^{(2)}$$
$$|N\rangle = |N\rangle^{(0)} \oplus g|N\rangle^{(1)} \oplus g^2|N\rangle^{(2)}$$

which leads to

$$\mathscr{H}^{(0)}|N\rangle^{(2)} \oplus \mathscr{H}'|N\rangle^{(1)} = E_N^{(0)}|N\rangle^{(2)} \oplus E^{(1)}|N\rangle^{(1)} \oplus E^{(2)}|N\rangle^{(0)}.$$

We expand the $|N\rangle^{(2)}$ and $|N\rangle^{(1)}$ in the usual way

$$|N\rangle^{(1)} = \sum_{n=1}^{\infty} a_n^{(1)}|n\rangle^{(0)}$$

$$|N\rangle^{(2)} = \sum_{m=1}^{\infty} a_m^{(2)}|m\rangle^{(0)}$$

and find, similar to equation (5.7),

$$\sum_{m=1}^{\infty} a_m^{(2)} E_m^{(0)}|m\rangle^{(0)} \oplus \sum_{n=1}^{\infty} a_n^{(1)} \mathscr{H}'|n\rangle^{(0)}$$

$$= E_N^{(0)} \sum_{m=1}^{\infty} a_m^{(2)}|m\rangle^{(0)} \oplus E^{(1)} \sum_{n=1}^{\infty} a_n^{(1)}|n\rangle^{(0)} \oplus E^{(2)}|N\rangle^{(0)}. \quad (5.14)$$

The scalar product of equation (5.14) and $|N_{k'}\rangle^{(0)}$ $(1 \leqslant k' \leqslant \beta)$ is

$$a_N^{(2)} E_N^{(0)} a_{k'}{}^N + \sum_{n=1}^{\infty} \sum_{k=1}^{\beta} a_n^{(1)} a_k{}^N H'_{N_{k'} n_k} = E_N^{(0)} a_N^{(2)} a_{k'}{}^N + E^{(2)} a_{k'}{}^N$$

since $a_N^{(1)} \equiv 0$, as defined in the first-order treatment. Because

$$a_n^{(1)} = \frac{H'_{nN}}{E_N^{(0)} - E_n^{(0)}} \qquad (n \neq N),$$

we can write the scalar product in the form

$$\sum_{k=1}^{\beta} a_k{}^N \left(\sum_{n=1}^{\infty}{}' \frac{H'_{nN} H'_{N_{k'} n_k}}{E_N^{(0)} - E_n^{(0)}} - E^{(2)} \delta_{k'k} \right) = 0.$$

This set of linear equations will have no solution for the $a_k{}^N$ save the trivial one unless

$$\det \left| \sum_{n=1}^{\infty}{}' \frac{H'_{nN} H'_{N_{k'} n_k}}{E_N^{(0)} - E_n^{(0)}} - E^{(2)} \delta_{k'k} \right| = 0 \qquad (k' = 1, \ldots, \beta). \quad (5.15)$$

Quantum Mechanical Approximation Methods

We see that there are once again β possible distinct values for the perturbation energy. The expansion coefficients for $|N\rangle^{(2)}$ are calculated by forming the scalar product of equation (5.14) and $|N'\rangle^{(0)}$, with the result

$$a_{N'}^{(2)} = \begin{cases} \displaystyle\sum_{n=1}^{\infty}{}' \frac{H'_{nN}(H'_{N'n} - E^{(1)}\delta_{N'n})}{(E_N^{(0)} - E_{N'}^{(0)})(E_N^{(0)} - E_n^{(0)})} & (N' \neq N) \\[3mm] \displaystyle -\frac{1}{2}\sum_{n=1}^{\infty}{}' \frac{|H'_{nN}|^2}{(E_N^{(0)} - E_n^{(0)})^2} & (N' = N). \end{cases} \tag{5.16}$$

The case $N' = N$ is found by insisting that the wavefunction $|N\rangle$ be normalized to terms of second order.

In the special case $\beta = 1$, we note that equation (5.15) reduces to

$$E^{(2)} = \sum_{n=1}^{\infty}{}' \frac{|H'_{Nn}|^2}{E_N^{(0)} - E_n^{(0)}}. \tag{5.17}$$

It is clear from our treatment of the second-order approximation that perturbation theory can get rather cumbersome as the degree of accuracy sought becomes greater. We can expect that the third-order expressions for the total energy and state vector will be very complicated in appearance and difficult to evaluate numerically. Fortunately, it is not always necessary to creep order-by-order to the exact solution of a perturbation problem. A generalized perturbation theory exists which is often quite useful and, when coupled with good physical insight, may give results which are exact or nearly so in a single operational stroke. Let us see how it works.

Suppose we have to solve the eigenvalue problem

$$\mathcal{H}\psi = E\psi. \tag{5.18}$$

By reference to a similar problem, physical intuition, or by just guessing, we choose a complete set of orthonormal polynomials ϕ_n $(n = 1, 2, \ldots)$ and expand ψ in terms of them:

$$\psi = \sum_{n=1}^{\infty} a_n \phi_n.$$

Equation (5.18) then becomes

$$\sum_{n=1}^{\infty} a_n(\mathcal{H} - E)\phi_n = 0$$

which, upon taking the scalar product with ϕ_m $(1 \leqslant m \leqslant \infty)$, may be re-written as

$$\sum_{n=1}^{\infty} a_n(H_{mn} - E\delta_{mn}) = 0 \qquad (m = 1, 2, \ldots) \tag{5.19}$$

where, as usual,

$$H_{mn} \equiv (\phi_m, \mathscr{H} \phi_n).$$

A very strong formal resemblance to the method of ordinary perturbation theory is obvious here. As before, the energy levels E are calculated by imposing the condition for nontrivial solutions of equations (5.19), namely,

$$\det |H_{mn} - E\delta_{mn}| = 0. \tag{5.20}$$

However, it must be remembered that the elements of the determinant in equation (5.20) are computed by using *arbitrary* orthonormal polynomials, *not* zeroth-order wavefunctions. Moreover, equation (5.20) will not be particularly easy to solve in general, because the determinant is of infinite order. This should be expected, of course, since a solution of equation (5.20) is equivalent to a solution of equation (5.18), which by hypothesis is not straightforward. On the other hand, the beauty and utility of the generalized method is that the polynomials ϕ_n *are* arbitrary and that a systematic way of reducing equation (5.20) is at hand. For example, if the ϕ_n were chosen very astutely, then we might have

$$\begin{vmatrix} H_{11} - E & 0 & 0 & \cdots \\ 0 & H_{22} - E & 0 & \cdots \\ 0 & 0 & H_{33} - E & \cdots \\ \vdots & \vdots & \vdots & \vdots \end{vmatrix} = 0$$

which implies, obviously, that

$$E_1 = H_{11}$$
$$E_2 = H_{22}$$

and so on. This represents an exact solution to the eigenvalue problem! If the ϕ_n were close in form to the solutions of equation (5.18), then we still might neglect the elements of the determinant which are not in the main diagonal, with the result that

$$E_N \equiv H_{NN} = (\phi_N, \mathscr{H} \phi_N) \equiv (\phi_N, [\mathscr{H}^{(0)} + \mathscr{H}']\phi_N)$$
$$= (\phi_N, \mathscr{H}^{(0)}\phi_N) + (\phi_N, \mathscr{H}'\phi_N) \equiv E_N^{(0)} + H'_{NN}$$

where the Hamiltonian operator $\mathscr{H}$ has been divided into a part for which the ϕ_N are essentially eigenfunctions and a remainder $\mathscr{H}'$. We see that neglecting the off-diagonal elements of the perturbation determinant is equivalent to using first-order perturbation theory for non-degenerate eigenvalues. (Note here that we have a unit coupling constant.)

Quantum Mechanical Approximation Methods

In certain applications we may find that some of the off-diagonal elements of the perturbation determinant are too large to be neglected. For example, we might have

$$\begin{vmatrix} H_{11} - E & 0 & 0 & 0 & \cdots \\ 0 & H_{22} - E & H_{23} & 0 & \cdots \\ 0 & H_{32} & H_{33} - E & 0 & \cdots \\ 0 & 0 & 0 & H_{44} - E & \cdots \\ \vdots & \vdots & \vdots & \vdots & \vdots \end{vmatrix} = 0,$$

which reduces to

$$H_{11} - E_1 = 0$$

$$\begin{vmatrix} H_{22} - E & H_{23} \\ H_{32} & H_{33} - E \end{vmatrix} = 0$$

$$H_{44} - E_3 = 0$$

and so on. The second expression—the unique one here—corresponds to first-order perturbation theory for a doubly degenerate energy state; for

$$H_{22} - E = [H_{22}^{(0)} + H_{22}'] - E^{(0)} - E^{(1)} = H_{22}' - E^{(1)}$$
$$H_{32} = [H_{32}^{(0)} + H_{32}'] = H_{32}',$$

and similarly for the remaining two determinant elements.

If the second order of approximation is desired, say, for (non-degenerate) E_2, then H_{22} should be substituted for the eigenvalues in equation (5.20) while all elements other than those corresponding to H_{n2} and H_{2n} are neglected:

$$\begin{vmatrix} H_{11} - H_{22} & H_{12} & 0 & 0 & \cdots \\ H_{21} & H_{22} - E_2 & H_{23} & H_{24} & \cdots \\ 0 & H_{32} & H_{33} - H_{22} & 0 & \cdots \\ \vdots & \vdots & \vdots & \vdots & \vdots \end{vmatrix} = 0.$$

The reduction of this determinant yields

$$(H_{22} - E_2)(H_{11} - H_{22})(H_{33} - H_{22}) - \cdots - H_{12}H_{21}(H_{33} - H_{22}) - \cdots$$
$$- H_{32}H_{23}(H_{11} - H_{22})(H_{44} - H_{22}) - \cdots - \cdots = 0$$

or

$$E_2 = H_{22} - \sum_{n=1}^{\infty} \frac{H_{n2}H_{2n}}{H_{nn} - H_{22}} \qquad (n \neq 2),$$

which is the same as equation (5.17) when $H_{nn} \equiv E_n^{(0)}$. The reduction process may be carried on this way indefinitely.

5.2. THE STARK EFFECT

As a first example of the use of perturbation theory, we shall investigate what happens to a hydrogen atom in the presence of a constant, uniform electric field of intensity **E**. The relative Hamiltonian function, in spherical polar space, for a hydrogen electron in such an external field is[3]

$$H(\mathbf{p}, \mathbf{r}) = \frac{p^2}{2\mu_e} - \frac{e^2}{r} - \boldsymbol{\mu} \cdot \mathbf{E} \tag{5.21}$$

where $\boldsymbol{\mu}$ is the *electric dipole moment vector* for the atom. We may choose the electric field vector to point along the z-axis of the atom, so that equation (5.21) may be written

$$H(\mathbf{p}, \mathbf{r}) = \frac{p^2}{2\mu_e} - \frac{e^2}{r} - (|e|r \cos \vartheta)|\mathbf{E}|$$

where $(|e|r \cos \vartheta)$ is the component of the dipole moment vector along the direction of the electric field. By the Schrödinger postulate, the Hamiltonian operator corresponding to $H(\mathbf{p}, \mathbf{r})$ is

$$\mathscr{H} = -\frac{\hbar^2}{2\mu_e} \nabla_{\mathbf{r}}^2 - \frac{e^2}{r} - (|e|r \cos \vartheta)|\mathbf{E}| \equiv \mathscr{H}^{(0)} + \mathscr{H}'$$

where

$$\mathscr{H}' \equiv -(|e|r \cos \vartheta)|\mathbf{E}|$$

is the perturbation Hamiltonian operator.

The zeroth-order state vector is an eigenfunction of $\mathscr{H}^{(0)}$ and, in general, has the form

$$\Psi_n(r, \vartheta, \varphi) = \sum_{l=0}^{n-1} a_l R_{nl}(r) \sum_{m'=-l}^{l} c_{m'} \Upsilon_l^{m'}(\vartheta, \varphi) \tag{5.22}$$

where a_l and $c_{m'}$ are constants and R_{nl} and $\Upsilon_l^{m'}$ have their previous meanings. As the constants $c_{m'}$ are quite arbitrary, we shall choose, for convenience,

$$c_{m'} \equiv \delta_{mm'}$$

so that equation (5.22) reduces to

$$\Psi_n^{(m)}(r, \vartheta, \varphi) = \sum_{l=|m|}^{n-1} a_l R_{nl}(r) \Upsilon_l^m(\vartheta, \varphi). \tag{5.23}$$

The zeroth-order energy is, by equations (5.1) and (5.23), and equation (4.48)

$$E_n^{(0)} = (\Psi_n^{(m)}, \mathscr{H}^{(0)} \Psi_n^{(m)}) = -\left(\frac{1}{n}\right)^2 \text{Rydbergs} \quad (n = 1, 2, 3, \ldots),$$

[3] For details concerning the perturbation term, see W. T. Scott, *The Physics of Electricity and Magnetism*, John Wiley and Sons, New York, 1966, pp. 105*ff*.

as expected. The first-order corrections to this energy, according to equation (5.9), are the solutions of the determinantal equation

$$\begin{vmatrix} H'^{(m)}_{n_{|m|}n_{|m|}} - E^{(1)} & \cdots & H'^{(m)}_{n_{|m|}n_{n-1}} \\ \vdots & & \vdots \\ H'^{(m)}_{n_{n-1}n_{|m|}} & \cdots & H'^{(m)}_{n_{n-1}n_{n-1}} - E^{(1)} \end{vmatrix} = 0 \tag{5.24}$$

where

$$H'^{(m)}_{nl'nl} = (R_{nl'}\Upsilon^m_{l'}, \mathscr{H}'R_{nl}\Upsilon^m_l)$$

$$= -|e|\,|\mathbf{E}|\int_0^{2\pi}\int_0^\pi\int_0^\infty R_{nl'}\overline{\Upsilon}^m_{l'}r\cos\vartheta R_{nl}\Upsilon^m_l r^2\,dr\,\sin\vartheta\,d\vartheta\,d\varphi.$$

We might ponder a moment on the physical significance of the determinantal elements, $H'^{(m)}_{nl'nl}$. First, it should be noted that *if the mean value of the z-component of the dipole moment is zero, then the unperturbed energy will suffer no first-order changes when the hydrogenic atom is in the presence of an electric field.* To see this, we need only compare

$$\langle\mu_z\rangle^{(m)}_n \equiv |e|\langle r\cos\vartheta\rangle^{(m)}_n = -\frac{1}{|\mathbf{E}|}\sum_{l'=|m|}^{n-1}\sum_{l=|m|}^{n-1}\bar{a}_{l'}a_l H'^{(m)}_{nl'nl}\Big/\sum_{l=|m|}^{n-1}|a_l|^2 \equiv 0$$

with the implication of equation (5.24), namely,

$$\sum_{l=|m|}^{n-1}a_l(H'^{(m)}_{nl'nl} - E^{(1)}\delta_{l'l}) = 0. \tag{5.25}$$

A vanishing $\langle\mu_z\rangle^{(m)}_n$ suggests that

$$\sum_{l'=|m|}^{n-1}\sum_{l=|m|}^{n-1}\bar{a}_{l'}a_l H'^{(m)}_{nl'nl} = \sum_{l'=|m|}^{n-1}\bar{a}_{l'}\sum_{l=|m|}^{n-1}a_l H'^{(m)}_{nl'nl}$$

vanishes also. However,

$$\sum_{\substack{l=|m|\\l\neq l'}}^{n-1}a_l H'^{(m)}_{nl'nl} = a_{l'}(E^{(1)} - H'^{(m)}_{nl'nl'})$$

by equation (5.25), so that

$$\sum_{l'=|m|}^{n-1}\bar{a}_{l'}\sum_{l=|m|}^{n-1}a_l H'^{(m)}_{nl'nl} = \sum_{l=|m|}^{n-1}|a_{l'}|^2[(E^{(1)} - H'^{(m)}_{nl'nl'}) + H'^{(m)}_{nl'nl'}]$$

$$= \left[\sum_{l'=|m|}^{n-1}|a_{l'}|^2\right]E^{(1)} \equiv 0$$

shows that $E^{(1)}$ vanishes. A zero value of $\langle\mu_z\rangle_n^{(m)}$ also suggests that in the state $|n\rangle^{(m)}$ all possible angles of orientation ϑ of the electron about the z-axis enjoy an equal probability of occurrence, such that $\langle\cos\vartheta\rangle_n^{(m)}$ receives equal and opposite contributions from integration over the two domains $(0 \leqslant \vartheta \leqslant \pi/2)$ and $(\pi/2 \leqslant \vartheta \leqslant \pi)$. It is clear that this can happen only for the state having a principal quantum number n equal to unity, because only for this state does $\Upsilon_l^m(\vartheta, \varphi)$ reduce to a value independent of ϑ. In all other states $\Upsilon_l^m(\vartheta, \varphi)$ depends on its arguments and there is vanishing expectation of finding the electron *randomly* oriented at any angle relative to the z-axis of the atom.

Explicitly, the determinant elements $H'^{(m)}_{nl'nl}$ are

$$H'^{(m)}_{nl'nl} = -\frac{|e|a_0|\mathbf{E}|}{8}\left\{\frac{(n-l'-1)!\,(2l'+1)(l'-|m|)!}{[(n+l')!]^3(l'+|m|)!}\right\}^{\frac{1}{2}}$$

$$\times \left\{\frac{(n-l-1)!\,(2l+1)(l-|m|)!}{[(n+l)!]^3(l+|m|)!}\right\}^{\frac{1}{2}}$$

$$\times \int_{-1}^{1} P_{l'}^m(z)zP_l^m(z)\,dz \int_0^\infty e^{-\rho}\rho^{l'+l}L_{n+l'}^{2l'+1}(\rho)\rho L_{n+l}^{2l+1}\rho^2\,d\rho \qquad (5.26)$$

where

$$\rho \equiv \left(\frac{2}{na_0}\right)r, \qquad z \equiv \cos\vartheta.$$

With the help of the recurrence relation [4]

$$zP_l^m(z) = (2l+1)^{-1}[(l+|m|)P_{l-1}^m(z) + (l-|m|+1)P_{l+1}^m(z)], \qquad (5.27)$$

equation (5.26) may be reduced to

$$H'^{(m)}_{nl'nl} = -\frac{|e|a_0|\mathbf{E}|}{4}\left[\left\{\frac{(l+|m|)(l-|m|)(n-l)!\,(n-l-1)!}{(2l+1)(2l-1)[(n+l-1)!\,(n+l)!]^3}\right\}^{\frac{1}{2}}I_n^{l-1,l}\delta_{l',l-1}\right.$$

$$+ \left\{\frac{(l+|m|+1)(l-|m|+1)(n-l-1)!\,(n-l-2)!}{(2l+3)(2l+1)[(n+l)!\,(n+l+1)!]^3}\right\}^{\frac{1}{2}}$$

$$\left. \times I_n^{l+1,l}\delta_{l',l+1}\right] \qquad (5.28)$$

where

$$I_n^{k',k} \equiv \int_0^\infty e^{-x}x^{k'+k+3}L_{n+k'}^{2k'+1}(x)L_{n+k}^{2k+1}(x)\,dx.$$

We see from this result that only the perturbation elements $H'^{(m)}_{nl-1nl}$ and $H'^{(m)}_{nl+1,nl}$ are not zero in equation (5.24). [Actually, because $\mathscr{H}'$ is a Hermitian operator,

[4] For a derivation of equation (5.27), see H. Margenau and G. M. Murphy, *The Mathematics of Physics and Chemistry*, D. Van Nostrand Co., Princeton, 1956, Vol. I, p. 106f.

the element $H'^{(m)}_{n_{l-1},n_l}$ can be obtained from $H'^{(m)}_{n_{l+1},n_l}$ by putting $l \to (l-1)$, taking the complex conjugate, and transposing indices.]

As a concrete example of the first-order theory, let us investigate qualitatively the perturbation of the hydrogen atom in its first and second excited states. Electronic transitions between these states give rise in the unperturbed atom to the 6562-Å (red) line in the Balmer series. In the first excited state, the electric field only partially removes the four-fold degeneracy; for, the first-order corrections to the energy vanish when $m = \pm 1$. When $m = 0$, equation (5.24) reduces to

$$\begin{vmatrix} -E^{(1)} & H'^{(0)}_{2_02_1} \\ H'^{(0)}_{2_12_0} & -E^{(1)} \end{vmatrix} = 0$$

which has two distinct solutions. Thus the second energy level in hydrogen is split by an electric field into *three* equally spaced levels. (See Figure 35.) The first and third levels are of multiplicity one, while the second level—whose magnitude is that of the unperturbed energy—is of multiplicity two. In the second excited state, the first-order corrections vanish when $m = \pm 2$. When $m = \pm 1$, we have

$$\begin{vmatrix} -E^{(1)} & H'^{(1)}_{3_13_2} \\ H'^{(1)}_{3_23_1} & -E^{(1)} \end{vmatrix} = 0,$$

giving two distinct solutions for $E^{(1)}$. When $m = 0$, the perturbation determinant becomes

$$\begin{vmatrix} -E^{(1)} & H'^{(0)}_{3_03_1} & 0 \\ H'^{(0)}_{3_13_0} & -E^{(1)} & H'^{(0)}_{3_13_2} \\ 0 & H'^{(0)}_{3_23_1} & -E^{(1)} \end{vmatrix} = 0$$

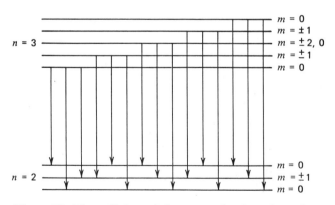

Figure 35. The splitting of the energy levels and resultant allowed transitions in the Stark effect for the 6562-Å line in the hydrogen spectrum.

Figure 36. The Stark effect for the 6562-Å line in the hydrogen spectrum. (From *Introduction to Atomic and Nuclear Physics*, by H. E. White, Copyright © 1964, by Litton Educational Publishing, Inc., by permission of Van Nostrand Reinhold Company).

which also possesses only two non-vanishing roots. Therefore, the third energy level splits into *five* levels. Two of the levels are of multiplicity one, two are of multiplicity two, while the remaining level is of multiplicity three. The electronic transitions between the eight levels corresponding to the first two excited states are diagrammed in Figure 35. The fifteen possible transitions correspond to fifteen different frequencies of radiation; *we therefore expect to see the red Balmer line split into that number of closely spaced components.* This splitting of a spectral line in an electric field is called the *Stark effect*, after J. Stark, who first reported it in 1913. In Figure 36, the splitting of the red Balmer line in an electric field of 10^5 volts/cm is shown. Nine distinct components are to be seen. The six lines missing here, it turns out,[5] are of far too small intensities to be easily detected.

Before closing our investigation of the Stark effect, we shall consider the behavior of a diatomic molecule—approximated as the so-called "rigid rotator"—in an external electric field. The unperturbed molecule may be thought of as two particles constrained to remain a distance r_e apart, such that the Hamiltonian function for their relative motion is

$$H(\mathbf{L}, \mathbf{r}_e) = \frac{L^2}{2\mu r_e^{\,2}}$$

[5] See L. Pauling and S. Goudsmit, *The Structure of Line Spectra*, McGraw-Hill Book Co., New York, 1930, p. 82*f*.

143

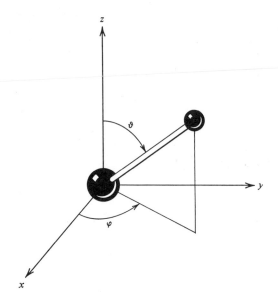

Figure 37. Angular coordinates for the rigid rotator.

where **L** is the relative angular momentum and μ is the reduced mass. The Hamiltonian operator corresponding to this function is, according to equation (4.19),

$$\mathcal{H}^{(0)} = \frac{-\hbar^2}{2\mu r_e^2} \left\{ \frac{1}{\sin \vartheta} \frac{\partial}{\partial \vartheta} \left(\sin \vartheta \frac{\partial}{\partial \vartheta} \right) + \frac{1}{\sin^2 \vartheta} \frac{\partial^2}{\partial \varphi^2} \right\} \tag{5.29}$$

where the angles ϑ and φ are defined in Figure 37. It is not difficult to see that the eigenfunctions of $\mathcal{H}^{(0)}$ are sums over the spherical harmonics, and that the energy levels permitted the rotating molecule are given by

$$E_l^{(0)} = l(l + 1)\frac{\hbar^2}{2I} \qquad (l = 0, 1, 2, \ldots)$$

where

$$I = \mu r_e^2$$

is the moment of inertia of the molecule. The energy levels $E_l^{(0)}$ are of multiplicity $(2l + 1)$.

In the electric field, the molecule is described by the eigenvalue problem

$$(\mathcal{H}^{(0)} - \mu|\mathbf{E}| \cos \vartheta)\psi(\vartheta, \varphi) = E\psi(\vartheta, \varphi)$$

where μ is now the electric dipole moment, $|\mathbf{E}|$ is the magnitude of the electric field, and $\psi(\vartheta, \varphi)$ is the new state vector. Of interest to our approximation scheme is the determinant element

$$H'_{l'm'lm} = (\Upsilon_{l'}^{m'}, \mathscr{H}'\Upsilon_l^m) = -\mu|\mathbf{E}| \int_0^{2\pi} \int_0^{\pi} \overline{\Upsilon_{l'}^{m'}} \cos \vartheta \Upsilon_l^m \sin \vartheta \, d\vartheta \, d\varphi.$$

We have, by equation (5.27) and the integral properties of the spherical harmonics,

$$H'_{l'm'lm} = -\mu|\mathbf{E}| \frac{\delta_{m'm}}{2} \left\{ \left[\frac{(l + |m|)(l - |m|)}{(2l + 1)(2l - 1)} \right]^{1/2} \delta_{l',l-1} \right.$$
$$\left. + \left[\frac{(l + |m| + 1)(l - |m| + 1)}{(2l + 3)(2l + 1)} \right]^{1/2} \delta_{l',l+1} \right\}. \tag{5.30}$$

We see that the determinant elements depend only upon the absolute value of the magnetic quantum number and that all the elements $H'_{lm'lm}$ of the first-order perturbation determinant vanish. Thus *there is no first-order correction to the energy.* The second-order corrections are, by equation (5.15), the solutions of

$$\det \left| \sum_{l'=0}^{\infty} \frac{H'_{l'l}H'_{lm'l'm}}{E_l^{(0)} - E_{l'}^{(0)}} - E^{(2)}\delta_{m'm} \right| = 0. \tag{5.31}$$

But, by equation (5.30), the perturbation determinant has non-vanishing elements only on the main diagonal. The reason for this result is that the perturbation Hamiltonian does not depend on the azimuthal angle φ. The second-order corrections are not all equal in value, however, since the determinant element $H'_{lml'm}$ does at least depend on the absolute value of m. We note that in this case of the Stark effect, as before, the applied electric field does not remove all of the degeneracy in the unperturbed state, there being two possible m-values for each value of the second-order correction.

Because of the simplification in the perturbation determinant we may, if we wish, assume the unperturbed state to be an eigenstate of $\mathscr{L}_z$. In that case, equation (5.31) becomes

$$E_{l|m|}^{(2)} = \sum_{l'=1}^{\infty} \frac{|H'_{lml'm}|^2}{E_l^{(0)} - E_{l'}^{(0)}} = \frac{|H'_{(l-1)mlm}|^2}{E_l^{(0)} - E_{l-1}^{(0)}} + \frac{|H'_{(l+1)mlm}|^2}{E_l^{(0)} - E_{l+1}^{(0)}}.$$

When the results in equation (5.30) are used, we find

$$E_{l|m|}^{(2)} = \frac{I\mu^2}{2\hbar^2} \left[\frac{(l^2 - |m|^2)}{(2l - 1)2l(2l + 1)} - \frac{(l + 1)^2 - |m|^2}{(2l + 1)(2l + 3)^2} \right] |\mathbf{E}|^2. \tag{5.32}$$

Now, $E^{(2)}$ may be construed physically as the energy attributable to the dipole moment induced in the molecule (rotator) by the electric field. This energy always is a quadratic function of the field intensity such that

$$\text{induced dipole energy} = -\tfrac{1}{2}\alpha|\mathbf{E}|^2$$

where α is called the induced dipole moment per unit field strength, or the *isotropic polarizability* of the molecule. Here we have, evidently,

$$\alpha_{l|m|} = -\frac{I\mu^2}{\hbar^2}\left[\frac{(l^2 - |m|^2)}{(2l - 1)2l(2l + 1)} - \frac{(l + 1)^2 - |m|^2}{(2l + 1)(2l + 3)^2}\right]. \qquad (5.33)$$

For all but the state corresponding to $l = 0$ the polarizability is a *negative* number. Physically, this result means that the electric field polarizes the molecule in a direction parallel to its own when the latter is in its (rotational) groundstate. Otherwise, the field tends to polarize the molecule in the anti-parallel direction. This is because in the groundstate the rotator's energy is insufficient to permit it to do more than oscillate; while in its excited states, however, the rotator can revolve freely and is slowed down only when its induced dipole moment vector points opposite the field direction.

5.3. THE GROUNDSTATES OF HELIUM-LIKE ATOMS

Time-independent perturbation theory is really quite a powerful device. In many physical applications (fortunately), a quantitative picture can be had from just the first- and second-order formalisms. As a case in point, let us investigate the two-electron atom in its groundstate, using the first-order theory. In this case the complete relative Hamiltonian operator

$$-\frac{\hbar^2}{2\mu_e}\{\nabla_1^2 + \nabla_2^2\} - \left\{\frac{Ze^2}{r_1} + \frac{Ze^2}{r_2} - \frac{e^2}{r_{12}}\right\} \equiv \mathscr{H}$$

factors into

$$\mathscr{H}^{(0)} \equiv -\frac{\hbar^2}{2\mu_e}\{\nabla_1^2 + \nabla_2^2\} - \left(\frac{Ze^2}{r_1} + \frac{Ze^2}{r_2}\right)$$

and

$$\mathscr{H}' \equiv \frac{e^2}{r_{12}} \qquad (5.34)$$

where $\mathbf{r}_1$ is the displacement of one electron from the nucleus, $\mathbf{r}_2$ is the displacement of the other one, and $\mathbf{r}_{12}$ is the vector separation of the two particles. (See Figure 38.) The zeroth-order state vector is an eigenfunction of $\mathscr{H}^{(0)}$ and so may be expressed

$$\Psi_1^{(0)}(\mathbf{r}_1, \mathbf{r}_2) = \psi_{100}(\mathbf{r}_1)\psi_{100}(\mathbf{r}_2) \qquad (5.35)$$

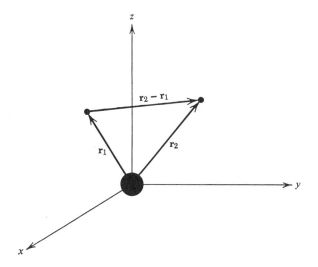

Figure 38. Vector coordinates for a two-electron atom.

where, by equation (4.39),

$$\psi_{100}(r) = (Z^3/\pi a_0^3)^{1/2} \exp(-Zr/a_0)$$

is the one-electron groundstate wavefunction. It is easy to see, then, that the zeroth-order energy is

$$E_1^{(0)} = -2Z^2 \text{ Rydbergs}.$$

Because the zeroth-order energy is of unit multiplicity, the first-order correction to the total electronic energy is expressed by equation (5.12) rather than (5.9). It follows from equations (5.34) and (5.35) that

$$E_1^{(1)} = (\Psi_1^{(0)}, \mathcal{H}'\Psi_1^{(0)})$$

$$= \frac{Z^6 e^2}{\pi^2 a_0^6} \int_0^{2\pi} \int_0^\pi \int_0^\infty \int_0^{2\pi} \int_0^\pi \int_0^\infty \exp\left[-2Z(r_1 + r_2)/a_0\right]$$

$$\times \frac{r_1^2}{r_{12}} dr_1 \sin\vartheta_1 \, d\vartheta_1 \, d\varphi_1 r_2^2 \, dr_2 \sin\vartheta_2 \, d\vartheta_2 \, d\varphi_2.$$

Now, the volume element

$$r_2^2 \, dr_2 \sin\vartheta_2 \, d\vartheta_2 \, d\varphi_2$$

may be replaced by

$$-r_{12}^2 \, dr_{12} \sin\psi_{12} \, d\psi_{12} \, d\varphi_{12}$$

where

$$\psi_{12} = \pi - \vartheta_{12}$$

147

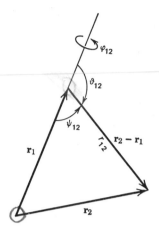

Figure 39. Relative coordinate
frames for a two-electron atom.

and r_{12}, ϑ_{12}, and φ_{12} are defined in Figure 39. To demonstrate the validity of this substitution we need show but three things:

(a) The volume element $dx\, dy\, dz$ in cartesian coordinates is equal to the volume element $r^2\, dr \sin \vartheta\, d\vartheta\, d\varphi$ in spherical polar coordinates.

(b) The volume element $d(x + a)d(y + b)d(z + c)$, where a, b, and c are constants, is equal to $dx\, dy\, dz$.

(c) The volume element $dx''\, dy''\, dz''$, defined through the second of the coordinate transformations shown in Figure 40, is equal to the volume element $dx_2\, dy_2\, dz_2$, where x_2, y_2, z_2 are the cartesian coordinates of the second electron (Figure 39).

In Figure 40 it is shown that the spherical polar coordinate system $(r_{12}, \vartheta_{12}, \varphi_{12})$ may be obtained from $(r_2, \vartheta_2, \varphi_2)$ by first converting to the cartesian coordinates x_2, y_2, z_2, rotating the reference frame through the angle ϑ_1 about the x_2-axis and then through φ_2 about the z''-axis, translating the reference frame along the z''-axis to the point $(0, 0, r_1)$, and, finally, converting to the spherical polar coordinates $(r_{12}, \vartheta_{12}, \varphi_{12})$. These operations must be shown to preserve the infinitesimal volume element $r_2{}^2\, dr_2 \sin \vartheta_2\, d\vartheta_2\, d\varphi_2$. The steps required in the proof are (a), (c), and (b) given above. The general method for carrying out these steps involves the use of the *Jacobian determinant*, defined by

$$ J(\mathbf{q'}, \mathbf{q}) \equiv \begin{vmatrix} \dfrac{\partial q_1'}{\partial q_1} & \dfrac{\partial q_1'}{\partial q_2} & \cdots & \dfrac{\partial q_1'}{\partial q_k} \\ \vdots & & & \vdots \\ \dfrac{\partial q_k'}{\partial q_1} & & & \dfrac{\partial q_k'}{\partial q_k} \end{vmatrix} \qquad (k = 1, 2, 3, \ldots). $$

The Groundstates of Helium-Like Atoms

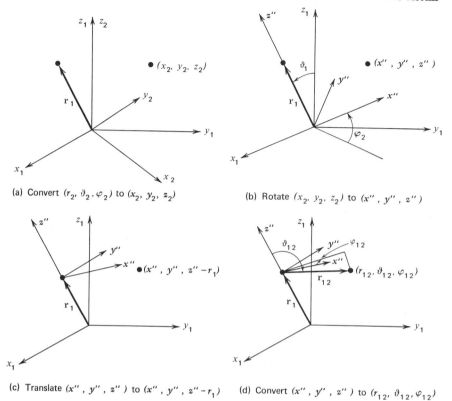

(a) Convert $(r_2, \vartheta_2, \varphi_2)$ to (x_2, y_2, z_2)

(b) Rotate (x_2, y_2, z_2) to (x'', y'', z'')

(c) Translate (x'', y'', z'') to $(x'', y'', z'' - r_1)$

(d) Convert (x'', y'', z'') to $(r_{12}, \vartheta_{12}, \varphi_{12})$

Figure 40. A four-step coordinate transformation from the spherical polar coordinates $(r_2, \vartheta_2, \varphi_2)$ to the relative coordinates $(r_{12}, \vartheta_{12}, \varphi_{12})$.

The importance of the Jacobian lies with the fact that[6]

$$dq_1' \, dq_2' \cdots dq_k' = J(\mathbf{q}', \mathbf{q}) \, dq_1 \, dq_2 \cdots dq_k \qquad (k = 1, 2, 3, \ldots). \qquad (5.36)$$

We see from equation (5.36) that two infinitesimal volume elements are equal if the Jacobian determinant is equal to one. As regards step (a) above, we have already shown implicitly in section 4.2 that the Jacobian determinant is of unit value. Step (b), which is the second to last step portrayed in Figure 40, may also be dealt with summarily, because $d(q + \alpha) = dq$ for any q and any constant α. It remains only to demonstrate that the set of coordinate transforms

$$x'' = \cos \varphi_2 x_2 + \sin \varphi_2 \cos \vartheta_1 y_2 + \sin \varphi_2 \sin \vartheta_1 z_2$$
$$y'' = -\sin \varphi_2 x_2 + \cos \varphi_2 \cos \vartheta_1 y_2 + \cos \varphi_2 \sin \vartheta_1 z_2$$
$$z'' = -\sin \vartheta_1 y_2 + \cos \vartheta_1 z_2$$

[6] H. Margenau and G. M. Murphy, *The Mathematics of Physics and Chemistry*, D. Van Nostrand Co., Princeton, 1956, Vol. I, pp. 17–20.

149

corresponds to a unit Jacobian determinant. We have

$$J(\mathbf{x}'', \mathbf{x}_2) = \begin{vmatrix} \cos \varphi_2 & \sin \varphi_2 \cos \vartheta_1 & \sin \varphi_2 \sin \vartheta_1 \\ -\sin \varphi_2 & \cos \varphi_2 \cos \vartheta_1 & \cos \varphi_2 \sin \vartheta_1 \\ 0 & -\sin \vartheta_1 & \cos \vartheta_1 \end{vmatrix}$$

$$= \cos^2 \varphi_2 \cos^2 \vartheta_1 + \cos^2 \varphi_2 \sin^2 \vartheta_1 + \sin^2 \varphi_2 \sin^2 \vartheta_1$$
$$+ \sin^2 \varphi_2 \cos^2 \vartheta_1 = \cos^2 \varphi_2 + \sin^2 \varphi_2 = 1.$$

The final result is now obtained by noting that

$$\sin \vartheta_{12} \, d\vartheta_{12} = -d(\cos \vartheta_{12}) = -d(\cos (\pi - \psi_{12}))$$
$$= -d(\cos \pi \cos \psi_{12}) = d(\cos \psi_{12}).$$

Just one more change in the volume element

$$-r_{12}{}^2 \, dr_{12} \sin \psi_{12} \, d\psi_{12} \, d\varphi_{12}$$

needs to be made before $E_1^{(1)}$ can be evaluated. From Figure 39 and the law of cosines we have

$$r_2{}^2 = r_1{}^2 + r_{12}{}^2 - 2r_1r_{12} \cos \psi_{12}$$

so that

$$r_2 \, dr_2 = r_1r_{12} \sin \psi_{12} \, d\psi_{12}.$$

When this result is incorporated into the volume element, we find

$$\frac{-r_{12}}{r_1} \, dr_{12}r_2 \, dr_2 \, d\varphi_{12}.$$

The expression for $E_1^{(1)}$ may now be written

$$E_1^{(1)} = -\frac{Z^6e^2}{\pi^2a_0{}^6} \int_0^{2\pi} \int_{r_2+r_1}^{|r_2-r_1|} \int_0^\infty \int_0^{2\pi} \int_0^\pi \int_0^\infty$$

$$\times \exp\left[-2Z(r_1 + r_2)/a_0\right]r_1 \, dr_1 \sin \vartheta_1 \, d\vartheta_1 \, d\varphi_1 r_2 \, dr_2 \, dr_{12} \, d\varphi_{12}$$

where it has been noted that

$$(r_2 + r_1) \geqslant r_{12} \geqslant |r_2 - r_1|$$

corresponds to

$$0 \leqslant \psi_{12} \leqslant \pi.$$

The integration over the angles can be done immediately with the result

$$E_1^{(1)} = \frac{8Z^6e^2}{a_0{}^6} \int_{|r_2-r_1|}^{r_2+r_1} \int_0^\infty \int_0^\infty \exp\left[-2Z(r_1 + r_2)/a_0\right]r_1 \, dr_1 r_2 \, dr_2 \, dr_{12}.$$

In doing the integration over r_{12}, we must distinguish the two cases $r_2 \geqslant r_1$ and $r_2 \leqslant r_1$. The triple integral above thus splits into a pair of triple integrals, such that

$$E_1^{(1)} = \frac{8Z^6 e^2}{a_0^6} \left\{ \int_{|r_2-r_1|}^{r_2+r_1} \int_{r_1}^{\infty} \int_0^{\infty} \exp\left[-2Z(r_1+r_2)/a_0 \right] r_1 \, dr_1 r_2 \, dr_2 \, dr_{12} \right.$$

$$\left. + \int_{|r_1-r_2|}^{r_1+r_2} \int_{r_2}^{\infty} \int_0^{\infty} \exp\left[-2Z(r_1+r_2)/a_0 \right] r_2 \, dr_2 r_1 \, dr_1 \, dr_{12} \right\}.$$

$r_2 \geqslant r_1$

$r_1 + r_2$

Upon making the change of variable

$$\rho = \left(\frac{2Z}{a_0} \right) r$$

and noting that

$$\int_A^{\infty} x e^{-x} = e^{-A}(A+1)$$

$$\int_0^{\infty} x^n e^{-ax} = \frac{n!}{a^{n+1}}$$

we find for $E_1^{(1)}$ the expression

$$E_1^{(1)} = \frac{5}{8} \frac{Ze^2}{a_0}.$$

The total electronic energy, correct to first order, is then

$$E_1 = -\left(2Z^2 - \frac{5Z}{4} \right) \text{Rydbergs}. \tag{5.37}$$

In Table 5.1 values of E_1 calculated using equation (5.37) have been compared with the appropriate experimental data compiled by J. C. Slater[7] for Z-values

Table 5.1. Theoretical and experimental values of the groundstate energies of several two-electron atoms.

Atom	$-E_1$ (calcd.) Rydbergs	$-E_1$ (obsd.) Rydbergs
He	5.50	5.81
Li^+	14.25	14.56
Be^{2+}	27.00	27.31
B^{3+}	43.75	44.07
C^{4+}	64.50	64.83
N^{5+}	89.25	89.60
O^{6+}	118.00	118.39

[7] J. C. Slater, *Quantum Theory of Atomic Structure*, McGraw-Hill Book Co., New York, 1960, pp. 339–342.

between two and eight. The agreement is seen to be very good, the estimates from first-order theory differing from the empirical data by only 0.3 Rydberg in each case.

5.4. TIME-DEPENDENT PERTURBATION THEORY

When a physical system is perturbed, there are actually two ways in which it may respond. The first possibility is that the permitted states of the system will be altered, such that an entirely new set of state vectors and energy eigenvalues are generated by the perturbation. The formalism associated with this kind of result has been the subject of our analysis thus far. The second possibility, which will be taken up now, is that the perturbation may cause a transition between whatever state the system may be found in and one other of the set of possible *unperturbed* states. This kind of perturbation effect will, in general, lead to physical situations quite different in character from those engendered by the first possibility. The theoretical alternative in a given instance, of course, should be decided on an empirical basis. As a rough guideline in this matter, it may be suggested that the alteration of state is encountered when the perturbing agent exerts its influence for periods of time comparable to the natural lifetime of the isolated, unperturbed system, and that the transition of state is found when the perturbing influence is of short enough duration to be labeled "time-dependent." It is not necessary, however, that in the second case the perturbation be *explicitly* time-dependent.

Now let us consider the time-dependent perturbation theory. We suppose to begin with that the Hamiltonian operator for a chosen system may be written in the form

$$\mathscr{H} = \mathscr{H}^{(0)} + \mathscr{H}'$$

where $\mathscr{H}'$ is to be regarded as a perturbation Hamiltonian operator dependent in some way on the time. The Hamiltonian operator $\mathscr{H}^{(0)}$ is then to be taken as the generator of the initial state of the system, in the sense that

$$\mathscr{H} = \begin{cases} \mathscr{H}^{(0)} & (t \leqslant t_i) \\ \mathscr{H}^{(0)} + \mathscr{H}', & (t_i \leqslant t \leqslant t_f) \end{cases}$$

where t_i and t_f are arbitrary values of the time variable. The Schrödinger equation constructed from $\mathscr{H}^{(0)}$ is

$$\mathscr{H}^{(0)}\Psi^{(0)}(\mathbf{r}, t) = i\hbar \frac{\partial \Psi^{(0)}}{\partial t} \tag{5.38}$$

which has as its general solution

$$\Psi^{(0)}(\mathbf{r}, t) = \sum_{k=1}^{n} c_k \psi_k^{(0)}(\mathbf{r}) \exp\left(-iE_k t/\hbar\right), \tag{5.39}$$

the c_k $(k = 1, \ldots, n)$ being arbitrary constants and the $\psi_k^{(0)}(\mathbf{r})$ being the n solutions of the Schrödinger eigenvalue problem corresponding to equation (5.38). When $t \leqslant t_i$, it is clear that $\Psi^{(0)}(\mathbf{r}, t)$ is the complete solution of

$$\mathscr{H}\Psi(r, t) = i\hbar \frac{\partial \Psi}{\partial t} \tag{5.40}$$

with the operator $\mathscr{H}$ defined as above. At all other times it is still possible to construct a formal solution for equation (5.40) because the functions

$$\Psi_k^{(0)}(\mathbf{r}, t) \equiv \psi_k^{(0)}(\mathbf{r}) \exp\left(-iE_k t/\hbar\right)$$

form a complete orthonormal set. The only condition on this formal solution is that the coefficients c_k $(k = 1, \ldots, n)$ must be functions of the time such that equation (5.40) is satisfied. Thus we have

$$\Psi(\mathbf{r}, t) = \sum_{k=1}^{n} c_k(t)\psi_k^{(0)}(\mathbf{r}) \exp\left(-iE_k t/\hbar\right) \qquad (t > t_i) \tag{5.41}$$

as the complete, formal solution of (5.40). By means of this device we have reduced the problem of what happens to the system after the perturbation $\mathscr{H}'$ is "switched on" to the deduction of the behavior of the $c_k(t)$. A differential equation for these functions may be derived by substituting equation (5.41) into equation (5.40), noting equation (5.38), and taking the scalar product of the result with any one of the $\Psi_k^{(0)}(\mathbf{r}, t)$ [say, $\Psi_l^{(0)}(\mathbf{r}, t)$]. The expressions for the $c_l(t)$ then turn out to be

$$\frac{dc_l}{dt} = -\frac{i}{\hbar} \sum_{k=1}^{n} c_k(t)H_{lk}' \exp\left(i\omega_{lk}t\right) \qquad (t > t_i) \tag{5.42}$$

where

$$H_{lk}' \equiv (\psi_l^{(0)}, \mathscr{H}'\psi_k^{(0)})$$
$$\omega_{lk} \equiv (E_l - E_k)/\hbar.$$

Equations (5.42) are a set of coupled, linear, homogeneous differential equations which cannot be solved in general by any straightforward means. It is for this reason *only* that approximation schemes find a place in time-dependent perturbation theory.

Suppose, now, that the system is in the state $\Psi_m^{(0)}(\mathbf{r}, t)$ when $t \leqslant t_i$. Then we must have

$$c_k(t) = \delta_{mk} \qquad (t \leqslant t_i). \tag{5.43}$$

After the perturbation has been switched on, we expect that $c_m(t)$ will decrease from its initial value and that some of the other $c_k(t)$ will grow in magnitude. However, during the very small time interval

$$t_i < t \leqslant t_i + \tau$$

we can without serious error substitute equation (5.43) into equations (5.42). The result is

$$\frac{dc_m}{dt} = -\frac{i}{\hbar} H'_{mm'} \qquad (t_i < t \leqslant t_i + \tau)$$

$$\frac{dc_l}{dt} = -\frac{i}{\hbar} H'_{lm} \exp(i\omega_{lm}t) \qquad (t_i < t \leqslant t_i + \tau), \tag{5.44}$$

the second expression being for all $l \neq m$. If the time interval τ is so short that the explicit time-dependence of H'_{lm} (if any) can be ignored, then the solutions of equations (5.44) are

$$c_m(t_i + \tau) = 1 - \frac{i}{\hbar} H'_{mm}\tau \tag{5.45}$$

$$c_l(t_i + \tau) = H'_{lm} \frac{\exp(i\omega_{lm}t_i)[\exp(i\omega_{lm}\tau) - 1]}{E_m - E_l} \qquad (l \neq m). \tag{5.46}$$

Equations (5.45) and (5.46) are the first-order solutions for the coefficients in equation (5.41). By analogy with the interpretation of the time-independent coefficients in equation (5.39), the quantity

$$\frac{|c_l(t)|^2}{\sum\limits_{l=1}^{n} |c_l(t)|^2}$$

is called *the probability of finding the system in the state* $\Psi_l^{(0)}(\mathbf{r}, t)$ *at the time t*, given that its initial state was $\Psi_m^{(0)}(\mathbf{r}, t_i)$. In the case of equation (5.46) the transition probability is

$$|c_l(t_i + \tau)|^2 = |H'_{lm}|^2 \frac{\sin^2(\omega_{lm}\tau/2)}{(\hbar\omega_{lm}/2)^2} \tag{5.47}$$

in the first-order approximation. We see that the probability is very small unless E_l is near in value to E_m. (*Cf.* Figure 41.) This result is consistent with the energy conservation principle.

To obtain the second-order expressions for the $c_l(t)$, we impose the condition that the transition from the state $\Psi_m^{(0)}$ to a certain one of the $\Psi_l^{(0)}$ must involve the *intermediate states* $\Psi_{l'}^{(0)}(l' \neq l, m)$, such that (for $t_i = 0$)

$$\frac{dc_{l'}}{dt} = -\frac{i}{\hbar} H'_{l'm} \exp(i\omega_{l'm}t) \qquad (0 < t \leqslant \tau) \tag{5.48}$$

is the analog of the first of equations (5.44). However, we do not require that the $c_{l'}(0)$ be equal to zero. The solution of equation (5.48) follows as

$$c_{l'}(t) = \frac{H'_{l'm}}{E_m - E_{l'}} \exp(i\omega_{l'm}t) \qquad (0 < t \leqslant \tau) \tag{5.49}$$

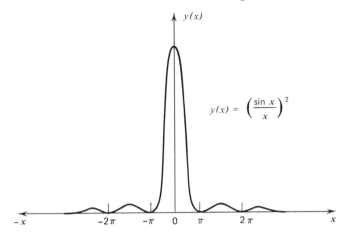

Figure 41. A plot of the function $y(x) = (\sin x/x)^2$.

where it is understood that $H'_{l'm}$ is virtually time-independent and that $c_{l'}(0) \equiv H'_{l'm}/(E_m - E_{l'})$. The differential equation for the $c_l(t)$ will be, in agreement with the second-order condition and the second of equations (5.44),

$$\frac{dc_l}{dt} = -\frac{i}{\hbar} \sum_{l'} c_{l'}(t) H'_{ll'} \exp(i\omega_{ll'}t)$$

$$= -\frac{i}{\hbar} \sum_{l'} \frac{H'_{ll'} H'_{l'm}}{E_m - E_{l'}} \exp(i\omega_{lm}t) \qquad (0 < t \le \tau) \qquad (5.50)$$

where equation (5.49) has been used in getting the second step. If we define

$$H_{lm} \equiv \sum_{l'} \frac{H'_{ll'} H'_{l'm}}{E_m - E_{l'}},$$

then the solution of equation (5.50) follows easily as

$$c_l(t) = H_{lm} \frac{\exp(i\omega_{lm}t) - 1}{E_m - E_l} \qquad (0 < t \le \tau). \qquad (5.51)$$

This is the second-order expression for the $c_l(t)$. It is not very difficult to see that the approximation scheme may be carried on indefinitely by mixing more and more intermediate states into the differential equation for the $c_l(t)$. One point should be clearly understood here. Although the transition probability computed, for example, from equation (5.51) and considered as a function of ω_{lm}, will be sharply peaked about zero, energy conservation is *not* implied for transitions from the initial state to any of the intermediate states. Indeed, we must have it that $E_m \ne E_{l'}$ for the quantity H_{lm} to be mathematically determinate. This contradiction of the energy conservation principle,

155

however, is of no physical importance because only the initial state $\Psi_m^{(0)}$ and the final state $\Psi_l^{(0)}$ are of concern in an experiment. For this reason, transitions from $\Psi_m^{(0)}$ to any of the $\Psi_l^{(0)}$ are called *virtual transitions* and have only a computational, not empirical, significance.

5.5. THE HYDROGENIC ATOM IN A PURE RADIATION FIELD

As an example of the very great utility of time-dependent perturbation theory in the first approximation, we shall investigate the behavior of a hydrogenic atom subjected to a pure radiation field. Out of this study we expect to obtain a derivation of Bohr's postulate concerning the relation between the energy levels of a hydrogenic atom and the frequencies of light it may absorb; that is, we expect to find

$$\nu = \frac{(E_f - E_i)}{h},$$

and to deduce the selection rules for electronic transitions which were stated in Chapter 4 without proof:

$$\Delta l = \pm 1$$
$$\Delta m = 0, \pm 1$$

l and m being the angular momentum and magnetic quantum numbers, respectively.

The Hamiltonian function for a single hydrogenic electron in a pure radiation field is[8]

$$H(\mathbf{p}, \mathbf{A}, \mathbf{r}) = \frac{1}{2\mu_e} \left(\mathbf{p} - \frac{e}{c}\mathbf{A}\right)^2 - \frac{Ze^2}{r} \qquad (e < 0)$$

where $\mathbf{A}(\mathbf{r}, t)$ is the vector potential, related to the magnetic and electric field vectors by

$$\mathbf{H} = \nabla \times \mathbf{A}$$
$$\mathbf{E} = -\frac{1}{c}\frac{\partial \mathbf{A}}{\partial t}.$$

The vector potential is in this case a solution of the wave equation

$$\nabla^2 \mathbf{A} - \frac{1}{c^2}\frac{\partial^2 \mathbf{A}}{\partial t^2} = 0$$

subject to

$$\nabla \cdot \mathbf{A} = 0.$$

[8] See, for example, T. C. Bradbury, *Theoretical Mechanics*, John Wiley and Sons, New York, 1968, pp. 212–215; 285–289, for details.

It follows from this that $A(\mathbf{r}, t)$ may be construed as a superposition of plane waves, each having a wavenumber vector $\mathbf{k}$, pointing in the direction of propagation, and a frequency ω. Thus

$$\mathbf{A}(\mathbf{r}, t) = \int_{-\infty}^{\infty} \mathbf{A}(\mathbf{k}, \omega) \exp(i\mathbf{k} \cdot \mathbf{r} - i\omega t)\, d\omega \tag{5.52}$$

where $\mathbf{A}(\mathbf{k}, \omega)$ is the Fourier transform of the vector potential. (It is permissible to use but one integration in equation (5.52) because

$$\mathbf{k} = \frac{2\pi}{\lambda}\hat{\mathbf{k}} = \frac{\omega}{c}\hat{\mathbf{k}}$$

where $\hat{\mathbf{k}}$ is a unit vector.)

The Hamiltonian operator corresponding to $H(\mathbf{p}, \mathbf{A}, \mathbf{r})$ may be written

$$\mathscr{H} = -\frac{\hbar^2}{2\mu_e}\nabla_{\mathbf{r}}^2 - \frac{Ze^2}{r} + \frac{i\hbar e}{\mu_e c}(\mathbf{A} \cdot \nabla_{\mathbf{r}})$$

where the squared term has been expanded, $e^2 A^2/2\mu_e c^2$ has been neglected because of its very small magnitude relative to the other terms, and the identity

$$\nabla \cdot (\mathbf{A}\psi) = \psi(\nabla \cdot \mathbf{A}) + \mathbf{A} \cdot (\nabla\psi)$$

(ψ being any function) has been involved in conjunction with the vanishing of the divergence of the vector potential. An obvious decomposition of the operator $\mathscr{H}$ is

$$\mathscr{H} = \mathscr{H}^{(0)} + \mathscr{H}'$$

$$\mathscr{H}^{(0)} = -\frac{\hbar^2}{2\mu_e}\nabla_{\mathbf{r}}^2 - \frac{Ze^2}{r}$$

$$\mathscr{H}' = +\frac{i\hbar e}{\mu_e c}(\mathbf{A} \cdot \nabla_{\mathbf{r}}) \qquad (e < 0). \tag{5.53}$$

Speaking physically, we expect the perturbation $\mathscr{H}'$ will cause the hydrogenic electron to absorb energy from the radiation field and thus undergo a transition from one eigenstate of $\mathscr{H}^{(0)}$ to another. A calculation of the transition probability should, for this reason, provide information about the conditions to be imposed on the absorption process.

With the help of equation (5.52), the perturbation Hamiltonian (5.53) may be expressed as

$$\mathscr{H}' = \frac{i\hbar e}{\mu_e c}\int_{-\infty}^{\infty} \exp(i\mathbf{k} \cdot \mathbf{r} - i\omega t)(\mathbf{A}(\mathbf{k}, \omega) \cdot \nabla_{\mathbf{r}})\, d\omega.$$

Because $\mathscr{H}'$ is here an explicit function of the time, the transition probability amplitude $c_n(t)$ will not have the form given it in equation (5.46). Instead, we must include the perturbation Hamiltonian under the integral, so that

$$c_n(t) = -\frac{i}{\hbar} \int_{-\infty}^{t} H'_{nn'} \exp(i\omega_{nn'}t')\,dt' \qquad (t_i = -\infty)$$

$$= \frac{e}{\mu_e c} \int_{-\infty}^{t} \int_{-\infty}^{\infty} (\psi_n(\mathbf{r}), \exp(i\mathbf{k}\cdot\mathbf{r})\nabla_\mathbf{r}\psi_{n'}(\mathbf{r}))$$

$$\cdot\mathbf{A}(\mathbf{k}, \omega)\,d\omega \exp[i(\omega_{nn'} - \omega)t']\,dt'$$

where $\psi_n(r)$ is a hydrogenic eigenfunction. We are interested in the value of $c_n(t)$ after a long period of time, when the atom may be once again characterized by constant values of this coefficient. Thus we have

$$c_n(+\infty) = \frac{e}{\mu_e c} \int_{-\infty}^{\infty} \int_{-\infty}^{\infty} (\psi_n, \exp(i\mathbf{k}\cdot\mathbf{r})\nabla_\mathbf{r}\psi_{n'})\cdot\mathbf{A}(\mathbf{k}, \omega)\,d\omega$$

$$\times \exp[i(\omega_{nn'} - \omega)t']\,dt'$$

$$= \frac{2\pi e}{\mu_e c} \int_{-\infty}^{\infty} \int_{-\infty}^{\infty} (\psi_n, \exp(i\mathbf{k}\cdot\mathbf{r})\nabla_\mathbf{r}\psi_{n'})\cdot\mathbf{A}(\mathbf{k}, \omega)\delta(\omega_{nn'} - \omega)\,d\omega$$

$$= \frac{2\pi e}{\mu_e c} (\psi_n, \exp(i\mathbf{k}_{nn'}\cdot\mathbf{r})\nabla_\mathbf{r}\psi_{n'})\cdot\mathbf{A}(\mathbf{k}_{nn'}, \omega_{nn'}) \tag{5.54}$$

where we have employed the definition of the delta-"function." Equation (5.54) shows that only *one* out of all the plane waves making up the radiation field is involved with the electronic transition from the eigenstate $\psi_{n'}$ to the eigenstate ψ_n in a hydrogenic atom. It is that wave of frequency $\omega_{nn'}$; this frequency, of course, is expressed by

$$\omega_{nn'} = \frac{E_n - E_{n'}}{\hbar} \tag{5.55}$$

where E_n and $E_{n'}$ are the energies of the final and initial states, respectively. [See equation (5.42).] Equation (5.55) is quite the same as the postulate Niels Bohr had to introduce to explain the line spectrum of hydrogen. We see now that Bohr's postulate is already contained, in a very general way, within the Schrödinger postulate.

Equation (5.54) may be written in the form of a series expansion:

$$c_n(+\infty) = \frac{2\pi e}{\mu_e c} \left[\left(\psi_n, \sum_{l=0}^{\infty} \frac{(i\mathbf{k}\cdot\mathbf{r})^l}{l!} \nabla_\mathbf{r}\psi_{n'} \right)\cdot\mathbf{A}(\mathbf{k}_{nn'}, \omega_{nn'}) \right].$$

If only the first term of the series is retained, we have

$$c_n(+\infty) \doteq \frac{2\pi e}{\mu_e c} (\psi_n, \nabla_\mathbf{r}\psi_{n'})\cdot\mathbf{A}(\mathbf{k}_{nn'}, \omega_{nn'}). \tag{5.56}$$

The procedure leading to equation (5.56) is called the *electric dipole approximation*. If the scalar product

$$(\psi_n, \nabla_r \psi_{n'})$$

does not vanish, the transition represented by the truncated $c_n(+\infty)$ is said to be of the "dipole type." The reason for this name may be seen from the identity

$$\frac{\partial}{\partial x} = \frac{1}{2}\left\{ x \frac{\partial^2}{\partial x^2} + \frac{\partial}{\partial x} + \frac{\partial}{\partial x} - x \frac{\partial^2}{\partial x^2} \right\} = \frac{1}{2}\left\{ \frac{\partial^2}{\partial x^2} x - x \frac{\partial^2}{\partial x^2} \right\}$$

which, in vector notation, may be expressed

$$\nabla_r = \tfrac{1}{2}[\nabla_r^2 \mathbf{r} - \mathbf{r}\nabla_r^2].$$

Because

$$\mathscr{H}^{(0)} = -\frac{\hbar^2}{2\mu_e} \nabla_r^2 - \frac{Ze^2}{r},$$

we may write

$$(\psi_n, \nabla_r \psi_{n'}) = \tfrac{1}{2}(\psi_n, \nabla_r^2 \mathbf{r}\psi_{n'}) - \tfrac{1}{2}(\psi_n, \mathbf{r}\nabla_r^2 \psi_{n'})$$

$$= -\frac{\mu_e}{\hbar^2}(\psi_n, \mathscr{H}^{(0)}\mathbf{r}\psi_{n'}) + \frac{\mu_e}{\hbar^2}(\psi_n, \mathbf{r}\mathscr{H}^{(0)}\psi_{n'})$$

$$= -\frac{\mu_e}{\hbar^2} E_n(\psi_n, \mathbf{r}\psi_{n'}) + \frac{\mu_e}{\hbar^2} E_{n'}(\psi_n, \mathbf{r}\psi_{n'})$$

$$= \frac{\mu_e}{\hbar^2}(E_{n'} - E_n)(\psi_n, \mathbf{r}\psi_{n'}).$$

Therefore, in the dipole approximation,

$$c_n(+\infty) = \frac{2\pi}{\hbar c}\,\omega_{nn'}\langle \mu \rangle_{nn'} \cdot \mathbf{A}(\mathbf{k}_{nn'},\,\omega_{nn'}) \tag{5.57}$$

where

$$\langle \mu \rangle_{nn'} = (\psi_n, |e|\mathbf{r}\psi_{n'}) \equiv (\psi_n \mathscr{M} \psi_{n'}),$$

the last term being introduced in Chapter 4.

Now, in general, the hydrogenic eigenfunction $\psi_n(\mathbf{r})$ will be of the form of $\Psi_n(r, \vartheta, \varphi)$ in equation (5.22). However, for the purpose of obtaining the selection rules for electric dipole transitions, we shall incur no loss of generality by putting $\Psi_n(\mathbf{r}) \equiv \psi_{nlm}(r, \vartheta, \varphi)$, where the latter function is given by equation (4.39) (with $a_0 \rightarrow a_0/Z$). Since

$$\langle \mu \rangle_{nn'} = \langle \mu_x \rangle_{nn'}\hat{\mathbf{x}} + \langle \mu_y \rangle_{nn'}\hat{\mathbf{y}} + \langle \mu_z \rangle_{nn'}\hat{\mathbf{z}}$$
$$= |e|[\{\langle r \sin \vartheta \cos \varphi \rangle_{nn'}\}\hat{\mathbf{x}} + \{\langle r \sin \vartheta \sin \varphi \rangle_{nn'}\}\hat{\mathbf{y}}$$
$$+ \{\langle r \cos \vartheta \rangle_{nn'}\}\hat{\mathbf{z}}],$$

159

the problem of calculating $\langle\mu\rangle_{nn'}$ reduces to the integrations

$$\int_0^\infty R_{nl} r R_{n'l'} r^2\, dr$$

$$\int_{-1}^1 \theta_l^m(z)(1 - z^2)^{\frac12}\theta_{l'}^{m'}(z)\, dz$$

$$\int_{-1}^1 \theta_l^m(z)z\theta_l^{m'}(z)\, dz$$

$$\int_0^{2\pi} \overline{\Phi}_m(\varphi)\Phi_{m'}(\varphi)\, d\varphi$$

$$\int_0^{2\pi} \overline{\Phi}_m \cos\varphi\Phi_{m'}\, d\varphi$$

$$\int_0^{2\pi} \overline{\Phi}_m \sin\varphi\Phi_{m'}\, d\varphi.$$

The last two integrals vanish, except where $m' = m \pm 1$; the next one, except when $m' = m$; and the next two, except when $l' = l \pm 1$, $m' = m$. The first one does not vanish. The *selection rules* for electric dipole transitions in a hydrogenic atom are, therefore,

$$\Delta l = \pm 1$$
$$\Delta m = 0,\ \pm 1$$

as stated earlier in our discussion of hydrogenic spectra.

5.6. THE VARIATIONAL PRINCIPLE

There may be instances when the Schrödinger eigenvalue problem cannot be solved rigorously and the methods of ordinary time-independent perturbation theory are of no use, simply because no practicable choice of an unperturbed-system Hamiltonian is possible. When faced with this situation, we can still employ the general perturbation theory developed in section 5.1. However, another method exists for attacking the problem which is entirely equivalent to general perturbation theory, but is often more manageable, especially if only the groundstate of a physical system is being investigated. The method is called the *variational method* and is founded upon a theorem which may be named the *Variational Principle*. The proof of the theorem is dependent only on the state postulate and the observable postulate. The consequences of the theorem are the Schrödinger eigenvalue problem and a host of successful applications to otherwise insoluble problems in quantum physics. An informal proof runs as follows.

The Variational Principle. If $\varphi(\mathbf{r})$ is an arbitrary, variable, complex function of the position coordinates, and if

$$\int |\varphi(\mathbf{r})|^2 \, d\mathbf{r}$$

exists, then the quantity

$$\frac{(\varphi, \mathcal{H}\varphi)}{(\varphi, \varphi)}$$

where $\mathcal{H}$ is the Hamiltonian operator, will be an extremum if and only if $\varphi(\mathbf{r})$ is an eigenfunction of $\mathcal{H}$; that is, a solution of the Schrödinger eigenvalue problem.

Proof. The theorem states that the quantity

$$\langle \mathcal{H} \rangle \equiv \frac{(\varphi, \mathcal{H}\varphi)}{(\varphi, \varphi)} \tag{5.58}$$

will be an extremum under certain conditions. This means that the first-order variation in $\langle \mathcal{H} \rangle$ will vanish:

$$\delta\langle \mathcal{H} \rangle \equiv 0. \tag{5.59}$$

Substituting equation (5.58) into equation (5.59) and carrying out the variations, we get

$$\langle \mathcal{H} \rangle + \delta\langle \mathcal{H} \rangle \equiv \frac{\int_{\mathbf{r}} (\bar{\varphi} + \delta\bar{\varphi})\mathcal{H}(\varphi + \delta\varphi) \, d\mathbf{r}}{\int_{\mathbf{r}} (\bar{\varphi} + \delta\bar{\varphi})(\varphi + \delta\varphi) \, d\mathbf{r}}$$

$$= \frac{(\varphi, \mathcal{H}\varphi) + \int_{\mathbf{r}} \bar{\varphi}\mathcal{H} \, \delta\varphi \, d\mathbf{r} + \int_{\mathbf{r}} \delta\bar{\varphi}\mathcal{H}\varphi \, d\mathbf{r}}{(\varphi, \varphi) + \int_{\mathbf{r}} \bar{\varphi} \, \delta\varphi \, d\mathbf{r} + \int_{\mathbf{r}} \delta\bar{\varphi}\varphi \, d\mathbf{r}}$$

or

$$(\varphi, \varphi)\delta\langle \mathcal{H} \rangle = (\delta\varphi, \mathcal{H}\varphi) + \overline{(\delta\varphi, \mathcal{H}\varphi)} - \langle \mathcal{H} \rangle\{(\delta\varphi, \varphi) + \overline{(\delta\varphi, \varphi)}\}$$

$$\equiv 0 = (\delta\varphi, [\mathcal{H} - \langle \mathcal{H} \rangle]\varphi) + \overline{(\delta\varphi, [\mathcal{H} - \langle \mathcal{H} \rangle]\varphi)} \tag{5.60}$$

neglecting all terms of higher than first order and noting that $\mathcal{H}$ is an Hermitian operator. The two terms in equation (5.60) must be equal to zero individually in order that the expression be valid in general. Each term, moreover, will not vanish unless the multiplier of $\delta\varphi$, which is an arbitrary function, also vanishes, whatever the values of the position coordinates. This requirement implies

$$(\mathcal{H} - \langle \mathcal{H} \rangle)\varphi(\mathbf{r}) = 0 \tag{5.61}$$

which, in turn, means that $\varphi(\mathbf{r})$ is an eigenfunction of $\mathcal{H}$, $\langle \mathcal{H} \rangle$ being stationary, by hypothesis. According to the observable postulate, $\langle \mathcal{H} \rangle$ is the mean value of the energy; the state postulate designates eigenfunctions of $\mathcal{H}$ as representatives of the states of a physical system. Under these conditions, equation (5.61) is, of course, the Schrödinger eigenvalue problem.

Quantum Mechanical Approximation Methods

The equivalence between the variational principle and general perturbation theory is easiest seen by noting that the variation function $\varphi(\mathbf{r})$ may always be written as a linear combination of a complete set of polynomials spanning the space to which $\varphi(\mathbf{r})$ belongs. Thus

$$\varphi(\mathbf{r}) = \sum_n a_n \phi_n(\mathbf{r}) \tag{5.62}$$

where the $\phi_n(\mathbf{r})$ form the complete set. When equation (5.62) is put into equation (5.58), the result is

$$\langle \mathcal{H} \rangle = \frac{\sum_m \sum_n \bar{a}_m a_n H_{mn}}{\sum_m |a_m|^2}$$

where

$$H_{mn} \equiv (\phi_m, \mathcal{H} \phi_n).$$

Proceeding as before, we vary the function $\varphi(\mathbf{r})$ by varying the coefficients a_m:

$$(\varphi, \varphi)\delta\langle \mathcal{H} \rangle = (\delta\varphi, \mathcal{H}\varphi) + \overline{(\delta\varphi, \mathcal{H}\varphi)} - \langle \mathcal{H} \rangle[(\delta\varphi, \varphi) + \overline{(\delta\varphi, \varphi)}]$$

$$= \sum_m \sum_n \delta\bar{a}_m a_n H_{mn} + \overline{\left(\sum_m \sum_n \delta\bar{a}_m a_n H_{mn}\right)}$$

$$- \langle \mathcal{H} \rangle \left[\sum_m \delta\bar{a}_m a_m + \overline{\left(\sum_m \delta\bar{a}_m a_m\right)}\right]$$

$$= \sum_m \delta\bar{a}_m \sum_n (H_{mn} - \langle \mathcal{H} \rangle\delta_{mn})a_n$$

$$+ \overline{\sum_m \delta\bar{a}_m \sum_n (H_{mn} - \langle \mathcal{H} \rangle\delta_{mn})a_n} \equiv 0.$$

Once again, the requirement that $\langle \mathcal{H} \rangle$ be stationary implies that

$$\sum_n a_n(H_{mn} - \langle \mathcal{H} \rangle\delta_{mn}) = 0$$

which is satisfied in general only if

$$\det |H_{mn} - \langle \mathcal{H} \rangle\delta_{mn}| = 0. \tag{5.63}$$

Equation (5.63) is identical with equation (5.20), the fundamental expression in general perturbation theory, once the observable postulate has been invoked to associate $\langle \mathcal{H} \rangle$ with the mean value of the energy.

When applying the variational method to the calculation of the ground-state energy of a physical system, use is made of the fact that equation (5.59), in this case, implies that $\langle \mathcal{H} \rangle$ is *minimized* by finding the correct variation function. Thus we choose some $\varphi(\mathbf{r})$, vary it so as to provide the lowest value of $\langle \mathcal{H} \rangle$, and compare the result with the experimentally determined groundstate energy. The result, in general, *will always be greater in magnitude*. Evidently, the choice of $\varphi(\mathbf{r})$ which provides the lowest value of $\langle \mathcal{H} \rangle$ is the one nearest in form to the correct wavefunction. An estimate of the error involved, for real-valued trial wavefunctions, may be had by calculating the mean-square deviation

$$(\Delta\varphi)^2 \equiv \int_{\mathbf{r}} (\varphi(\mathbf{r}) - \psi_1(\mathbf{r}))^2 \, d\mathbf{r}$$

where $\psi_1(\mathbf{r})$ is the true (and real-valued) groundstate wavefunction. Now,

$$(\varphi(\mathbf{r}) - \psi_1(\mathbf{r}))^2 = |\varphi(\mathbf{r})|^2 - 2\psi_1(\mathbf{r}) \sum_n a_n \psi_n(\mathbf{r}) + |\psi_1(\mathbf{r})|^2$$

since the $\psi_n(\mathbf{r})$ form a complete orthonormal set. The mean-square deviation is then

$$(\Delta\varphi)^2 = (\varphi, \varphi) - 2 \sum_n a_n(\psi_1, \psi_n) + (\psi_1, \psi_1) = 2(1 - a_1)$$

where it has been assumed that the variation function is normalized. Because

$$\langle \mathcal{H} \rangle - E_1 = \sum_n |a_n|^2(\psi_n, \mathcal{H}\psi_n) - E_1 = \sum_n |a_n|^2(E_n - E_1)$$

$$= \sum_{n=2} |a_n|^2(E_n - E_1) \geqslant \sum_{n=2} |a_n|^2(E_2 - E_1)$$

$$= (1 - a_1^2)(E_2 - E_1),$$

owing to the normality of $\varphi(\mathbf{r})$, we can write

$$(1 - a_1^2) = (\Delta\varphi)^2 + \tfrac{1}{4}(\Delta\varphi)^4 \leqslant \left(\frac{\langle \mathcal{H} \rangle - E_1}{E_2 - E_1} \right)$$

or

$$(\Delta\varphi)^2 < \left(\frac{\langle \mathcal{H} \rangle - E_1}{E_2 - E_1} \right), \tag{5.64}$$

assuming that $(\Delta\varphi)^4$ is negligible compared with $(\Delta\varphi)^2$.

PROBLEMS

1. Calculate the first-order correction to the energy of the first excited state in hydrogen when the atom is in an electric field of magnitude $|\mathbf{E}|$. Show that the component of the permanent dipole moment along the direction of the field (the z-axis) is given by

$$\langle \mu_z \rangle_2^{(0)} = 3|e|a_0.$$

163

2. The Schrödinger eigenvalue problem for a diatomic molecule which can rotate only in a plane is

$$-\frac{\hbar^2}{2I}\frac{d^2}{d\varphi^2} = E\psi(\varphi).$$

(a) Solve the equation, obtaining an expression for E in terms of the magnetic quantum number.

(b) Calculate the first-order correction to E when the rotator is placed in an electric field of magnitude E. (The interaction energy is

$$\mathscr{H}' = -\mu|\mathbf{E}|\cos\varphi.)$$

(c) Calculate the second-order correction to E for the same circumstances. Show that the polarizability of the molecule is

$$\alpha_m = \frac{-2\mu^2 I}{\hbar^2(4m^2 - 1)}\qquad (m = 0, \pm 1, \ldots),$$

and interpret its sign for different values of the magnetic quantum number.

3. Suppose that a hydrogen atom in its groundstate is placed in a uniform, constant electric field E directed along the z-axis.

(a) Write down the appropriate complete Hamiltonian operator for the electron.

(b) Using the variation function

$$\varphi(\mathbf{r}) = \psi_{100}(\mathbf{r})(1 + Ar\cos\vartheta),$$

where A is the variable parameter, calculate $\langle\mathscr{H}\rangle$.

(c) Minimize $\langle\mathscr{H}\rangle$ with respect to the variable parameter, neglecting powers of $|E|$ greater than two. Calculate A and the minimized $\langle\mathscr{H}\rangle$.

(d) Compute the polarizability of the groundstate hydrogen atom. The accepted value is 0.67×10^{-24} cm^3.

4. Repeat Problem 3, using the variation function

$$\varphi(\mathbf{r}) = \psi_{100}(\mathbf{r})(1 + Ar\cos\vartheta + Br^2\cos\vartheta),$$

where A and B are both variable parameters.

5. A free electron moving in a one-dimensional cube of length $2L$ is subjected to the "coulomb potential" $V(q)$:

$$V(q) = -e^2\,\delta(q)\qquad (-L \leqslant q \leqslant L).$$

If the wavefunction representing the free particle is

$$\psi_n(q) = (L)^{-\frac{1}{2}}\cos\left(\frac{n\pi q}{2L}\right)\qquad (n = 1, 3, 5, \ldots),$$

what is the first-order correction to the energy caused by the perturbation $V(q)$? Over what range of cube dimension is the first-order theory likely to be accurate?

6. If, in the case of the helium atom, the perturbation Hamiltonian operator were of the form

$$\mathscr{H}' = \frac{e^2}{r_{12}} \delta(r_{12}),$$

what would be the first-order correction to the groundstate energy of the atom? Compare this result to

$$E_1^{(1)} = \tfrac{5}{8} Z \frac{e^2}{a_0}$$

and explain the discrepancy.

7. There are transitions, such as when an atom emits radiation, that do not end with the system in a single final state. In this case, the total transition probability is a sum over the transition probabilities from the initial state to each of the final states. If the final states form a continuum, as they do in the case of atomic radiation (atom + emitted photons), we can write

$$\sum_l |c_l(t)|^2 = 4 \int_{-\infty}^{\infty} |H'_{lm}|^2 \frac{\sin^2(\omega_{lm}t/2)}{(E_l - E_m)^2} \rho(E_l) \, dE_l, \qquad 0 < t$$

where $\rho(E_l)$ is the number of final states per unit energy.. Show that *the transition rate*, defined by

$$W \equiv \frac{d}{dt} \sum_l |c_l(t)|^2$$

is a *constant* in the first approximation. (*Hint*: Assume that $|H'_{lm}|^2$ and the density of states $\rho(E_l)$ are independent of E_l when the latter is near in value to E_m. Why are the E_l that are radically different from E_m not important here?) Enrico Fermi has called the fact that W is time-independent "the Golden Rule of time-dependent perturbation theory" because of its great importance in applications.

8. Show that the Golden Rule in the second approximation is expressed by

$$W = \frac{2\pi |H'_{lm}|^2 \rho(E_l)}{\hbar}$$

where

$$H'_{lm} = \sideset{}{'}\sum_{l'} \frac{H'_{ll'} H'_{l'm}}{E_m - E_{l'}}.$$

9. In time-independent perturbation theory, the correct groundstate wavefunction is approximated by

$$\varphi_1(\mathbf{r}) = \varphi_1^{(0)}(\mathbf{r}) + g\varphi_1^{(1)}(\mathbf{r}) + g^2\varphi_1^{(2)}(\mathbf{r}) + \cdots.$$

If the series is terminated at the nth term, the function $\varphi_1(\mathbf{r})$ should differ from the true wavefunction $\psi_1(\mathbf{r})$ by terms of order $(n + 1)$. Now, if the function $\varphi_1(\mathbf{r})$ is used as a variation function, the value of $\langle \mathcal{H} \rangle$ calculated should differ from E_1, the true groundstate energy, by terms of order $2(n + 1)$. Thus, the nth-order variation function should be correct to the $(2n + 1)$th order of approximation. Verify this conjecture by calculating the groundstate energy corresponding to the Hamiltonian operator $\mathcal{H} = \mathcal{H}^{(0)} + \mathcal{H}'$, using the zeroth-order variation function

$$\varphi_1(\mathbf{r}) = \varphi_1^{(0)}(\mathbf{r})$$

where $\varphi_1^{(0)}(\mathbf{r})$ is an eigenfunction of $\mathcal{H}^{(0)}$. Assume that $E_1^{(0)}$ is non-degenerate and that $\varphi_1^{(0)}(\mathbf{r})$ is normalized.

10. Calculate the groundstate energy of the hydrogen atom by using the variation function

$$\varphi(r) = \exp\left(-Ar/a_0\right)$$

where A is the variable parameter. What value must A have to give the lowest energy? (Remember that $a_0 = \hbar^2/\mu_e e^2$.)

CHAPTER 6

Matrix Quantum Mechanics

Beauty is in the eye of the beholder.

6.1. STATE VECTORS AND OBSERVABLES AS MATRICES

In Chapter 3 it was necessary to prescribe a representation of the state vector in order to construct an equation of motion from the state postulate and the observable postulate. The prescription made was to consider the state vector as a continuous, twice-differentiable, complex-valued function of real variables lying in some ray of Hilbert space. This representation is analogous to the depiction of a vector in ordinary space by

$$\mathbf{A} = A\hat{\mathbf{A}}$$

where A is the magnitude of $\mathbf{A}$, and $\hat{\mathbf{A}}$ is a unit vector pointing in the direction of $\mathbf{A}$. On the other hand, it is well known that vectors in ordinary space may be represented in a way entirely equivalent to the foregoing method by referring them to an appropriate set of three basis vectors, $\hat{\mathbf{x}}$, $\hat{\mathbf{y}}$, and $\hat{\mathbf{z}}$. Thus we may write

$$\mathbf{A} = A_x\hat{\mathbf{x}} \oplus A_y\hat{\mathbf{y}} \oplus A_z\hat{\mathbf{z}},$$

where A_x, A_y, and A_z are the components of $\mathbf{A}$ along the three principal axes (one-dimensional subspaces!) of the space spanned by $\hat{\mathbf{x}}$, $\hat{\mathbf{y}}$, and $\hat{\mathbf{z}}$. In accordance with this possibility of portraying ordinary vectors in two ways, we should expect that a state vector may be written in terms of an *arbitrary* set of basis vectors spanning Hilbert space:

$$|g\rangle = \sum_{n=1}^{\infty} \langle n|g\rangle |n\rangle \tag{6.1}$$

where the coefficients $\langle n|g\rangle$ are now to be considered as the *components* of $|g\rangle$ along the subspaces of the $|n\rangle$. Just as the components A_x, A_y, and A_z specify the vector $\mathbf{A}$ in Euclidean space, so should the components $\langle n|g\rangle$

167

specify the state vector $|g\rangle$ in Hilbert space. In this way we may picture a state vector as an *array* of its components:

$$(|g\rangle) \equiv \begin{pmatrix} \langle 1|g\rangle \\ \langle 2|g\rangle \\ \vdots \end{pmatrix} \quad \text{or} \quad (\langle g|) \equiv (\langle \overline{1|g}\rangle \langle \overline{2|g}\rangle \cdots).$$

In the first instance above, we call $|g\rangle$ a *column vector*; in the second case $\langle g|$ is a *row vector*. So much for state vectors; but what about observables? By definition, we can write

$$|II\rangle = O|I\rangle \tag{6.2}$$

where O is some observable (linear Hermitian operator) and $|I\rangle$ and $|II\rangle$ are two state vectors having the components $\{\langle n|I\rangle\}$ and $\{\langle n|II\rangle\}$, respectively, in terms of the basis vectors $\{|n\rangle\}$. If the scalar product of both sides of equation (6.2) with $\langle m|$ ($1 \leqslant m \leqslant \infty$) is taken, we find, applying equation (6.1) to $|I\rangle$,

$$\langle m|II\rangle = \sum_{n=1}^{\infty} \langle m|O|n\rangle\langle n|I\rangle \quad (m = 1, 2, \ldots). \tag{6.3}$$

Equations (6.3) form a set of simultaneous equations for the $\langle m|II\rangle$, which may be expressed

$$\begin{pmatrix} \langle 1|II\rangle \\ \langle 2|II\rangle \\ \vdots \end{pmatrix} = \begin{pmatrix} O_{11} & O_{12} & \cdots \\ O_{21} & & \\ \vdots & \vdots & \end{pmatrix} \times \begin{pmatrix} \langle 1|I\rangle \\ \langle 2|I\rangle \\ \vdots \end{pmatrix}, \tag{6.4}$$

where

$$O_{mn} \equiv \langle m|O|n\rangle \quad (m = 1, \ldots, \infty; n = 1, \ldots, \infty),$$

provided that a rule has been given for multiplying the two arrays on the right-hand side of the equation, such that the equality is satisfied. The rules which we shall prescribe are the rules of *matrix algebra*, so that the arrays portraying state vectors and observables may be called *matrices*. The observable O is thus pictured as the matrix

$$(O) \equiv \begin{pmatrix} O_{11} & O_{12} & \cdots \\ O_{21} & & \\ \vdots & \vdots & \end{pmatrix} \tag{6.5}$$

[equation (6.4) being the matrix form of equation (6.2)] where the O_{mn} are called the *matrix elements* of the observable O. Because all observables are Hermitian operators, we must have it that

$$\langle III|O|I\rangle = \overline{\langle I|O|III\rangle}$$

or, by equation (6.1),

$$\sum_{k,l} \langle III|k\rangle\langle k|O|l\rangle\langle l|I\rangle = \sum_{l,k} \overline{\langle I|l\rangle\langle l|O|k\rangle\langle k|III\rangle}$$

$$= \sum_{l,k} \overline{\langle I|l\rangle}\,\overline{\langle l|O|k\rangle}\,\overline{\langle k|III\rangle}$$

which implies the condition

$$O_{mn} = \overline{O}_{nm} \qquad (m = 1,\ldots,\infty;\, n = 1,\ldots,\infty) \qquad (6.6)$$

for the matrix elements of observables.

Now to the rules of matrix algebra. We can differentiate two kinds of operation here: conjugation and purely algebraic manipulation. There are three types of conjugation, namely, *complex conjugation,*

$$(C) \equiv (\overline{A}) \qquad \text{if} \qquad C_{mn} = \overline{A}_{mn}; \qquad (6.7)$$

transposition,

$$(T) \equiv (\tilde{A}) \qquad \text{if} \qquad T_{mn} = A_{nm}; \qquad (6.8)$$

and *Hermitian conjugation,*

$$(H) \equiv (A)^\dagger \qquad \text{if} \qquad H_{mn} = \overline{A}_{nm} \qquad (6.9)$$

where (A) is an arbitrary matrix and (C), (T), and (H) are its complex conjugate, transposed conjugate, and Hermitian conjugate, respectively. There are four kinds of algebraic operations of importance. The first three are *scalar multiplication,*

$$(M) \equiv c(A) \qquad \text{if} \qquad M_{mn} = cA_{mn}; \qquad (6.10)$$

addition,

$$(S) \equiv (A) + (B) \qquad \text{if} \qquad S_{mn} = A_{mn} + B_{mn}; \qquad (6.11)$$

and *the product from the left,*

$$(P) \equiv (A) \times (B) \qquad \text{if} \qquad P_{mn} = \sum_{k=1}^{N} A_{mk}B_{kn} \qquad (6.12)$$

where (A) is a matrix having M_A rows and N_A columns (an "M_A by N_A matrix," we say), (B) is a matrix having M_B rows and N_B columns, and

$$N_A = M_B \equiv N.$$

The product from the right, $(B) \times (A)$, may be defined as the product from the left of (B) and (A). In general, both products do not exist as the conditions

$$N_A = M_B, \qquad N_B = M_A$$

would have to be fulfilled. There are certain cases of importance when the two products are defined, however; for example, when (A) and (B) are square matrices.

If (A) and (B) are square matrices and it is possible to write

$$\sum_{k=1}^{N} B_{mk} A_{kn} = \delta_{mn},$$

the matrix (B) is said to be the *inverse* of the matrix (A): $(B) \equiv (A)^{-1}$. The matrix whose elements are δ_{mn} is thus the representation of the *identity operator*:

$$(I) = \begin{pmatrix} 1 & 0 & 0 & 0 & \cdots \\ 0 & 1 & 0 & 0 & \cdots \\ 0 & 0 & 1 & 0 & \cdots \\ \vdots & \vdots & \vdots & \vdots & \end{pmatrix}.$$

(I) is an example of what is termed a *diagonal matrix*; that is, a matrix having non-zero elements only on the principal diagonal. Another important special case which arises when the product from the left is a commutative operation may be expressed

$$(U) \times (U)^{\dagger} = (U)^{\dagger} \times (U) = (I) \tag{6.13}$$

which defines the (square) *unitary matrix* (U).

The fourth algebraic operation involving matrices is the *direct* (or *tensor*) *product*:

$$(D) \equiv (A) \otimes (B).$$

The operation $\otimes$ is to be understood as follows. The matrix (B) is juxtaposed with each matrix element of (A). This creates an M_A by N_A matrix whose elements are scalar multiples of matrices, rather than numbers. Then the multiplication of each element of (A) with the elements of (B) yields an $M_A M_B$ by $N_A N_B$ matrix—the direct product of (A) and (B). For example, if

$$(A) = \begin{pmatrix} 0 & 2 \\ 4 & 0 \end{pmatrix}, \qquad (B) = \begin{pmatrix} 0 & 1 \\ -1 & 0 \end{pmatrix},$$

then we have

$$(A) \otimes (B) = \begin{pmatrix} O & 2(B) \\ 4(B) & O \end{pmatrix} = \begin{pmatrix} 0 & 0 & 0 & 2 \\ 0 & 0 & -2 & 0 \\ 0 & 4 & 0 & 0 \\ -4 & 0 & 0 & 0 \end{pmatrix},$$

where O is a two by two null matrix, *not* the number zero. A direct product matrix can be thought of as matrices, in the space of the basis vectors of (A), whose elements are *matrices* in the space of the basis vectors of (*B*). Alternatively, the matrix elements of the direct product are *numbers* possessing as many pairs of indices as there are factors in the direct product:

$$(D)_{mm';nn'} = ((A) \otimes (B))_{mm';nn'} = (A)_{mn}(B)_{m'n'}.$$

Coming back to quantum physics, we see that a column vector is an *M* by one matrix, a row vector is a one by *N* matrix, and an observable is an *M* by *N* matrix. (Both *M* and *N* can be infinite.) The scalar product of a vector with itself is thus written, according to equation (6.12), as

$$\langle g|g \rangle = (\overline{\langle 1|g\rangle}\,\overline{\langle 2|g\rangle}\cdots) \times \begin{pmatrix} \langle 1|g \rangle \\ \langle 2|g \rangle \\ \vdots \end{pmatrix} = |\langle 1|g\rangle|^2 + |\langle 2|g\rangle|^2 + \cdots \quad (6.14)$$

which agrees with the usual definition. The eigenvalue problem is not difficult to formulate in matrix language, either. If the basis chosen to represent an Hermitian operator *H* as a matrix is its set of eigenvectors, then we must have

$$(H) = \begin{pmatrix} \lambda_1 & 0 & 0 & \cdots \\ 0 & \lambda_2 & 0 & \cdots \\ 0 & 0 & \lambda_3 & \cdots \\ \vdots & \vdots & \vdots & \end{pmatrix} \quad (6.15)$$

where, obviously,

$$H_{nm} = \langle n|H|m \rangle = \begin{cases} \lambda_n & n = m \\ 0 & n \neq m \end{cases}$$

and $\{|n\rangle\}$ is the set of orthonormal eigenstates of *H*:

$$H|n\rangle = \lambda_n|n\rangle.$$

We see that, if the set $\{|n\rangle\}$ is orthonormal, *an Hermitian operator is pictured as a diagonal matrix in the eigenvalue problem*. Moreover, since

$$H_{nn} = \langle n|H|n \rangle \equiv \langle H \rangle,$$

we can say that the diagonal matrix elements of (*H*) are the mean values of the operator *H* in the states $|n\rangle$, in accordance with the interpretation given in Chapter 3.

If the eigenvalues λ_n are degenerate, then we have in general

$$H|nl\rangle = \lambda_n|nl\rangle \qquad (l = 1,\ldots,L_n) \quad (6.16)$$

where $\{|nl\rangle\}$ is a set of not necessarily orthogonal eigenvectors. In this case, (*H*) would no longer have to be a diagonal matrix. Equation (6.16) states that there are L_n vectors $|nl\rangle$ corresponding to the eigenvalue λ_n; which is to say that the L_n basis vectors span an L_n-dimensional invariant subspace of H. It follows that, for a given n, the matrix expression analogous to (6.15) is

$$(H^{(n)}) = \lambda_n \begin{pmatrix} \langle n1|n1\rangle & \cdots & \langle n1|nL_n\rangle \\ \vdots & & \vdots \\ \langle nL_n|n1\rangle & \cdots & \langle nL_n|nL_n\rangle \end{pmatrix}$$

which, then, does *not* mean

$$(H^{(n)}) = \lambda_n(I)$$

unless the $|nl\rangle$ form an orthonormal set. The complete matrix for (H) is composed of all the matrices $(H^{(n)})$ and (H) is called the *direct sum* of the $(H^{(n)})$:

$$(H) = (H^{(1)}) \boxplus (H^{(2)}) \boxplus \cdots .$$

As an example of how the direct sum looks in matrix representation, suppose $H^{(1)}$ possessed doubly degenerate eigenvalues and $H^{(2)}$ possessed triply degenerate eigenvalues. Then we would write

$$(H) = (H^{(1)}) \boxplus (H^{(2)}) \boxplus \cdots = \begin{pmatrix} H^{(1)} & & O \\ & H^{(2)} & \\ O & & \ddots \end{pmatrix}$$

$$= \begin{pmatrix} H_{11}^{(1)} & H_{12}^{(1)} & 0 & 0 & 0 & \cdots \\ H_{21}^{(1)} & H_{22}^{(1)} & 0 & 0 & 0 & \cdots \\ 0 & 0 & H_{11}^{(2)} & H_{12}^{(2)} & H_{13}^{(2)} & \cdots \\ 0 & 0 & H_{21}^{(2)} & H_{22}^{(2)} & H_{23}^{(2)} & \cdots \\ 0 & 0 & H_{31}^{(2)} & H_{32}^{(2)} & H_{33}^{(2)} & \cdots \end{pmatrix}$$

If the eigenvectors of $H^{(1)}$ and $H^{(2)}$ had been made mutually orthonormal, the last matrix would look like this:

$$(H) = \begin{pmatrix} H_{11}^{(1)} & 0 & 0 & 0 & 0 & \cdots \\ 0 & H_{22}^{(1)} & 0 & 0 & 0 & \cdots \\ 0 & 0 & H_{11}^{(2)} & 0 & 0 & \cdots \\ 0 & 0 & 0 & H_{22}^{(2)} & 0 & \cdots \\ 0 & 0 & 0 & 0 & H_{33}^{(2)} & \cdots \end{pmatrix} .$$

6.2. THE EQUATION OF MOTION FOR OBSERVABLES

Throughout our discussion of quantum physics it has been apparent that the content of the theory, insofar as the results of experiments are concerned, revolves about the matrix elements of observables. For it is in terms of these quantities that mean values are defined:

$$\langle 0 \rangle = \frac{\langle g|O|g \rangle}{\langle g|g \rangle} = \sum_n \frac{|\langle n|g \rangle^2|}{\langle g|g \rangle} \langle n|O|n \rangle = \sum_n \omega_n O_n$$

Matrix quantum mechanics emphasizes this role in that the matrix elements of an observable play a central part in the formalism. Indeed, the matrix elements *are* the observable in this formalism. It follows that in the matrix formulation the observable is given a greater significance than is the state vector. This is most clearly demonstrated by the fact that *arbitrary* basis vectors in Hilbert space are used to construct observables. It may have appeared that such was not the case in our discussion of the eigenvalue problem, since there a set of basis vectors which were eigenvectors of the Hermitian operator H were employed. But this need not be done. For example, suppose the operator H were known only in terms of a certain set of basis vectors $\{\varphi_n\}$ in the coordinate representation. The eigenfunctions of H, say $\{\phi_n\}$, can still be found because every vector in Hilbert space can be reached by a suitable mapping. Thus, we can write

$$\phi_n \equiv U\varphi_n \qquad (n = 1, 2, 3, \ldots)$$

where U is a linear operator which maps φ_n into ϕ_n. To preserve the normality of the φ_n, the operator U will have to be unitary:

$$(\phi_n, \phi_n) = (U\varphi_n, U\varphi_n) = (\varphi_n, U^\dagger U\varphi_n) \equiv (\varphi_n, \varphi_n).$$

Moreover, we see that the matrix elements of U are just the projections of the ϕ_n onto the φ_m:

$$U_{mn} \equiv (\varphi_m, U\varphi_n) = (\varphi_m, \phi_n).$$

The matrix elements of H in the basis $\{\phi_n\}$ are

$$(\phi_n, H\phi_m) = \lambda_n \delta_{nm} = (U\varphi_n, HU\varphi_m) = (\varphi_n, U^\dagger H U\varphi_m),$$

and these may be construed as matrix elements of the operator $U^\dagger H U$ in the basis $\{\varphi_n\}$. *This latter operator, unlike H, is represented by a diagonal matrix in the basis $\{\varphi_n\}$.* The eigenvalue problem has thus become

$$U^\dagger H U\varphi_n = \lambda_n \varphi_n \tag{6.17}$$

or, of course,

$$HU\varphi_n = \lambda_n U\varphi_n. \tag{6.18}$$

Matrix Quantum Mechanics

Equation (6.18) written in terms of matrix elements in the basis $\{\varphi_n\}$ is, according to equation (6.12),

$$\sum_k H_{mk} U_{kn} = \lambda_n U_{mn} = \sum_k \lambda_n \delta_{mk} U_{kn}$$

or

$$\sum_k (H_{mk} - \lambda_n \delta_{mk}) U_{kn} = 0 \qquad (6.19)$$

which is a set of linear, homogeneous equations for the U_{kn}. As we know very well by now, equations (6.19) have no solutions but the trivial ones unless

$$\det |H_{mk} - \lambda_n \delta_{mk}| = 0. \qquad (6.20)$$

The roots of equation (6.20) are the λ_n. When these have been found, equations (6.19) may be solved for the U_{kn}, which, in turn, permit the construction of U and the computation of the ϕ_n.

We see that the problem in matrix quantum mechanics is to determine the forms of observables. This can be done (a) if the commutators for the observables are known, and (b) if an equation of motion has been prescribed. As has been pointed out earlier, no amount of study of empirical data will provide this information. Matrix quantum mechanics, like wave quantum mechanics, rests upon a postulate. The founding axiom was first conceived by Werner Heisenberg, who in 1925 wrote down the matrix formulation of quantum mechanics.

The Heisenberg Postulate. The commutators for the position and momentum operators are

$$[q_m, q_n] = [p_m, p_n] = 0, \qquad [q_m, p_n] = \begin{cases} 0 & m \neq n \\ i\hbar I & m = n \end{cases}$$

where q_m is any position operator in one dimension and p_m is the momentum operator in one dimension. The equation of motion for these and all other observables is

$$\frac{dO_t}{dt} = \frac{\partial O_t}{\partial t} - \frac{i}{\hbar} [O_t, H]$$

where the observable O_t may depend explicitly on the time and

$$H(t) \equiv \exp\left[i\mathscr{H}(t - t_i)/\hbar\right]\mathscr{H} \exp\left[-i\mathscr{H}(t - t_i)/\hbar\right] \qquad (t \geqslant t_i),$$

$\mathscr{H}$ being the Hamiltonian operator.

Using the commutation relations for the position and momentum, we can easily deduce those for all other observables. In general, for any observable $O(\mathbf{q}, \mathbf{p})$,

$$[q_m, O(\mathbf{q}, \mathbf{p})] = i\hbar \frac{\partial O}{\partial p_m} \tag{6.21}$$

$$[p_m, O(\mathbf{q}, \mathbf{p})] = -i\hbar \frac{\partial O}{\partial q_m} \tag{6.22}$$

For example, if $O = O(q, p)$ and is expressible as a power series in either of the two observables q and p, we can make use of the commutator identity

$$[q, p^2] = [q, p]p + p[q, p] = 2i\hbar p$$

to get equation (6.21). By direct generalization of the foregoing expression we have

$$[q, p^n] = ni\hbar p^{n-1}$$

which suffices to prove (6.21).

The equation of motion for observables, like its counterpart for state vectors, may in a certain sense be derived. Let us see how this comes about. In the matrix formulation, the state vectors are time-independent; they may be reached by operating on the time-dependent state vectors in the wave formulation with the unitary operator

$$U_t^\dagger = \exp\left[i\mathcal{H}(t - t_i)/\hbar\right] \qquad (t \geq t_i).$$

Thus

$$\begin{aligned}
|M\rangle &= U_t^\dagger |W\rangle_t = \exp\left[i\mathcal{H}(t - t_i)/\hbar\right]|W\rangle_{t_i} \exp\left[-iE(t - t_i)/\hbar\right] \\
&= |W\rangle_{t_i} \exp\left[iE(t - t_i)/\hbar\right] \exp\left[-iE(t - t_i)/\hbar\right] \\
&= |W\rangle_{t_i}
\end{aligned}$$

where $|M\rangle$ denotes a state vector in the Heisenberg (matrix) picture, $|W\rangle_t$ is a time-dependent state vector in the Schrödinger (wave) picture, $|W\rangle_{t_i}$ is an eigenvector of $\mathcal{H}$, and $U_t^\dagger$ is defined by the series

$$U_t^\dagger \equiv \sum_{k=0}^{\infty} \frac{[i\mathcal{H}(t - t_i)/\hbar]^k}{k!} \qquad (t \geq t_i).$$

Matrix elements of observables in the two pictures are therefore related by

$$_t\langle W|O|W\rangle_t = \langle M|U_t^\dagger O U_t|M\rangle \equiv \langle M|O_t|M\rangle$$

so that

$$O_t = \exp\left[i\mathcal{H}(t - t_i)/\hbar\right]O \exp\left[-i\mathcal{H}(t - t_i)/\hbar\right] \tag{6.23}$$

175

in the matrix picture, where O is the time-independent observable in the wave picture. When equation (6.23) is differentiated termwise with respect to the time, we have

$$\frac{dO_t}{dt} = \frac{i}{\hbar} \exp\left[i\mathcal{H}(t - t_i)/\hbar\right]\mathcal{H}O \exp\left[-i\mathcal{H}(t - t_i)/\hbar\right]$$

$$+ \exp\left[i\mathcal{H}(t - t_i)/\hbar\right]\frac{\partial O}{\partial t} \exp\left[-i\mathcal{H}(t - t_i)/\hbar\right]$$

$$- \frac{i}{\hbar} \exp\left[i\mathcal{H}(t - t_i)/\hbar\right]O\mathcal{H} \exp\left[-i\mathcal{H}(t - t_i)/\hbar\right]$$

$$\equiv \frac{\partial O_t}{\partial t} - \frac{i}{\hbar}[O_t, H]$$

which is the equation of motion stated in the Heisenberg postulate.

When the observable under consideration is a time-independent Hamiltonian operator, the equation of motion becomes

$$\frac{dH}{dt} = 0,$$

showing that the total energy is a constant of the motion. This result is certainly expected. However, it is also to be seen that any observable whose commutator with the Hamiltonian operator is zero and which does not depend explicitly on the time is also a constant of the motion. Observables that commute with the Hamiltonian operator are said to be *compatible* because they possess the same set of eigenvectors as does the latter operator.[1] Thus, *observables which are compatible with the total energy are well-defined constants.*

Because the state vectors in the matrix picture are not time-dependent, the equation of motion for the mean value of an observable may be written

$$\frac{d}{dt}\langle O_t \rangle = \left\langle \frac{\partial O_t}{\partial t} \right\rangle - \frac{i}{\hbar}\langle [O_t, H] \rangle \tag{6.24}$$

---

[1] That this is so may not be quite obvious. We shall not prove the statement, but suffice to say that

$$O\mathcal{H} = \mathcal{H}O$$

implies that

$$EO|\varphi\rangle = \mathcal{H}(O|\varphi\rangle),$$

which can hold true only if $|\varphi\rangle$ is an eigenvector of O as well as $\mathcal{H}$.

where

$$\langle O_t \rangle \equiv \langle M|O_t|M \rangle = {}_t\langle W|O|W \rangle_t$$

and so on. Two special cases of equation (6.24) are of interest. If the observable is chosen to be the position, we have in one dimension

$$\frac{d}{dt}\langle q_t \rangle = -\frac{i}{\hbar}\langle [q_t, H] \rangle = -\frac{i}{\hbar}\left\langle i\hbar \frac{\partial H}{\partial p_t} \right\rangle = \left\langle \frac{\partial H}{\partial p_t} \right\rangle$$

where the last step comes from equation (6.21). Similarly, if we choose the momentum,

$$\frac{d}{dt}\langle p_t \rangle = -\frac{i}{\hbar}\langle [p_t, H] \rangle = -\frac{i}{\hbar}\left\langle -i\hbar \frac{\partial H}{\partial q_t} \right\rangle = -\left\langle \frac{\partial H}{\partial q_t} \right\rangle$$

the last step coming from equation (6.22). Because the mean values of observables must be the same in the wave picture as in the matrix picture, the foregoing equations may be rewritten as

$$\frac{d\langle q \rangle}{dt} = \left\langle \frac{\partial \mathscr{H}}{\partial p} \right\rangle$$

$$\frac{d\langle p \rangle}{dt} = -\left\langle \frac{\partial \mathscr{H}}{\partial q} \right\rangle$$

(6.25)

where now the means are over observables in the wave picture. Equations (6.25) express the time-development of the mean values of position and momentum; they are known as *Ehrenfest's equations*. Their counterparts in classical mechanics are

$$\frac{dq}{dt} = \frac{\partial H}{\partial p}, \qquad \frac{dp}{dt} = -\frac{\partial H}{\partial q}$$

(6.26)

where $H(p, q)$ is the Hamiltonian function. These relations are known as *Hamilton's equations*.[2] Equations (6.25) and (6.26) are examples of the correspondence postulate at work. Indeed, knowing equations (6.26) we might have invoked the postulate and immediately written down equations (6.25). With a lot of ingenuity, perhaps, then, we might have been led to the Heisenberg postulate!

The equation of motion (6.24) also has a profound bearing upon the root-mean-square deviation of observables from their average values. To see this,

---

[2] Those interested in learning more of Hamilton's equations might consult W. A. Hauser, *Introduction to the Principles of Mechanics*, Addison-Wesley Publ. Co., Reading, Mass., 1965, pp, 195–198.

let us note at the outset that, by the Schwarz inequality (defined in problems 8 and 11 of Chapter 3),

$$(\Delta O)^2(\Delta \mathcal{H})^2 \equiv \langle (O - \langle O \rangle)^2 \rangle \langle (\mathcal{H} - \langle \mathcal{H} \rangle)^2 \rangle$$
$$\geq |\langle (O - \langle O \rangle)(\mathcal{H} - \langle \mathcal{H} \rangle) \rangle|^2$$
$$= \left| \frac{\langle (O - \langle O \rangle)(\mathcal{H} - \langle \mathcal{H} \rangle) + (\mathcal{H} - \langle \mathcal{H} \rangle)(O - \langle O \rangle) \rangle}{2} \right.$$
$$\left. + \frac{\langle [(O - \langle O \rangle), (\mathcal{H} - \langle \mathcal{H} \rangle)] \rangle}{2} \right|^2$$
$$\geq \tfrac{1}{4} |\langle [(O - \langle O \rangle), (\mathcal{H} - \langle \mathcal{H} \rangle)] \rangle|^2 = \tfrac{1}{4} |\langle [O, \mathcal{H}] \rangle|^2,$$

or

$$(\Delta O)(\Delta \mathcal{H}) \geq \tfrac{1}{2} |\langle [O, \mathcal{H}] \rangle|. \qquad (6.27)$$

According to equation (6.24), we can write

$$\langle [O_t, H] \rangle = \langle [O, \mathcal{H}] \rangle = i\hbar \frac{d\langle O \rangle}{dt},$$

assuming O is not explicitly time-dependent, so that equation (6.27) becomes

$$(\Delta O)(\Delta \mathcal{H}) \geq \frac{\hbar}{2} \left| \frac{d\langle O \rangle}{dt} \right|$$

or

$$\frac{(\Delta O)}{|d\langle O \rangle/dt|} (\Delta \mathcal{H}) \geq \frac{\hbar}{2}.$$

The quantity

$$\frac{(\Delta O)}{|d\langle O \rangle/dt|} \equiv \Delta \tau$$

is of the order of magnitude of the time required for the observable O to displace from its mean value $\langle O \rangle$ by an amount (ΔO). In this sense, $\Delta \tau$ is a characteristic time of evolution for the system whose property O has the value $\langle O \rangle$: At times greater than $\Delta \tau$, the system has significantly changed with respect to the property $\langle O \rangle$. Since $(\Delta \mathcal{H}) = \Delta E$, the equation above may be written

$$\Delta \tau \, \Delta E \geq \frac{\hbar}{2} \qquad (6.28)$$

which is a *time-energy uncertainty relation*. Equation (6.28) states that the spread in energy will always be greater than the product of Planck's constant and the frequency of serious departure from the initial state of the system. Conversely, the smallest time interval which has any physical meaning is of the order of $(\hbar/(\Delta E))$.

6.3. AN EXAMPLE: THE HARMONIC OSCILLATOR

The relative Hamiltonian function for a particle oscillating with an angular frequency ω about some point free to move through space is

$$H(p, q) = \frac{p^2}{2\mu} + \tfrac{1}{2}\mu\omega^2 q^2 \tag{6.29}$$

in one dimension, where μ is the reduced mass of the oscillator, p is its momentum, and q is its displacement from the resting position. The Hamiltonian operator constructed from (6.29) is

$$\mathscr{H} = \frac{p_{op}^2}{2\mu} \neq \tfrac{1}{2}\mu\omega^2 q_{op}^2, \tag{6.30}$$

where p_{op} and q_{op} are defined by the commutation relations

$$[q_{op}, q_{op}] = [p_{op}, p_{op}] = 0, \qquad [q_{op}, p_{op}] = i\hbar I. \tag{6.31}$$

The eigenvalue problem following from equation (6.30) may be solved by an operator method invented by P. A. M. Dirac. We commence the method by defining the *construction operators*

$$a = \left(\frac{\mu\omega}{2\hbar}\right)^{\frac{1}{2}}\left(q_{op} + \frac{i}{\mu\omega}p_{op}\right)$$

$$a^\dagger = \left(\frac{\mu\omega}{2\hbar}\right)^{\frac{1}{2}}\left(q_{op} - \frac{i}{\mu\omega}p_{op}\right) \tag{6.32}$$

so that equation (6.30) becomes

$$\mathscr{H} = (aa^\dagger + \tfrac{1}{2}I)\hbar\omega. \tag{6.33}$$

The commutation relations for the construction operators are, according to equations (6.31) and (6.32),

$$[a, a] = [a^\dagger, a^\dagger] = 0, \qquad [a, a^\dagger] = I. \tag{6.34}$$

These relations replace equations (6.31) and must be used to deduce the eigenvalues of $\mathscr{H}$.

Suppose the vector $|k\rangle$ ($k = 0, 1, 2, \ldots$) is an eigenvector of $aa^\dagger$. Then, by equation (6.33),

$$\langle k|aa^\dagger|k\rangle = \lambda_k \langle k|k\rangle = \left\langle k\left|\left(\frac{\mathscr{H}}{\hbar\omega} - \frac{1}{2}I\right)\right|k\right\rangle$$

$$= \left(\frac{E_k}{\hbar\omega} - \frac{1}{2}\right)\langle k|k\rangle,$$

179

where λ_k is the kth eigenvalue belonging to $aa^\dagger$ and E_k is an eigenvalue of $\mathcal{H}$. The matrix element

$$\langle k|aa^\dagger|k\rangle$$

may be interpreted as the magnitude of the vector $a^\dagger|k\rangle$, which, by definition, is always positive-definite. It follows, then, because $\langle k|k\rangle$ is always positive on physical grounds, that

$$\lambda_k \geqslant 0 \qquad (k = 0, 1, 2, \ldots) \tag{6.35}$$

which, in turn, means

$$E_k \geqslant \tfrac{1}{2}\hbar\omega \qquad (k = 0, 1, 2, \ldots). \tag{6.36}$$

Equations (6.35) and (6.36) are restrictions on the lower limits of the eigenvalues of $aa^\dagger$ and $\mathcal{H}$, respectively. The equalities hold only if

$$a|k\rangle = 0 \qquad \text{or} \qquad a^\dagger|k\rangle = 0 \qquad (k = 0, 1, 2, \ldots).$$

Now, $|k\rangle$ is not the only eigenvector of $aa^\dagger$. For, by equations (6.34),

$$(aa^\dagger)a|k\rangle = a(a^\dagger a)|k\rangle = a(aa^\dagger - I)|k\rangle = (\lambda_k - 1)a|k\rangle$$

and

$$(aa^\dagger)a^\dagger|k\rangle = (a^\dagger a + I)a^\dagger|k\rangle = a^\dagger(aa^\dagger)|k\rangle + a^\dagger|k\rangle = (\lambda_k + 1)a^\dagger|k\rangle.$$

The vectors $a|k\rangle$ and $a^\dagger|k\rangle$ are seen to correspond to the eigenvalues $(\lambda_k - 1)$ and $(\lambda_k + 1)$, respectively. Moreover, it is evident from this calculation that repeated application of a upon $|k\rangle$ generates eigenvectors corresponding to $(\lambda_k - 1)$, $(\lambda_k - 2)$, etc., while repeated application of $a^\dagger$ upon $|k\rangle$ generates eigenvectors corresponding to $(\lambda_k + 1)$, $(\lambda_k + 2)$, etc. As these two operators comprise all the dynamical information about the system [recall equations (6.32)], the set of eigenvectors they generate must be complete. We conclude, then, that *each λ_k differs from its predecessor by one.* It follows that we need only know what is the value of λ_0 in order to calculate the spectrum of $aa^\dagger$. The former quantity must lie between zero and one, however, because a value greater than one would imply that $a|0\rangle$ is an eigenvector corresponding to an eigenvalue smaller than that for $|0\rangle$, which is impossible, while a value lower than zero would be in violation of equation (6.35). It must be that

$$a|0\rangle \equiv 0|0\rangle \tag{6.37}$$

and, therefore, that λ_0 is zero. The eigenvalues of $aa^\dagger$, then, are seen to be the set of all positive integers, so that

$$E_k = (k + \tfrac{1}{2})\hbar\omega \qquad (k = 0, 1, \ldots) \tag{6.38}$$

180

are the eigenvalues of $\mathcal{H}$ (in agreement with Planck's postulate!). We notice that in the groundstate the total energy of the oscillator is not zero, but is given by

$$E_0 = \tfrac{1}{2}\hbar\omega$$

which is often called the *zero-point energy*. That E_k does not vanish in the groundstate is not surprising because

$$E_0 = \left\langle \frac{p_{\text{op}}^2}{2\mu} \right\rangle_0 + \tfrac{1}{2}\mu\omega^2 \langle q_{\text{op}}^2 \rangle_0.$$

A zero value of E_0 would suggest that the mean deviations of p_{op} and q_{op} were simultaneously zero, in violation of the momentum-position uncertainty relations.

We can get a look at the coordinate representatives of the $|k\rangle$ by noting that the last of equations (6.34) implies [see the discussion following equation (6.22)]

$$aa^{\dagger k} - a^{\dagger k}a = ka^{\dagger k-1} \qquad (k \geqslant 1)$$
$$a^{\dagger 0} \equiv I$$

so that

$$aa^{\dagger k}|0\rangle = ka^{\dagger k-1}|0\rangle$$

because of equation (6.37). It follows that

$$\langle 0|a^k a^{\dagger k}|0\rangle = k\langle 0|a^{k-1}a^{\dagger k-1}|0\rangle$$
$$= k(k-1)\langle 0|a^{k-2}a^{\dagger k-2}|0\rangle = \cdots = k!$$

with $\langle 0|0\rangle \equiv 1$. With this result we can normalize $a^{\dagger k}|0\rangle$ and write the normalized $|k\rangle$ as

$$|k\rangle = (k!)^{-1/2}a^{\dagger k}|0\rangle \qquad (k \geqslant 0). \tag{6.39}$$

All that remains is to find the functional form of the groundstate wavefunction. This we do by noting that equation (6.37), when $|0\rangle$ has been projected onto a member of the coordinate basis $\{|q\rangle\}$, is a differential equation:

$$\left(\frac{\mu\omega}{2\hbar}\right)^{1/2}\left(q + \frac{\hbar}{\mu\omega}\frac{d}{dq}\right)\langle q|0\rangle = 0$$

or

$$\frac{d\psi_0}{dq} + q\left(\frac{\mu\omega}{\hbar}\right)\psi_0(q) = 0, \tag{6.40}$$

where $\psi_0(q) \equiv \langle q|0\rangle$ is the wavefunction in the groundstate. The solution of equation (6.40) is

$$\psi_0(q) = c\exp\left[-(\mu\omega/2\hbar)q^2\right] \qquad (-\infty \leqslant q \leqslant +\infty)$$

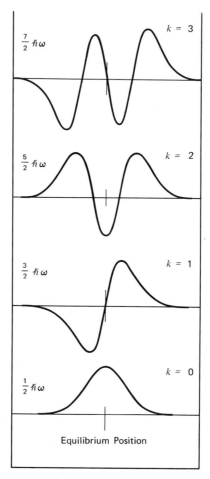

Figure 42. The first four eigenstates for the linear harmonic oscillator. The alternating symmetry and asymmetry of the wavefunction is the result of the invariance of the Hamiltonian operator under reflections about the equilibrium position of the oscillator (If this seems like magic, see Chapter 9!)

where c is a normalization constant. The normalized solution is

$$\psi_0(q) = \left(\frac{\mu\omega}{\hbar}\right)^{\frac{1}{2}} \exp\left[-(\mu\omega/2\hbar)q^2\right] \qquad (-\infty \leqslant q \leqslant +\infty) \qquad (6.41)$$

from which it follows that

$$\psi_k(q) = (k!)^{-\frac{1}{2}}\left(\frac{\mu\omega}{\hbar}\right)^{\frac{1}{4}}\left(\frac{\mu\omega}{2\hbar}\right)^{k/2}\left(q + \frac{\hbar}{\mu\omega}\frac{d}{dq}\right)^k \exp\left[-(\mu\omega/2\hbar)q^2\right]. \qquad (6.42)$$

The $\psi_k(q)$ are known as Hermite functions. They satisfy an equation of the Sturm-Liouville type:

$$\frac{d^2\psi_k}{dq^2} + \left(\frac{\mu\omega}{\hbar}\right)\left[(2k + 1) - \left(\frac{\mu\omega}{\hbar}\right)q^2\right]\psi_k(q) = 0 \qquad (6.43)$$

and are in form always the product of a polynomial in q (the *Hermite polynomial*) and $\psi_0(q)$. The first four of the $\psi_k(q)$ are shown in Figure 42.

6.4. THE MEASUREMENT OF OBSERVABLES

The matrix element of an observable represented in a basis composed of its eigenstates was defined in Chapter 3 to be a component of the mean value of the observable:

$$\langle O \rangle = \sum_{k=1}^{\infty} \omega_k \langle k|O|k \rangle \qquad (6.44)$$

where ω_k is the probability that the system is in the eigenstate $|k\rangle$. We have assumed along with this concomitant of the observable postulate that $\langle k|O|k \rangle$ can, in some way, be connected logically with the process of measurement. It is worthwhile now to discuss this possibility in detail.

The measurement process in quantum physics, as it is understood at present,[3] may be summarized in the following postulate, due essentially to John von Neumann.

The Measurement Postulate. The result of measuring the value of an observable for a physical system is always one of the eigenvalues of the observable.

It is to be understood, of course, that "measurement" here is considered in an ideal sense: an eigenvalue results only when the measurement is devoid of empirical error. In view of equation (6.44), the measurement postulate seems to mean the following. If the system to be investigated is in a superposition state prior to the measurement, then we can only ascribe a set of probabilities

---

[3] For a careful review of the measurement problem, see Chapter IV in K. Gottfried, *Quantum Mechanics*, W. A. Benjamin Co., New York, 1966, Vol. I.

$\{\omega_k\}$ to all the possible results of experiment. After the measurement has occurred, however, all of the ω_k but one reduces to zero; the single non-vanishing probability has now become equal to unity—it corresponds to the eigenvalue actually determined. On the other hand, if a system is in an eigenstate, it is clear that the measurement will have no effect on the set $\{\omega_k\}$. In other words, there appears to be an uncontrollable aspect of measurement, such that the measuring device exerts a discontinuous, "reducing influence" upon the wavefunction of the system being measured.

Thus, knowledge of a state vector is in general *not* the same as knowledge of the exact result of measurement, because the state vector, through equation (6.44), permits only the specification of the *probability* that a certain property will have a certain measured value. We have had an example of this behavior in our discussion, in Chapter 3, of Young's double-slit experiment with photons. In that consideration it was stated that prior to contact with the detecting screen the photon in transit could be thought of as in a superposition state composed of all the possible "trajectory states" connected with the interference pattern. At the moment of impact, however, the photon must be in just *one* of these states—the one measured. The measuring process has therefore "filtered" the state vector in the sense that it reduces all but one of the ω_k to zero. It is not possible or necessary to state deterministically how the "filtering" was done because (a) during the measurement the photon and the screen detecting it form but a single, inseparable quantum-physical system, and (b) the interference phenomenon can be predicted from a knowledge of the photon's state vector alone.

The filtering of the state vector is tantamount to the interaction between the system and the measuring device which is necessary for a reproducible experiment. For it is clear that no filtering will take place unless the state of the measuring device is disturbed sufficiently by the system measured, and that reproducible experimental data are the results of a well-filtered state vector. On the other hand, if the interaction is *too* strong, so as to completely modify the state of the measuring device, the system will filter the state vector of the measuring instrument and erratic data will be recorded. (For example, a too-intense beam of light in the double-slit experiment might cause damage to the detecting screen.) It follows that both a minimum and a maximum limit exist on the strength of the interaction during measurement. In principle, the interaction strength could be controlled by limiting the time period of interaction to an appropriate length. This fact implies a lower limit on any estimate of the total energy of the system via the measuring device, since, by equation (6.28),

$$\Delta E \geqslant \frac{\frac{1}{2}\hbar}{\Delta\tau} \qquad (6.45)$$

184

where $\Delta\tau$ is the time period of interaction during measurement. Equation (6.45) is simply another way of saying that a deterministic knowledge of the filtering process is impossible to obtain.

There is another aspect of the process of measurement, which lends itself very naturally to the branch of physics known as quantum statistical mechanics. Suppose our knowledge of a given system is not dynamically complete, such that it is impossible to assign a wavefunction to represent the state of the system. In this case, the best we can do is to say that the system could be in any one of several, or perhaps many, different physically possible pure states. The state of the system is therefore represented by a *set* of wavefunctions, each of which is associated with a probability that it is the "right" state. This kind of state is called a *statistical mixture* or *mixed state*. We see that it is quite different from a superposition state, which is a linear combination of basis vectors. Now, if $\langle O \rangle^{(j)}$ is the mean value of the observable O in the jth pure state,

$$\langle\langle O \rangle\rangle \equiv \sum_{j=1}^{N} p_j \langle O \rangle^{(j)}$$

is the mean value of O in a mixed state represented by N pure states, p_j being the probability that the mixed state is the jth pure state. It follows from equation (6.44) that

$$\langle\langle O \rangle\rangle = \sum_{j=1}^{N} \sum_{k=1}^{\infty} p_j \omega_k^{(j)} \langle k|O|k \rangle, \tag{6.46}$$

assuming each of the N pure states has been represented in terms of the basis $\{|k\rangle\}$. Equations (6.44) and (6.46) are formally quite similar, but refer to completely different physical situations. In the former case, the coefficients of $\langle k|O|k \rangle$ $(k = 1, \ldots, \infty)$ refer to quantities whose values can be determined from first principles, while in the latter case the coefficients cannot be so determined. Indeed, it must be understood clearly that the $\{p_j\}$ are *not* scalar products of the mixed state vector with one of its constituent pure states. Rather, this set of probabilities is determined from *a priori* physical considerations. (Usually the assignment of probabilities stems from an assumption of randomness in the likelihood that the mixed state is a given pure state.)

A prerequisite for any measurement is the partitioning of the system from the measuring assembly: the measurement must always be "objective." However, it is certainly possible in the case of a physical measurement (as opposed to biological or psychological measurements) to separate the experimenter from the experiment, as all findings can be reduced to digital information. Therefore, the "objectivity" of a measurement has really to do with the system and the measuring device, and not with the experimenter. We have noted that a partitioning of these two systems is not possible on the microscopic level because on that scale they form a single quantum-physical system. Stated alternatively, the observables for the pair cannot be represented as a sum of observables pertaining to the properties of one system or the other. The partitioning, then, must be done at the *macroscopic* level, where it is permissible to ignore the quantum-physical links between a system and its surroundings.

We conclude from this that *the measurements made on any physical system will always be expressed in macroscopic terms by the measuring device, no matter what is the microscopic nature of the system or its interaction with the surroundings.* In particular, the macroscopic concepts "particle" and "wave" must be attached to any measuring device, such that it must be called a "particle-sensor" or a "wave-sensor." Because of this, a measuring device can only give evidence of either wave or particle phenomena, and microscopic behavior, which is not so divisible, cannot be fully described by the results of any one class of experiments. Instead, each kind of experimental result has to be regarded as but one aspect of the totality of information that is possible to gain about some physical system. The results of different experiments performed on the system are, in this sense, always *complementary*.

The foregoing analysis comprises what has been called by Niels Bohr the *Complementarity Principle*. We do not elevate it to the level of a postulate (as in the case of the correspondence postulate) because it may be deduced from the Heisenberg uncertainty relations. Let us see how this is so. In the wave picture, the momentum of a particle is directly connected with its de Broglie wave. It follows that a measuring device capable of sensing wave phenomena will be useable for measuring the momentum of a particle to an arbitrarily high degree of precision. However, this capability prevents the device from every recording the *position* of a particle with any precision at all, since the property of localizability is completely inimical to the wave behavior, as we saw in Chapter 2. Momentum and position are therefore *complementary* observables, because different experimental arrangements are necessary in order to measure each with great precision. Stated another way, if, for example, a particle is represented by $|k\rangle \Rightarrow \exp(ikq)$, a state vector which is an eigenvector of $p_{op} = -i\hbar\, d/dq$, then $\langle p_{op}\rangle$ is equal to some eigenvalue of p_{op}, but $\langle q_{op}\rangle$ is certainly not equal to an eigenvalue of q_{op}. We cannot obtain *simultaneous* high-precision measurements of these two mean values.

Now let us make a more exact statement about complementary observables. In the case of position and momentum, each observable possesses a different set of eigenvectors. It follows in consequence that, for the state vector $|k\rangle$ given above,

$$p_{op}(q_{op}|k\rangle) = -i\hbar|k\rangle + \hbar k q_{op}|k\rangle \neq q_{op}(p_{op}|k\rangle) = \hbar k q_{op}|k\rangle.$$

This inequality is nothing more than a special case of the commutation relation

$$[q_{op}, p_{op}] = i\hbar I$$

which lies at the root of the uncertainty relations and from which we conclude that *complementary observables do not commute*. It is simply not possible to have a state vector for which complementary observables both possess dispersion-free mean values. (Note that all of our arguments apply to functions of p_{op}, such as the kinetic energy, and to functions of q_{op}, such as the potential energy.)

In summary, we may say that:

(a) An experiment cannot be devised for measuring the values of two non-commuting observables simultaneously with great precision.

(b) The state of a physical system is fully characterized by its set of compatible observables; these can, in principle, be measured simultaneously with arbitrarily high precision.

It is to be noted that unless the complete set of compatible observables has been measured, we do not possess full dynamical knowledge of a physical system.

PROBLEMS

1. Show that complex conjugation, transposition, and Hermitian conjugation are linear operations:

$$\overline{(O + P)} = (\bar{O}) + (\bar{P})$$
$$(O \overset{\sim}{+} P) = (\tilde{O}) + (\tilde{P})$$
$$(O + P)^\dagger = (O)^\dagger + (P)^\dagger$$

where (O) and (P) are arbitrary matrices.

2. The sum of the elements making up the principal diagonal of a square matrix is called the *trace* of the matrix. It is denoted by

$$\text{Trace}(A) \equiv Tr(A) \equiv \sum_k A_{kk}.$$

Show that the trace of a product of matrices is invariant under cyclic permutation of the matrices, that is,

$$\text{Tr}[(A) \times (B) \times (C)] = \text{Tr}[(C) \times (A) \times (B)] = \text{Tr}[(B) \times (C) \times (A)].$$

Matrix Quantum Mechanics

3. Show that one N by N matrix that commutes with all other N by N matrices is a constant matrix. (A constant matrix has elements $c\delta_{ij}$, where c is an arbitrary constant.)

4. Show that

$$\text{Tr}\,(|g\rangle) \times (\langle g|) = 1$$

where

$$(|g\rangle) = \begin{pmatrix} \langle 1|g\rangle \\ \langle 2|g\rangle \\ \vdots \end{pmatrix} \qquad (\langle g|) = (\overline{\langle 1|g\rangle}\,\overline{\langle 2|g\rangle}\cdots)$$

are normalized state vectors.

5. Prove, using matrix elements, that the eigenvalues of an Hermitian matrix are real.

6. Show that the eigenvalues of a unitary matrix are of unit absolute value.

7. An Hermitian matrix has the representation

$$(H) = \begin{pmatrix} \alpha & \beta \\ \beta & \alpha \end{pmatrix}$$

in the orthonormal basis composed of the vectors $|1\rangle$ and $|2\rangle$.

(a) What are the eigenvalues of (H)?

(b) What is the form of the matrix representing the unitary operator mapping $|1\rangle$ and $|2\rangle$ into $|I\rangle$ and $|II\rangle$, the basis in which (H) is diagonal?

(c) What are the expressions for $|I\rangle$ and $|II\rangle$ in terms of $|1\rangle$ and $|2\rangle$?

8. Show that the Schrödinger definitions of p_{op} and q_{op} are consistent with Heisenberg commutation relations for these operators.

9. Discuss the relationship between the Heisenberg prescription for the eigenvalue problem in quantum mechanics and the general perturbation theory introduced in Chapter 4.

10. Verify the commutation relations

$$[q, O(q, p)] = i\hbar\,\frac{\partial O}{\partial p}$$

$$[p, O(q, p)] = -i\hbar\,\frac{\partial O}{\partial q}$$

where q and p are position and momentum operators in one dimension and O is an arbitrary operator expressible as a Taylor series in q and p.

11. Show that

$$U_t\psi(\mathbf{r}) = \Psi(\mathbf{r}, t)$$

where

$$U_t = \exp\left(-i\mathcal{H}t/\hbar\right) \qquad (t \geqslant 0)$$

$\psi(\mathbf{r})$ is an eigenfunction of $\mathcal{H}$, and

$$\mathcal{H}\Psi(\mathbf{r}, t) = i\hbar \frac{\partial \Psi}{\partial t}.$$

12. Show that equation (6.24) may be rewritten

$$\frac{d}{dt} \langle O \rangle = \left\langle \frac{\partial O}{\partial t} \right\rangle - \frac{i}{\hbar} \langle [O, \mathcal{H}] \rangle$$

where O and $\mathcal{H}$ are observables in the Schrödinger picture.

13. Suppose we were to define a pair of operators by

$$t_{op}\psi(t) \equiv t\psi(t)$$

$$E_{op}\psi(t) \equiv i\hbar \frac{\partial \psi}{\partial t}$$

(a) Calculate $[t_{op}, E_{op}]$. How does it compare with $[q_{op}, p_{op}]$?

(b) The eigenvalues of q_{op} and p_{op} range over the infinite interval $(-\infty, +\infty)$. Evidently the same is true for t_{op} and E_{op}. Does this pose any problem for a physical interpretation of these operators? (For a rigorous discussion, see K. Gottfried, *Quantum Mechanics*, W. A. Benjamin Co., New York, Vol. I, p. 248.)

14. Derive the commutation relations for the *Pauli matrices*

$$\sigma_x = \begin{pmatrix} 0 & 1 \\ 1 & 0 \end{pmatrix} \qquad \sigma_y = \begin{pmatrix} 0 & -i \\ i & 0 \end{pmatrix} \qquad \sigma_z = \begin{pmatrix} 1 & 0 \\ 0 & -1 \end{pmatrix}$$

The mathematical significance of these matrices is that any Hermitian two by two matrix can be expressed as a superposition of Pauli matrices and the unit matrix. Show also that the Pauli matrices *anticommute*:

$$\sigma_x\sigma_y + \sigma_y\sigma_x = \sigma_y\sigma_z + \sigma_z\sigma_y = \sigma_z\sigma_x + \sigma_x\sigma_z = 0.$$

15. Show that the operators defined by equations (6.32) are not Hermitian. Why is their product Hermitian?

16. Let us attempt another interpretation of the operators a and $a,^\dagger$ defined in equations (6.32). Suppose we define

$$\hat{N} \equiv a^\dagger a$$

$$\hat{H} \equiv \mathcal{H} - \tfrac{1}{2}\hbar\omega$$

where $\hat{N}$ is to be called the *number operator*. We imagine that $\hat{H}$ is the Hamiltonian operator for an assembly of indistinguishable particles, each of which has the total energy $\hbar\omega$. Then $|0\rangle$ represents the state of a system with no particles (the *vacuum state*), $|1\rangle$ represents the state with one particle, and so on.

(a) Show that $\langle \hat{N} \rangle = \langle k|\hat{N}|k\rangle$ can be identified with the mean number of particles in the state $|k\rangle$.

(b) Show that it is reasonable to call $a^{\dagger}$ a *creation operator* and a, an *annihilation operator*.

(c) Interpret physically the mean value of $\hat{H}$ in the vacuum state. This interpretation of the harmonic oscillator problem is widely used in quantum field theory and statistical physics.

17. For the harmonic oscillator, show that

(a) $q_{op} = (\hbar/2\mu\omega)^{\frac{1}{2}}(a + a^{\dagger})$

$p_{op} = (-\hbar\mu\omega/2)^{\frac{1}{2}}(a - a^{\dagger})$

(b) $\langle k|q_{op}|k\rangle = \langle k|p_{op}|k\rangle = 0 \qquad (k = 0, 1, \ldots)$

(c) $\langle k|q_{op}^2|k\rangle = E_k/\mu\omega^2$

$\langle k|p_{op}^2|k\rangle = \mu E_k \qquad (k = 0, 1, \ldots)$

(d) $(\Delta q)_k(\Delta p)_k = (k + \frac{1}{2})\hbar \qquad (k = 0, 1, \ldots)$

where

$$(\Delta q)_k^2 = \langle k|(q_{op} - \langle k|q_{op}|k\rangle)^2|k\rangle$$

and so on. How do the results in (b) and (c) show that E_0 cannot be zero if the uncertainty relations are to be obeyed? (Remember the form of $\mathscr{H}$!)

18. Write down the Schrödinger eigenvalue problem for the harmonic oscillator in one dimension. Show that the normalized solution of the problem is

$$\psi_k(z) = \left(\frac{(\mu\omega/\pi\hbar)^{\frac{1}{2}}2^{-k}}{k!}\right)^{\frac{1}{2}} H_k(z) \exp\left(-z^2/2\right)$$

where

$$z = \left(\frac{\mu\omega}{\hbar}\right)^{\frac{1}{2}} q,$$

q being the oscillator position coordinate, and $H_k(z)$ is a solution of the differential equation

$$\frac{d^2 H_k}{dz^2} - 2z\frac{dH_k}{dz} + 2kH_k(z) = 0.$$

The function $H_k(z)$ is called a *Hermite polynomial* and has the form

$$H_k(z) = \sum_{l=0}^{[k/2]} \frac{(-1)^l k! (2z)^{k-2l}}{l! (k - 2l)!}$$

the sum over l extending to the largest integer less than or equal to $k/2$.

19. Suppose an harmonic oscillator in one dimension is subjected to the constant dragging force

$$F(q) = -(\mu\hbar\omega^3)^{1/2}.$$

(a) Write down the Hamiltonian operator for the oscillator in terms of the operators a and $a^\dagger$. What are the eigenvalues?

(*Hint:* Express the position coordinate in terms of the new variable $q' \equiv q + \alpha$, where α is a constant to be determined by the condition that all terms linear in q in the Hamiltonian function must vanish. Then express a and $a^\dagger$ in terms of q'_{op} instead of q_{op}.)

(b) Using the recurrence relation for Hermite polynomials,

$$2zH_n(z) = H_{n+1}(z) + 2nH_{n-1}(z),$$

calculate the second-order correction to the groundstate energy of the oscillator caused by the force $F(q)$. Compare the result with the exact value. (See problem 18 for the relation between $H_n(z)$ and the harmonic oscillator wavefunction.)

20. An atom in one of its excited states may be characterized by a *mean lifetime* $\Delta\tau$ which represents the time required for the atom to decay into a state of lower energy. Show that an excited atom is therefore not in a well-defined energy level, but must be regarded as being in a range of energy-values. Conversely, show that atoms in well-defined energy states will never decay.

PART III

ATOMIC and MOLECULAR STRUCTURE

CHAPTER 7

Many-Electron Atoms

A. The Helium Atom

The whole is greater than the sum of its parts.

7.1. ELEMENTARY EXCITATIONS IN HELIUM

We should like now to complete our picture of the helium atom. It has proven correct thus far to view this system as composed of two electrons interacting with one another and with a central nucleus through coulomb forces. The first-order perturbation treatment given in Chapter 5 showed that these interactions lead to a total groundstate energy about one-third greater in magnitude than what is expected for a hydrogenic atom containing two electrons. It may be recalled that the first-order calculation was relatively accurate, the estimated groundstate energy differing by just 5 per cent from the accepted empirical value. This result encourages us to apply the first-order theory to the lower-lying excited states of the atom in an effort to predict a portion of its emission spectrum. Such an application, if successful, would do much to pave the way for a comprehensive theory of many-electron atoms and would serve to strengthen our confidence in the fundamental theoretical structure of quantum physics.

Consider a helium atom possessing one electron in the $1s$ state and the other in some low-lying excited state denoted by the principal quantum number n. We should like to know what is the state vector for this system and what is its total energy. By analogy with the analysis given in section 5.3, we specify the zeroth-order wavefunctions, the eigenfunctions of

$$\mathscr{H}^{(0)} = -\frac{\hbar^2}{2\mu_e}\{\nabla_a{}^2 + \nabla_b{}^2\} - \left(\frac{2e^2}{r_a} + \frac{2e^2}{r_b}\right)$$

to be

$$\varphi_1^{(0)}(\mathbf{r}_a, \mathbf{r}_b) = \Psi_1(\mathbf{r}_a)\Psi_n(\mathbf{r}_b),$$
$$\varphi_2^{(0)}(\mathbf{r}_b, \mathbf{r}_a) = \Psi_1(\mathbf{r}_b)\Psi_n(\mathbf{r}_a) \tag{7.1}$$

where $\Psi_m(\mathbf{r})$ $(m = 1, n)$ is the hydrogenic wavefunction expressed by equation (4.43) (with a_0 replaced by $a_0/2$) and the subscripts a and b refer to the two electrons. According to first-order perturbation theory, the correct zeroth-order wavefunction is the linear combination

$$\varphi_N^{(0)} = \sum_{k=1}^{2} a_k{}^N \varphi_k^{(0)} \qquad (N = n + 1)$$

whose coefficients are found by solving

$$\sum_{k=1}^{2} a_k{}^N [(\varphi_{k'}^{(0)}, \mathscr{H}'\varphi_k^{(0)}) - E^{(1)}\delta_{k'k}] = 0 \qquad (k' = 1, 2) \qquad (7.2)$$

where

$$\mathscr{H}' = \frac{e^2}{r_{ab}}$$

is the perturbation Hamiltonian operator. Equations (7.2) have no useful solutions unless

$$\begin{vmatrix} H_{11}' - E^{(1)} & H_{12}' \\ H_{21}' & H_{22}' - E^{(1)} \end{vmatrix} = 0 \qquad (7.3)$$

where

$$H_{k'k}' = (\varphi_{k'}^{(0)}, \mathscr{H}'\varphi_k^{(0)}) \qquad (k', k = 1, 2).$$

By expanding the determinant in equation (7.3) we find

$$(E^{(1)})^2 - (H_{11}' + H_{22}')E^{(1)} + (H_{11}'H_{22}' - H_{21}'H_{12}') = 0$$

which has the roots

$$E^{(1)} = \begin{cases} H_{11}' + H_{12}' \\ H_{11}' - H_{12}' \end{cases} \qquad (7.4)$$

because

$$H_{12}' = H_{21}'$$

by the Hermitian nature of $\mathscr{H}'$ and

$$H_{11}' = H_{22}'$$

since both of these matrix elements are independent of the labeling of the two electrons. The total electronic energy of the helium atom is to first order

$$E_N = \begin{cases} E_1 + E_n + H_{11}' + H_{12}' \\ E_1 + E_n + H_{11}' - H_{12}'. \end{cases} \qquad (7.5)$$

196

The term H'_{11} in E_N is the analog of the correction term computed for the groundstate energy in Chapter 5 and is often termed a *coulomb integral*. It represents the effect on the total energy of the coulomb repulsion between electrons. The term H'_{12} has no direct analog. It represents the contribution to the total energy from a quantum-physical coupling between the states $\varphi_1^{(0)}$ and $\varphi_2^{(0)}$. The coupling arises because both $\varphi_1^{(0)}$ and $\varphi_2^{(0)}$ correspond to the same energy $(E_1 + E_n) = E_N$ and so the atom may, in a certain sense, *resonate* as do coupled oscillators in Newtonian mechanics.[1] The name *exchange integral* is usually attached to the term H'_{12} to emphasize that the basis vectors involved differ in the exchange of electron labels.

Returning to equations (7.2), we see that there will be four distinct a_k^N— two corresponding to the larger value of $E^{(1)}$ and two, to the lower value. These coefficients must satisfy

$$-a_1^{N+} + a_2^{N+} = 0$$
$$a_1^{N-} + a_2^{N-} = 0$$
$$|a_1^{N+}|^2 + |a_2^{N+}|^2 = |a_1^{N-}|^2 + |a_2^{N-}|^2 = 1$$

where the first equation is obtained by putting $k' = 1$ and $E^{(1)} = H'_{11} + H'_{12}$, in equations (7.2), the second is derived by doing the same with $E^{(1)} = H'_{11} - H'_{12}$, and the third ensures that $\varphi_N^{(0)}$ is normalized. The solutions for the a_k^N are then readily seen to be

$$a_1^{N+} = a_2^{N+} = a_1^{N-} = 2^{-\frac{1}{2}}$$
$$a_2^{N-} = -2^{-\frac{1}{2}}$$

so that

$$\varphi_{N+}^{(0)} \equiv 2^{-\frac{1}{2}}[\varphi_1^{(0)} + \varphi_2^{(0)}]$$
$$\varphi_{N-}^{(0)} \equiv 2^{-\frac{1}{2}}[\varphi_1^{(0)} - \varphi_2^{(0)}]. \tag{7.6}$$

Equations (7.6) tell us that the correct zeroth-order wavefunctions for excited helium are antisymmetric and symmetric linear combinations of the product wavefunctions. Because

$$(\varphi_{N-}^{(0)}, \mathcal{H}\varphi_{N-}^{(0)}) = \tfrac{1}{2}[(\varphi_1^{(0)}, \mathcal{H}\varphi_1^{(0)}) - (\varphi_1^{(0)}, \mathcal{H}\varphi_2^{(0)}) - (\varphi_2^{(0)}, \mathcal{H}\varphi_1^{(0)}) + (\varphi_2^{(0)}, \mathcal{H}\varphi_2^{(0)})]$$
$$= E_1 + E_n + H'_{11} - H'_{12}$$

we must associate the antisymmetric wavefunction with the *lower* first-order energy level as given in equation (7.5).

Now it is time to compare the results of our theoretical work with the accepted spectral data on helium. In Figure 43 are shown portions of familiar

---

[1] See L. Pauling and E. B. Wilson, Jr., *Introduction to Quantum Mechanics*, McGraw-Hill Book Co., New York, 1935, §41, for a discussion of resonance.

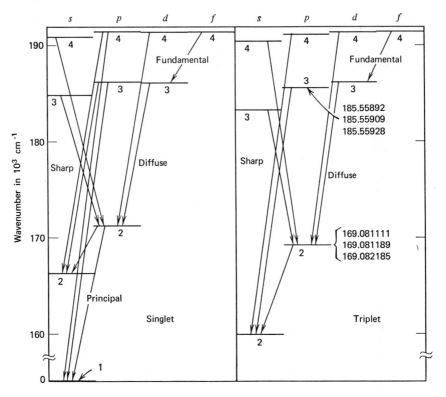

Figure 43. The spectral series for helium gas. Whole numbers placed near the levels indicate the value of the principal quantum number. The energy of the singlet groundstate corresponds to 198,305 cm^{-1}.

spectral series (principal, sharp, diffuse, and fundamental). Four attributes of these series remain to be explained:

(a) The energy levels depend on the angular quantum number as well as the principal quantum number.

(b) There are twice as many spectral series as are observed for "one-electron" atoms (such as hydrogen, lithium, and sodium). The two sets of spectra do not appear to combine with one another in any way.

(c) No electronic transitions into or from the groundstate are observed for one of the spectra.

(d) The spectral lines in the series are singlets in one case and triplets in the other.

The dependence of the energy levels upon the angular quantum number was masked in our treatment of the helium atom because we took the zeroth-order state vectors to be products of the state vectors $\Psi_n(\mathbf{r})$, which do not correspond to well-defined angular momenta. If we had chosen the hydrogenic wavefunction $\psi_{nlm}(\mathbf{r})$ (equation 4.39) instead, we would have had $2n^2$ different $\varphi^{(0)}$ as zeroth-order basis vectors making up $\varphi_N^{(0)}$. In that case the perturbation Hamiltonian operator would be a $2n^2$ by $2n^2$ matrix and there would have been $2n^2$ solutions of the determinantal equation for $E^{(1)}$—two solutions for each value of the term $(l + m)$, where l and m are angular and magnetic quantum numbers, respectively. Of these, only those corresponding to different l-values would be unequal.[2]

That there are two sets of spectral lines for helium is, of course, the principal result of our theory. The singlet levels in Figure 43 are displaced above the triplet levels and, therefore, must be associated with the larger of the two values possible for each E_N. Accordingly, the atom in these states must be represented by the symmetric wavefunction $\varphi_{N+}^{(0)}$. It follows that the triplet levels must be assigned to the asymmetric wavefunction $\varphi_{N-}^{(0)}$. Spectroscopists before the advent of quantum mechanics differentiated these states by the names "parahelium" for the symmetric state and "orthohelium" for the antisymmetric state. The lack of combination between parahelium and orthohelium spectra is not hard to understand because the probability of a transition between the states $\varphi_{N+}^{(0)}$ and $\varphi_{N-}^{(0)}$ is zero. To see this, we have only to compute the integral

$$(\varphi_{N+}^{(0)}, \mathbf{r}\varphi_{N-}^{(0)}) \qquad (\mathbf{r} = \mathbf{r}_a + \mathbf{r}_b)$$

which is related to the transition probability amplitude for dipole radiation, given in equation (5.57). Now,

$$(\varphi_{N+}^{(0)}, \mathbf{r}\varphi_{N-}^{(0)}) = \tfrac{1}{2} \int \{[\varphi_1^{(0)}(\mathbf{r}_a, \mathbf{r}_b)]^2 - [\varphi_2^{(0)}(\mathbf{r}_b, \mathbf{r}_a)]^2\}\mathbf{r}\, d\mathbf{r}_a\, d\mathbf{r}_b$$

but, upon exchanging a for b, we have

$$\tfrac{1}{2} \int \{[\varphi_2^{(0)}(\mathbf{r}_b, \mathbf{r}_a)]^2 - [\varphi_1^{(0)}(\mathbf{r}_a, \mathbf{r}_b)]^2\}\mathbf{r}\, d\mathbf{r}_a\, d\mathbf{r}_b = -(\varphi_{N+}^{(0)}, \mathbf{r}\varphi_{N-}^{(0)}).$$

This implies that the value of $(\varphi_{N+}^{(0)}, \mathbf{r}\varphi_{N-}^{(0)})$ depends on the labeling of the electrons, which is not possible. We conclude that the integral must be zero.

The remarkable absence of transitions involving the groundstate triplet level is a direct consequence of the asymmetry of its associated state vector. For, if we put $n = 1$ in $\varphi_{N-}^{(0)}$, we find

$$\varphi_{2-}^{(0)} = (\sqrt{2})^{-1}\{\Psi_1(\mathbf{r}_a)\Psi_1(\mathbf{r}_b) - \Psi_1(\mathbf{r}_a)\Psi_1(\mathbf{r}_b)\} = 0.$$

[2] A detailed analysis is given by L. Pauling and E. B. Wilson, Jr. in reference 1, pp. 210–214.

The groundstate simply vanishes! However, this is not as good a result as it seems. If we assume, reasonably, that all helium atoms first must lie in their groundstates, then it follows that no orthohelium existed initially. But if the probability of a transition from a parahelium state to an orthohelium state is zero, as shown above, then no orthohelium should ever be found! This unhappy conclusion certainly contradicts the facts; but we have no choice in the matter.

To compound the injury, we must face the fact that our theory offers no explanation for the appearance of triplets, rather than singlets, in the spectrum of orthohelium. It is doubtful, moreover, that a second-order treatment would alleviate the problem, since the first-order calculations of the energy levels for both kinds of helium turn out to be quite accurate on the average. It would seem that our picture of helium does not ascribe a sufficient number of attributes to the system or that the basic formalism of quantum mechanics is somehow faulty. If either case be true, unfortunately, we shall not be able to do without another fundamental postulate.

Nov. 17

7.2. ELECTRON SPIN

The character of the line spectrum of helium is not the only evidence for imposing a fundamental change on the quantum theory. (Indeed, if it were, we should be on safer ground to avoid such drastic action and look for other, perhaps *ad hoc* means of explaining the data.) A careful scrutiny of the behavior of hydrogenic atoms also reveals some discrepancies between our current picture of the electron and the results of experiment. For example, it is expected that a beam of hydrogen atoms passing through an inhomogeneous magnetic field (a Stern-Gerlach experiment) will separate into an *odd* number of parts, the exact number depending on the angular momentum of the atoms. This follows because the separation of the beam is proportional to the z-component of the magnetic moment of the atoms which, in turn, is proportional to the z-component of their angular momentum. The latter quantity, of course, can take on $(2l + 1)$ values for each value of l, the angular momentum quantum number. Thus, we expect a beam of groundstate hydrogen atoms to remain unseparated upon passing through an inhomogeneous field. It has been observed, on the other hand, that such a beam splits into *two* components, in direct contradiction with theory. (The same is true for the groundstate silver atom, as discussed in Chapter 1. That result is also unwanted, since silver atoms are "effectively one-electron" atoms.) But it is not only our concept of the *structure* of hydrogenic atoms that is faulty; for, upon taking a close look, we should find that the line spectrum of hydrogen consists not of singlets, but of *doublets*. (See Figure 44.)

6562.8 Å

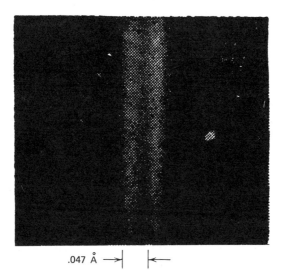

.047 Å →| |←

Figure 44. Doublet structure of the red Balmer line in the spectrum of hydrogen gas. (From *Introduction to Atomic and Nuclear Physics*, by H. E. White, Copyright © 1964, by Litton Educational Publishing, Inc., by permission of Van Nostrand Reinhold Company.)

It seems clear now that the theory must be modified. We certainly do not wish to abandon the general formalism, as contained in the equations of motion, but would rather modify our notion of "state" to account for the inexplicable data. This is indeed possible to do; for, our statement of the probability postulate, it turns out, includes an unnecessary assumption. Recall that we chose a state vector, in the wave picture, to be a complex, twice-differentiable function of the space coordinates and time. Doing this, we have made no allowance for a dependence of the state vector upon purely quantum-physical variables. This oversight may be the reason for the lack of correspondence between experiment and theory. Of course, it is not obvious that this is so, nor is it clear just which quantum-physical variable has been neglected; physical intuition is necessary once more for the proper extension of our theory. The correct quantum-physical variable was suggested by S. Goudsmit and G. E. Uhlenbeck in 1926. We may state their hypothesis in a revised form of the probability postulate, as:

The Probability Postulate. The square of the absolute value of the wavefunction, that is, $|\psi(\mathbf{r}, s, t)|^2$, is to be interpreted as the intensity distribution function for a matter wave. Here **r** refers to the space coordinates, t is the time, and s represents the direction of the intrinsic angular momentum of the particle represented by $\psi(\mathbf{r}, s, t)$. Rigorously, $|\psi(\mathbf{r}, s, t)|^2 \, d\mathbf{r}$ is the probability that the particle it represents occupies a point in $d\mathbf{r}$, with intrinsic angular momentum direction s, at a time t.

The variable s refers, in a classical-mechanical sense, to the angular momentum a particle has when spinning about its own axis—an intrinsic angular momentum, or *spin* angular momentum. For this reason, s is called the *spin variable*. We may deduce the properties of the spin variable in a natural way by imposing the restriction that its inclusion within the quantum theory elucidate the previously unexplained empirical data. In order to explain the Stern-Gerlach experiment with groundstate hydrogen atoms, we must say that the spin magnetic moment is responsible for the beam splitting, and, therefore, that the spin variable can have but two values. (Actually, we have anticipated this requirement—that the spin variable take on only a finite number of values—in our statement of the probability postulate.) Thus we shall assign to s the convenient values $+1$ and -1, corresponding to "spin up" and "spin down," respectively. Accordingly, the eigenfunction for an electron must be written $\psi(\mathbf{r}, s)$ and normalized in the sense that

$$\sum_{s=-1}^{1}{}' \int_{\mathbf{r}} \bar{\psi}(\mathbf{r}, s)\psi(\mathbf{r}, s) \, d\mathbf{r} \equiv 1 \qquad (7.7)$$

where the prime means the sum is to exclude $s = 0$. Equation (7.7) contains two terms on its left-hand side, which implies that there are two solutions to every Schrödinger eigenvalue problem: one for spin up and one for spin down. Therefore, we should now write the Schrödinger equation of motion in the matrix form

$$\mathscr{H}\begin{pmatrix} \psi(\mathbf{r}, 1, t) \\ \psi(\mathbf{r}, -1, t) \end{pmatrix} = i\hbar \frac{\partial}{\partial t}\begin{pmatrix} \psi(\mathbf{r}, 1, t) \\ \psi(\mathbf{r}, -1, t) \end{pmatrix}. \qquad (7.8)$$

If $\mathscr{H}$ does not depend upon the spin variable, equation (7.8) factors into a pair of equations of the usual kind. The eigenvalue problem may then be written

$$\mathscr{H}\psi(\mathbf{r}, s) = E\psi(\mathbf{r}, s) \qquad (s = \pm 1).$$

Because

$$\mathscr{H}\psi(\mathbf{r}) = E\psi(\mathbf{r})$$

is also true in the present case by hypothesis, it follows that $\psi(\mathbf{r}, s)$ *must be a constant multiple of* $\psi(\mathbf{r})$. We shall write

$$\psi(\mathbf{r}, s) = \begin{cases} \psi(\mathbf{r})\alpha(s) & (s = 1) \\ \psi(\mathbf{r})\beta(s) & (s = -1) \end{cases} \qquad (7.9)$$

where

$$\alpha(s) = \begin{cases} 1 & (s = 1) \\ 0 & (s = -1) \end{cases}$$

$$\beta(s) = \begin{cases} 0 & (s = 1) \\ 1 & (s = -1). \end{cases}$$

The functions $\alpha(s)$ and $\beta(s)$ are called *spin wavefunctions*. They may be postulated to be eigenfunctions of the operator to be associated with the square of the spin angular momentum:

$$S^2 \begin{pmatrix} \alpha(s) \\ \beta(s) \end{pmatrix} = l_s(l_s + 1)\hbar^2 \begin{pmatrix} \alpha(s) \\ \beta(s) \end{pmatrix} \equiv \frac{3\hbar^2}{4} \begin{pmatrix} \alpha(s) \\ \beta(s) \end{pmatrix} \qquad (7.10)$$

where l_s is the spin quantum number, equal in magnitude—as we shall show—to $\frac{1}{2}$. The spin operator S^2 acts only upon the two-dimensional vector space spanned by the spin wavefunctions and so is an identity operator with regard to the coordinate wavefunction. Conversely, the angular momentum operators are identity operators with regard to the spin wavefunctions. Therefore, S^2 and each of its components must commute with $\mathscr{L}^2$, $\mathscr{L}_x$, $\mathscr{L}_y$, and $\mathscr{L}_z$:

$$[S_q, \mathscr{L}_q] = [S^2, \mathscr{L}^2] = [S^2, \mathscr{L}_q] = [S_q, \mathscr{L}^2] = 0 \qquad (q = x, y, z). \qquad (7.11)$$

Moreover, because we have imparted to S^2 the structure of an angular momentum operator, it and its components may be postulated to obey angular momentum commutation relations, as given in the problems at the end of Chapter 3. For example,

$$[S_x, S_y] = i\hbar S_z, \qquad [S_y, S_z] = i\hbar S_x \qquad (7.12)$$

and so on.

The magnitude of the z-component of the spin angular momentum could, in principle, take on the values $-\frac{1}{2}, 0, \frac{1}{2}$ (because $l_s = \frac{1}{2}$). However, the results of the Stern-Gerlach experiment restrict us to choosing only the extremes $\pm\frac{1}{2}$. This means that $\alpha(s)$ and $\beta(s)$ are simultaneous eigenfunctions of S^2 and S_z, the latter yielding the eigenvalue equations

$$S_z \begin{pmatrix} \alpha(s) \\ \beta(s) \end{pmatrix} = m_s \hbar \begin{pmatrix} \alpha(s) \\ \beta(s) \end{pmatrix} \qquad (m_s = \pm\frac{1}{2}) \qquad (7.13)$$

where m_s may be called the magnetic spin quantum number. If the Hamiltonian operator is independent of the spin variable,

$$[S^2, \mathscr{H}] = [S_q, \mathscr{H}] = 0 \qquad (q = x, y, z). \qquad (7.14)$$

We conclude from equations (7.11) through (7.14) that, in a spherically symmetric system, the total energy, the orbital angular momentum and its *z*-component, and the spin angular momentum and its *z*-component are a set of simultaneously measurable observables.

As regards a hydrogenic atom, the existence of spin requires an additional quantum number for the specification of each electronic state. The wavefunction then becomes

$$\psi_{nlmm_s}(\mathbf{r}, s) = \psi_{nlm}(\mathbf{r})\begin{pmatrix}\alpha(s)\\ \beta(s)\end{pmatrix}.$$

For each value of the principal quantum number, n, there are now $2n^2$, rather than n^2, distinct eigenfunctions: each value of the total energy is now of multiplicity $2n^2$. The effect of electron spin is to leave the value of the total energy unchanged while doubling its multiplicity. It would appear that this result is the key to the doublets in the hydrogen spectrum, once some mechanism is discovered for a spin-induced splitting of the energy levels.

In the case of the helium atom, we should like to obtain a three-fold increase in the multiplicity of the energy of the antisymmetric state and no increase in that of the energy of the symmetric state. This result can be had by putting

$$\varphi_N^{(3)} \equiv \varphi_{N-}^{(0)}\begin{cases}\alpha(s_1)\alpha(s_2)\\ (2)^{-\frac{1}{2}}[\alpha(s_1)\beta(s_2) + \alpha(s_2)\beta(s_1)]\\ \beta(s_1)\beta(s_2)\end{cases}$$

$$\varphi_N^{(1)} \equiv \varphi_{N+}^{(0)}(2)^{-\frac{1}{2}}[\alpha(s_1)\beta(s_2) - \alpha(s_2)\beta(s_1)] \tag{7.15}$$

where $\varphi_{N+}^{(0)}$ and $\varphi_{N-}^{(0)}$ are expressed in equations (7.6) and the subscripts on the spin variables refer to the two electrons. The reason for choosing just these combinations of the spin and coordinate wavefunctions will be made apparent later on. Here again the multiplicity, not the magnitude, of each energy level has been altered. In order to get singlet and triplet lines into the spectrum of helium we shall have to consider a spin-dependent Hamiltonian operator.

nov. 19

7.3. TOTAL ANGULAR MOMENTUM

The helium atom, as we see it now, consists of two spinning electrons stationed in orbits about a central nucleus. Each of these particles interacts with the other, but yet may be construed to possess distinct angular momenta (in the quantum-theoretical sense) by virtue of its orbital and spin motions. We can compose a useful picture of this system by regarding the orbital and spin angular momenta as *vectors* of magnitudes $\sqrt{l(l + 1)}\,\hbar$ and $\sqrt{l_s(l_s + 1)}\,\hbar$, respectively. The angular momentum vectors of each type, we shall suppose, can be added (or "coupled") to form vectors $\mathbf{L}$ and $\mathbf{S}$ of length $\sqrt{L(L + 1)}\,\hbar$

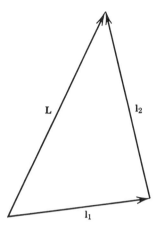

Figure 45. The total orbital angular momentum vector for a two-electron atom.

and $\sqrt{S(S+1)}\,\hbar$. In the case of a two-electron system such as helium, the quantum number L must be restricted to the values

$$L = l_1 + l_2, l_1 + l_2 - 1, \ldots, |l_1 - l_2| + 1, |l_1 - l_2|$$

where the subscripts refer to the two electrons. (See Figure 45.) The quantum number S, in a similar way, is restricted to the values 0 or 1, corresponding to sets of antiparallel and parallel spin directions.

With the two new resultant vectors we can define yet a third vector, **J**, which shall be called the *total angular momentum*. The total angular momentum is of length $\sqrt{J(J+1)}\,\hbar$, where J, the total angular quantum number, must be restricted to values lying in the range

$$(L + S) \geqslant J \geqslant |L - S| \qquad (7.16)$$

each value differing from its predecessor by unity. (See Figure 46.) In general, if the angular momentum of an atom can be spoken of at all, it is J, and not L or S, that is a "good quantum number." However, except for atoms with large numbers of electrons (eighty or more), both L and S are approximately constants of the motion, and the vector method just introduced may be used with confidence. The procedure itself is called the *Russell-Saunders coupling scheme*, after the two men who first employed it, in 1925, to interpret the emission spectrum of calcium. Russell-Saunders coupling is but one aspect of the *vector model of the atom*, which we shall use extensively in this chapter to elucidate the orgins of the spectrum of helium.

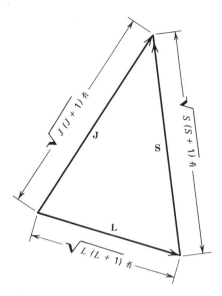

Figure 46. The total angular momentum vector **J**.

The concept of total angular momentum makes it possible to introduce a notation for multielectron states analogous with that for single-electron states. Those states of total orbital angular momentum $\sqrt{L(L+1)}\,\hbar$ are to be designated S for $L = 0$, P for $L = 1$, D for $L = 2$, and so on. The multiplicity of these states, however, is not $(2L + 1)$ unless L is smaller than S; rather, the multiplicity is $(2S + 1)$ and is denoted by a superscript before the L-symbol. The total angular quantum number for a state is made apparent by a subscript after the L-symbol. Thus, the triplet ($S = 1$, $L = 0$, $J = 1$) and singlet ($S = L = J = 0$) states for helium are given the symbols $^3S_1$ and $^1S_0$, respectively.

The state vector representing the state of two electrons may be written, in the Russell-Saunders approximation, as a linear combination of products of the basis vectors representing the individual electrons. In Dirac notation, we have:

$$|L, S; J, M_J\rangle = \sum_{M_L=-L}^{L} \sum_{M_S=-S}^{S} \langle L, S; M_L, M_S|L, S; JM_J\rangle|L, M_L\rangle|S, M_S\rangle$$

$$(7.17)$$

where M_L is the magnetic quantum number corresponding to L, and M_S is that corresponding to S. $|L, M_L\rangle$ is an eigenstate of total orbital angular momentum $\sqrt{L(L+1)}\,\hbar$ and z-axis projection $M_L\hbar$ and $|S, M_S\rangle$ is an eigenstate of total spin angular momentum $\sqrt{S(S+1)}\,\hbar$ and z-axis projection $M_S\hbar$. These eigenstates are, in turn, superpositions of single-electron eigenstates of $\mathscr{L}^2$, $\mathscr{L}_z$, S^2, and S_z:

$$|L, M_L\rangle = \sum_{m_1 = -l_1}^{l_1} \sum_{m_2 = -l_2}^{l_2} \langle l_1, l_2; m_1, m_2|L, M_L\rangle|l_1, m_1\rangle|l_2, m_2\rangle$$

$$|S, M_S\rangle = \sum_{m_s = -l_s}^{l_s}{}' \sum_{m_s' = -l_s'}^{l_s'}{}' \langle l_s, l_s'; m_s, m_s'|S, M_S\rangle|l_s, m_s\rangle|l_s', m_s'\rangle \quad (7.18)$$

The coefficients in equations (7.17) and (7.18) are known as *vector-coupling coefficients*, *Clebsch-Gordon coefficients*, and *Wigner coefficients*. They are of great importance to the detailed study of complex atoms and have been tabulated extensively.[3]

We are now in a position to understand the choice of spin wavefunctions made in equations (7.15). The second of equations (7.18) tells us that the state vector corresponding to $S = M_S = 1$ (both spins up, triplet state) is

$$|1, 1\rangle = \sum_{m_s = -\frac{1}{2}}^{\frac{1}{2}}{}' \sum_{m_s' = -\frac{1}{2}}^{\frac{1}{2}}{}' \langle \tfrac{1}{2}, \tfrac{1}{2}; m_s, m_s'|1, 1\rangle|\tfrac{1}{2}, m_s\rangle|\tfrac{1}{2}, m_s'\rangle$$

$$= \langle \tfrac{1}{2}, \tfrac{1}{2}; \tfrac{1}{2}, \tfrac{1}{2}|1, 1\rangle|\tfrac{1}{2}, \tfrac{1}{2}\rangle|\tfrac{1}{2}, \tfrac{1}{2}\rangle$$

$$\Rightarrow \alpha(s_1)\alpha(s_2)$$

since, on physical grounds, all of the vector coupling coefficients but one must vanish. The remaining triplet spin wavefunctions can be derived similarly. For $S = 1$, $M_S = 0$ we have

$$|1, 0\rangle = \langle \tfrac{1}{2}, \tfrac{1}{2}; \tfrac{1}{2}, -\tfrac{1}{2}|1, 0\rangle|\tfrac{1}{2}, \tfrac{1}{2}\rangle|\tfrac{1}{2}, -\tfrac{1}{2}\rangle \oplus \langle \tfrac{1}{2}, \tfrac{1}{2}; -\tfrac{1}{2}, \tfrac{1}{2}|1, 0\rangle|\tfrac{1}{2}, -\tfrac{1}{2}\rangle|\tfrac{1}{2}, \tfrac{1}{2}\rangle$$

$$\Rightarrow (2)^{-\frac{1}{2}}(\alpha(s_1)\beta(s_2) + \beta(s_1)\alpha(s_2))$$

if, as above, we choose the coupling coefficients so as to normalize the wavefunction. For $S = 1$, $M_S = -1$ we have

$$|1, -1\rangle = \langle \tfrac{1}{2}, \tfrac{1}{2}; -\tfrac{1}{2}, -\tfrac{1}{2}|1, -1\rangle|\tfrac{1}{2}, -\tfrac{1}{2}\rangle|\tfrac{1}{2}, -\tfrac{1}{2}\rangle$$

$$\Rightarrow \beta(s_1)\beta(s_2).$$

The singlet wavefunction corresponds to $S = 0$, $M_S = 0$. Therefore,

$$|0, 0\rangle = \langle \tfrac{1}{2}, \tfrac{1}{2}; \tfrac{1}{2}, -\tfrac{1}{2}|0, 0\rangle|\tfrac{1}{2}, \tfrac{1}{2}\rangle|\tfrac{1}{2}, -\tfrac{1}{2}\rangle \oplus \langle \tfrac{1}{2}, \tfrac{1}{2}; -\tfrac{1}{2}, \tfrac{1}{2}|0, 0\rangle|\tfrac{1}{2}, -\tfrac{1}{2}\rangle|\tfrac{1}{2}, \tfrac{1}{2}\rangle$$

$$\Rightarrow (2)^{-\frac{1}{2}}(\alpha(s_1)\beta(s_2) - \beta(s_1)\alpha(s_2))$$

where the last step comes from the fact that $|0, 0\rangle$ must be orthogonal to $|1, 0\rangle$ as well as normalized.

[3] For a good discussion without lethal mathematics, see B. W. Shore and D. H. Menzel, *Principles of Atomic Spectra*, John Wiley and Sons, New York, 1968, Chapter 6.

7.4. THE ANOMALOUS ZEEMAN EFFECT

The vector model of helium, with Russell-Saunders coupling, may be used to extend the theory of the Zeeman effect developed in section 4.6. By analogy with equation (4.53), we postulate that the perturbation Hamiltonian function for the interaction between a group of spinning electrons and a weak[4] magnetic field is of the form

$$H' = \frac{[\mu_B(\mathbf{L} \cdot \mathbf{H}_e) + 2\mu_B(\mathbf{S} \cdot \mathbf{H}_e)]}{\hbar} \tag{7.19}$$

where μ_B is the Bohr magneton and $\mathbf{H}_e$ is the applied field. It is to be noted that we have *assumed*, in consonance with experiment, that the absolute magnitude of the spin magnetic moment is $2\mu_B$, rather than μ_B, as might be expected on the basis of equation (4.52). The vectors $\mathbf{L}$ and $\mathbf{S}$ are related to their corresponding magnetic moments by

$$\boldsymbol{\mu}_\mathbf{L} = \frac{-\mu_B \mathbf{L}}{\hbar}, \qquad \boldsymbol{\mu}_\mathbf{S} = \frac{-2\mu_B \mathbf{S}}{\hbar}. \tag{7.20}$$

Following this tack, we shall assume that

$$\boldsymbol{\mu}_\mathbf{J} = \frac{-g\mu_B \mathbf{J}}{\hbar} \tag{7.21}$$

where g is a constant factor yet to be determined. Now, if the $\mathbf{J}$, $\mathbf{L}$, and $\mathbf{S}$ are coupled as suggested in section 7.3, their corresponding magnetic moments must also be coupled. Indeed, we have

$$\boldsymbol{\mu}_\mathbf{J} = \boldsymbol{\mu}_\mathbf{L} \cos\theta + \boldsymbol{\mu}_\mathbf{S} \cos\varphi \tag{7.22}$$

where θ is the angle between $\mathbf{L}$ and $\mathbf{J}$ and φ is that between $\mathbf{S}$ and $\mathbf{J}$. (See Figure 47.)

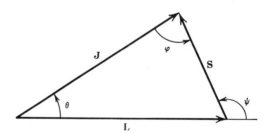

Figure 47. A geometrical picture of Russell-Saunders coupling of angular momenta.

[4] By "weak" we mean of low enough field strength to produce level splitting small compared to the separation between triplet levels.

Now, the wavefunctions for the unperturbed electrons are also eigen-functions of H', according to the basic hypothesis of the vector model. If the magnetic field is taken to point along the z-axis, the correction to the unperturbed energy level is then

$$\langle \mathcal{H}' \rangle = \langle \mu_{\mathbf{H}_e} \rangle H_e$$

where $\mu_{\mathbf{H}_e}$ is the component of $\mu_\mathbf{J}$ along the z-axis. Thus we have

$$\langle \mathcal{H}' \rangle = g\mu_B H_e M_J \qquad (M_J = -J, \ldots, +J) \qquad (7.23)$$

where g may be computed from

$$\langle |\mu_\mathbf{J}| \rangle = -\frac{g}{\hbar} \mu_B \langle |\mathbf{J}| \rangle = -g\mu_B \sqrt{J(J+1)}$$

$$= \langle \mu_\mathbf{L} \rangle \langle \cos\theta \rangle + \langle \mu_\mathbf{S} \rangle \langle \cos\varphi \rangle.$$

$$= -\mu_B \sqrt{L(L+1)} \langle \cos\theta \rangle - 2\mu_B \sqrt{S(S+1)} \langle \cos\varphi \rangle. \qquad (7.24)$$

The mean values of the cosines may be found by applying the law of cosines to the vector diagram in Figure 47. The results are

$$\langle \cos\theta \rangle = \frac{J(J+1) + L(L+1) - S(S+1)}{2\sqrt{J(J+1)L(L+1)}}$$

$$\langle \cos\varphi \rangle = \frac{J(J+1) + S(S+1) - L(L+1)}{2\sqrt{J(J+1)S(S+1)}}.$$

Upon putting these results into equation (7.24) and solving for g we get

$$g = 1 + \frac{J(J+1) - L(L+1) + S(S+1)}{2J(J+1)}. \qquad (7.25)$$

The factor g is known as the *Landé g-factor*, after the man who first derived expression (7.25). The correction to the unperturbed energy level is then

$$\langle \mathcal{H}' \rangle = M_J \mu_B H_e \left\{ 1 + \frac{J(J+1) - L(L+1) + S(S+1)}{2J(J+1)} \right\}. \qquad (7.26)$$

In the special case $S = 0$, the term in brackets reduces to unity and M_J plays a role analogous to that played by the orbital magnetic quantum number. Equation (7.26) in that case expresses the normal Zeeman effect. We conclude, then, that *all the singlet states of helium will exhibit normal Zeeman splitting in an applied magnetic field.* Conversely, all the triplet states of helium will show a more complicated splitting in accordance with equation (7.26), provided the applied magnetic field is of the appropriate strength. This splitting is known as the *anomalous Zeeman effect.* Now, equation (7.26), and the conclusions just drawn from it, are in good agreement with experiment[5] only if it is assumed that the spin quantum number l_s is equal to one-half. This fact corroborates the abundance of empirical data on other spin-dependent properties of atoms and is strong evidence for the postulated form of equation (7.10). We must for now be content with this foundation for our theory of spin; a more general formalism will be necessary in order to give a purely theoretical backing for our choice of spin eigenvalue.

Nov. 22

7.5. THE SPIN-ORBIT INTERACTION

It has been suggested that the appearance of singlets and triplets in the line spectrum of helium comes about because of spin-dependent terms in the Hamiltonian operator for the atom. That this is indeed the case may be seen, in terms of the vector model, as follows. A tenet of elementary electromagnetic theory has it that a moving charge may be associated with a magnetic field. We expect, therefore, that a magnetic field is engendered by the orbital motion of an atomic electron, and that this field interacts with the spin magnetic moment. Because the induced field is proportional to the vector product of the coulomb field of the nucleus with the electron momentum, it is also proportional to the orbital angular momentum. The Hamiltonian function for a two-electron system is, accordingly,

$$H(\mathbf{p}_1, \mathbf{p}_2, \mathbf{r}_1, \mathbf{r}_2, \mathbf{r}_{12}, \mathbf{L}, \mathbf{S}) = H^{(0)}(\mathbf{p}_1, \mathbf{p}_2, \mathbf{r}_1, \mathbf{r}_2, \mathbf{r}_{12}) + K(\mathbf{S} \cdot \mathbf{L})$$

where the first term on the right-hand side is the usual spin-independent Hamiltonian and K is a constant depending upon the details of the nuclear potential field interacting with the electrons. The second term on the right-hand side is intended to be analogous to the terms in equation (7.19). As was the case when we dealt with the latter expression, the eigenstates of $\mathscr{H}^{(0)}$ (constructed from $H^{(0)}$ in the foregoing equation) will be taken to be the representation in which the perturbation Hamiltonian operator is diagonal. Thus we may write

$$\langle \mathscr{H}' \rangle = K\hbar^2 \sqrt{L(L+1)} \sqrt{S(S+1)} \langle \cos \psi \rangle \tag{7.27}$$

where ψ is the angle between $\mathbf{L}$ and $\mathbf{S}$. (See Figure 47.) Once more we employ the law of cosines, to obtain

$$\langle \cos \psi \rangle = \frac{J(J+1) - L(L+1) - S(S+1)}{2\sqrt{L(L+1)S(S+1)}}.$$

[5] A good discussion of the Zeeman effect at all magnetic field strengths is given by J. C. Van den Bosch, The Zeeman-effect, *Handbuch der Physik* **XXVIII**: 296–332 (1957).

Equation (7.27) can then be rewritten

$$\langle \mathscr{H}' \rangle = \frac{K\hbar^2}{2}[J(J+1) - L(L+1) - S(S+1)]. \qquad (7.28)$$

The spin-orbit correction disappears from singlet states because for these states $S = 0$ and $J = L$. For the triplet states $S = 1$ and we have

$$J = \begin{cases} L + 1 \\ L \\ L - 1. \end{cases}$$

We see that there are three possible values for $\langle \mathscr{H}' \rangle$ in a two-electron system. This means that *each triplet energy level $\langle \mathscr{H}^{(0)} \rangle$ splits into three levels because of the spin-orbit interaction.* The amount of the splitting is proportional to the difference between successive values of $\langle \mathscr{H}' \rangle$ for a given L and S; that is,

$$\text{triplet interval} = \langle \mathscr{H}' \rangle_{(J+1)} - \langle \mathscr{H}' \rangle_{(J)} \propto J + 1. \qquad (7.29)$$

Equation (7.29) provides an accurate description of many two-electron spectra. We see now that the spin-orbit interaction is the key to the fine structure of the spectral lines for triplet states.[6]

7.6. THE EXCLUSION PRINCIPLE

Our study of the helium atom has led us to the conclusion that the zeroth-order wavefunctions for each low-lying energy state are of the forms

$$\begin{aligned}
\varphi_{N_1}^{(3)} &= (\varphi_1^{(0)} - \varphi_2^{(0)})\alpha(s_1)\alpha(s_2)/\sqrt{2} \\
\varphi_{N_2}^{(3)} &= \tfrac{1}{2}(\varphi_1^{(0)} - \varphi_2^{(0)})[\alpha(s_1)\beta(s_2) + \alpha(s_2)\beta(s_1)] \\
\varphi_{N_3}^{(3)} &= (\varphi_1^{(0)} - \varphi_2^{(0)})\beta(s_1)\beta(s_2)/\sqrt{2} \\
\varphi_N^{(1)} &= \tfrac{1}{2}(\varphi_1^{(0)} + \varphi_2^{(0)})[\alpha(s_1)\beta(s_2) - \alpha(s_2)\beta(s_1)]
\end{aligned} \qquad (7.30)$$

where

$$\begin{aligned}
\varphi_1^{(0)} &= \Psi_1(\mathbf{r}_1)\Psi_n(\mathbf{r}_2) \\
\varphi_2^{(0)} &= \Psi_1(\mathbf{r}_2)\Psi_n(\mathbf{r}_1)
\end{aligned}$$

and the Ψ_m are hydrogenic wavefunctions. Equations (7.30) are the results of perturbation theory, the spin hypothesis, and the operators defined by equations (7.19). These devices are, in turn, the products of empirical data

[6] However, the triplet interval as expressed in equation (7.29) is *not* the one observed in the case of helium. A correct estimate is obtained only when the magnetic field engendered by the spin motion also is taken into account. This induced field interacts with the spin magnetic moment, giving rise to a spin-spin interaction and leading to an accurate prediction of the triplet separation.

and physical intuition. We have as yet found no underlying general principle that can elucidate just why we must have the particular combinations of coordinate and spin wavefunctions shown. It is this problem that we must turn to now.

The equations (7.30) have a common property first pointed out by Wolf-gang Pauli in an elementary way and then embellished by Heisenberg and Dirac. If the labeling of the position and spin coordinates is interchanged—that is, if we put $\mathbf{r}_1, s_1 \rightarrow \mathbf{r}_2, s_2$ and vice versa—then each of the wavefunctions (7.30) changes sign:

$$\varphi_q^{(m)}(\mathbf{r}_1, s_1; \mathbf{r}_2, s_2) = -\varphi_q^{(m)}(\mathbf{r}_2, s_2; \mathbf{r}_1, s_1)$$

where $q = N_1, N_2, N_3, N$ and $m = 3, 1$. We say, therefore, that these wave-functions are *antisymmetric* with respect to the interchange of position and spin coordinates. Evidently, it is only the antisymmetric wavefunction which has a role to play in the description of the helium atom. In keeping with the attitude we have adopted toward the definition of "state" in the quantum theory, we shall not ask why symmetric wavefunctions are prohibited, but simply accept the fact. Moreover, although we certainly are not fully justified in doing so at this point, we shall generalize our discovery concerning helium and formulate it as a postulate concerning all electron systems.

The Exclusion Postulate. The wavefunction representing a system of particles having half odd-integral spin quantum numbers, such as electrons, must always be antisymmetric with respect to the interchange of coordinates.

The foregoing statement is often called the *Pauli Exclusion Principle*; the particles to which it refers are called *fermions*, after Enrico Fermi, who studied some of their general properties.

The exclusion postulate has great significance for hydrogenic atoms. It may be recast, for these systems, into the statement: *no two electrons in a hydrogenic atom can possess one and the same set of quantum numbers n, l, m, and m_s.* To see this, consider the wavefunction for two non-interacting atomic electrons. The wavefunction must be a linear combination of the form

$$\phi_\pm = [\psi_m(1)\psi_n(2) \pm \psi_m(2)\psi_n(1)],$$

aside from normalization factors, where m and n refer to the sets of quantum numbers $\{nlmm_s\}$ for the two electrons and

$$\psi_m(1) = \psi_m(\mathbf{r}_1)\begin{pmatrix} \alpha(s_1) \\ \beta(s_1) \end{pmatrix}$$

$$\psi_n(2) = \psi_n(\mathbf{r}_2)\begin{pmatrix} \alpha(s_2) \\ \beta(s_2) \end{pmatrix}$$

as pointed out in section 7.2. The exclusion postulate requires that we drop the symmetric ϕ_+ as being without physical significance. There remains only

$$\phi_- = [\psi_m(1)\psi_n(2) - \psi_m(2)\psi_n(1)]$$

which, it is not difficult to see, vanishes whenever $m = n$. We expect that this result holds true in the general case, where an arbitrary number of electrons is considered.

It is clear that the exclusion postulate will place definite restrictions upon the maximum number of electrons in a given hydrogenic energy state. For example, the $1s$ state contains at most two electrons since $n = 1, l = m = 0$, and $m_s = \pm\frac{1}{2}$ in this instance. Similarly, there are just two electrons in a $2s$ state and six in a $2p$ state; *in general, a given (nl) state can hold at most $2(2l + 1)$ electrons.* When a hydrogenic atom possesses a full complement of electrons in a certain (nl) state, it is said to have a *completed group* or a *closed shell.* Conversely, where less than $2(2l + 1)$ electrons are in one of its (nl) states, it is said to have an *incomplete group.* The number of electrons in an (nl) state is designated by a superscript in front of the state symbol; the number of electrons in an atom may be denoted by giving the number of electrons in each (nl) state in succession. Hydrogen, then, is $(1s)$, helium is $(1s)^2$, lithium is $(1s)^2(2s)$, carbon is $(1s)^2(2s)^2(2p)^2$, and sodium is $(1s)^2(2s)^2$ $(2p)^6(3s)$. Groups of (nl) states arranged in this way are called *configurations.* The configuration for an atom, of course, is strictly accurate only in the hydrogenic approximation. On the other hand, it provides an excellent starting point for the interpretation of chemical behavior on the microscopic level.

As regards our model of the helium atom, the exclusion postulate gives the reason for the forms of the singlet and triplet wavefunctions and leads immediately to the disappearance of the triplet groundstate. Moreover, in the language of hydrogenic atom theory, we see that helium atoms represent complete groups. It is this fact, as we shall see, that underlies the well-known chemical lethargy of helium: atoms possessing only complete groups do not interact well with other atoms.

nov. 24

PROBLEMS

1. Suppose a helium atom is in its first excited state. This state is of multiplicity eight, corresponding to the zeroth-order wavefunctions (in the absence of spin)

$$\psi_{100}(1)\psi_{200}(2)$$
$$\psi_{200}(1)\psi_{100}(2)$$

$$\psi_{100}(1)\psi_{211}(2)$$
$$\psi_{211}(1)\psi_{100}(2)$$
$$\psi_{100}(1)\psi_{210}(2)$$
$$\psi_{210}(1)\psi_{100}(2)$$
$$\psi_{100}(1)\psi_{21-1}(2)$$
$$\psi_{21-1}(1)\psi_{100}(2)$$

where the numerical arguments refer only to position coordinates.

(a) Show that determinantal equation for the first-order correction to the energy is composed of subdeterminants along the principal diagonal, which have the form

$$\begin{vmatrix} J_l - E^{(1)} & K_l \\ K_l & J_l - E^{(1)} \end{vmatrix}$$

where

$$J_l = \left(\psi_{100}\psi_{2lm}, \frac{e^2}{r_{12}} \psi_{100}\psi_{2lm} \right) \quad \text{(coulomb integral)}$$

$$K_l = \left(\psi_{100}\psi_{2lm}, \frac{e^2}{r_{12}} \psi_{2lm}\psi_{100} \right) \quad \text{(exchange integral)}$$

for each appropriate l, m pair.

(b) Show that

$$E^{(1)} = \begin{cases} J_0 + K_0 \\ J_0 - K_0 \\ J_1 + K_1 \\ J_1 - K_1 \end{cases}$$

and draw a diagram of the resultant splitting of the zeroth-order energy level. Which values of $E^{(1)}$ correspond to triplet states? What has happened to the degeneracy of the zeroth-order state?

2. Why is it not essential, in carrying out the analysis in problem 1, to include the spin wavefunctions?

3. The *Hund rule*, which has it that states of higher multiplicity have lower energy, appears to be accurate for most atoms. Verify this rule for helium.

4. Show that the presence of spin-orbit coupling makes the argument against parahelium-orthohelium transitions, given in section 7.1, only approximate.

5. Show that the spin wavefunctions are orthogonal in the sense that

$$\sum_{s=-1}^{1} {}' \alpha(s)\beta(s) = 0.$$

6. Consider the operators defined by the commutation relations

$$[S_z, a^\dagger] = \hbar a^\dagger \qquad [S_z, a] = -\hbar a.$$

Show that $a^\dagger \beta(s) = \alpha(s)$ and $a\alpha(s) = \beta(s)$ by considering the effect of S_z on the right-hand sides of these equations. What is the relation between these operators and those introduced in section 6.3?

7. Consider the Pauli matrices, defined in problem 6 of Chapter 14:

$$\sigma_x = \begin{pmatrix} 0 & 1 \\ 1 & 0 \end{pmatrix} \qquad \sigma_y = \begin{pmatrix} 0 & -i \\ i & 0 \end{pmatrix} \qquad \sigma_z = \begin{pmatrix} 1 & 0 \\ 0 & -1 \end{pmatrix}$$

Show that, if we represent the spin state vectors $|\alpha\rangle$ and $|\beta\rangle$ by the column matrices

$$(|\alpha\rangle) = \begin{pmatrix} 1 \\ 0 \end{pmatrix} \qquad (|\beta\rangle) = \begin{pmatrix} 0 \\ 1 \end{pmatrix},$$

the Pauli matrices can represent the spin angular momentum operators as

$$(S_x) = \frac{1}{2}\hbar\sigma_x,$$

and so on.

8. Show that the projection of the spin angular momentum vector onto the xy-plane in spin-variable space is of length $\hbar/\sqrt{2}$.

9. Consider a helium atom in the excited configuration $(2p)^2$.

(a) How many zeroth-order product wavefunctions of the form $\psi_{21m}\psi_{21m'} \alpha(s_1)\alpha(s_2)$, etc., are there for this system? How many antisymmetric combinations are there?

(b) What are the L-symbol designations for the antisymmetric states?

10. Show that, in the case of the helium atom,

$$\mathscr{H}(1, 2) = \mathscr{H}(2, 1)$$

and

$$\|\varphi_N^{(0)}(1, 2)\|^2 = \|\varphi_N^{(0)}(2, 1)\|^2.$$

These are examples of the quantum-physical indistinguishability of electrons. Demonstrate how it follows from the foregoing equations that mean values, especially of the energy, are independent of particle coordinate labels.

11. Consider the hydrogen atom in a weak magnetic field. Draw a diagram illustrating the splitting of the states $^2P_{3/2}$, $^2P_{1/2}$, and $^2S_{1/2}$. By applying the selection rules for hydrogenic atoms, show the allowed transitions between the two P-states and the S-state. Calculate the separation of each level (in units of $\mu_B H_e$) according to Landé's expression.

12. When a strong magnetic field is applied to a helium atom, it is expected that the interactions between **L** and **S** and the field will be stronger than that which provides coupling between the two vectors.

(a) Show that, neglecting any spin-orbit coupling, we have in this case

$$\langle \mathcal{H}' \rangle = \mu_B H_e (M_L + 2M_S)$$

where the M's are z-axis projections corresponding to their subscript quantum numbers.

(b) How many $\langle \mathcal{H}' \rangle$ are there for a given M_J and L? Compare this result with what occurs in the normal Zeeman effect.

13. Suppose a hydrogen atom in the $^2P_{1/2}$ state is subjected to a weak, time-varying magnetic field.

(a) Calculate the separation of the two energy levels into which the unperturbed energy level is split, per oersted of applied field strength.

(b) Show that electric dipole transitions between the two energy levels are virtually impossible.

(c) *If the field frequency is equivalent to the energy difference between the levels, magnetic dipole transitions can occur.* This phenomenon is called *electron paramagnetic resonance*. What is the field frequency necessary for electron paramagnetic resonance to occur for the split $^2P_{1/2}$ state, assuming the field strength is one oersted?

14. For a hydrogen atom the spin-orbit Hamiltonian is

$$H' = \frac{he^2}{2\mu_e^2 c^2} \left(\frac{1}{r^3} \right) (\mathbf{L} \cdot \mathbf{S})$$

so that

$$\langle \mathcal{H}' \rangle = \frac{\hbar^2 e^2}{4\mu_e^2 c^2} \langle r^{-3} \rangle [J(J+1) - L(L+1) - S(S+1)]$$

$$= \frac{\hbar^2 e^2}{4\mu_e^2 c^2} (n^3 a_0^3 L(L+\tfrac{1}{2})(L+1))^{-1}$$

$$\times [J(J+1) - L(L+1) - S(S+1)] \qquad (L \neq 0)$$

$$= \alpha^2 \frac{|E_n|}{n} \frac{J(J+1) - L(L+1) - S(S+1)}{2[L(L+\tfrac{1}{2})(L+1)]} \qquad (L \neq 0)$$

where

$$\alpha = \frac{e^2}{\hbar c} = 7.2972 \pm 0.0001 \times 10^{-3}$$

is called the *fine structure constant* and E_n is a hydrogen energy eigenvalue.

(a) Write down expressions for the spin-orbit energy for the appropriate values of J. (It is possible to show that the spin-orbit energy vanishes for $L = 0$.)

(b) Calculate the separation of the hydrogen doublet line due to spin-orbit coupling for the transition $3p \rightarrow 2s$ (the red Balmer line). The empirical value is 0.047 Å.

15. The perturbation Hamiltonian for the spin-spin interaction in helium is directly proportional to

$$A \equiv [2(\mathbf{S}_1 \cdot \mathbf{S}_2) + \tfrac{3}{2}\hbar^2]|\mathbf{L}|^2 - \tfrac{3}{2}(\mathbf{S} \cdot \mathbf{L})\hbar^2 - 3(\mathbf{S} \cdot \mathbf{L})^2.$$

where $\mathbf{S} = \mathbf{S}_1 + \mathbf{S}_2$.

(a) Show that

$$2\langle \mathbf{S}_1 \cdot \mathbf{S}_2 \rangle = [S(S + 1) - \tfrac{3}{2}]\hbar^2$$

(b) Prove that the spin-spin interaction is unimportant for parahelium and that, for orthohelium, we have

$$\frac{\langle A \rangle}{\hbar^4} = -\frac{1}{2}\begin{cases} L(2L - 1) & J = L + 1 \\ (2L - 3)(1 - 2L) & J = L \\ (2L + 3)(L + 1) & J = L - 1. \end{cases}$$

CHAPTER 8

Many-Electron Atoms

B. The Structures of Complex Atoms

E pluribus unum

8.1. A VARIATIONAL CALCULATION ON THE HELIUM ATOM

Our initial efforts have been directed toward the discovery of a microscopic basis for many-electron spectra, beginning with the simplest, with only casual interest being given atomic structure. It is time to remedy this deficiency and turn to the problem of determining multielectron wavefunctions. The general method we shall employ for this task has the advantages of a simple physical interpretation and of a convergent approximation scheme. Its essential tenets are best illustrated by a fresh consideration of the groundstate of helium.

Let us suppose that electron-electron interactions, spin, and the exclusion postulate may be neglected, so that the groundstate wavefunction for helium is expressible as

$$\psi_1(\mathbf{r}_1, \mathbf{r}_2) = \frac{Z'^3}{\pi a_0{}^3} \exp\left[-Z'(r_1 + r_2)/a_0\right] \qquad (8.1)$$

where Z' is not the atomic number, but is to be a parameter capable of variation. Except for being a variation function, $\psi_1(\mathbf{r}_1, \mathbf{r}_2)$ is just a product of hydrogenic wavefunctions and so is a natural starting point for our calculation. The means for converting $\psi_1(\mathbf{r}_1, \mathbf{r}_2)$ into an acceptable groundstate wavefunction is, evidently, the variational principle, stated in Chapter 5. We may begin by computing the first diagonal matrix element of

$$\mathscr{H} = -\frac{\hbar^2}{2\mu_e}(\nabla_{\mathbf{r}_1}{}^2 + \nabla_{\mathbf{r}_2}{}^2) - 2e^2\left(\frac{1}{r_1} + \frac{1}{r_2}\right) + \frac{e^2}{r_{12}}.$$

This is

$$(\psi_1, \mathscr{H}\psi_1) = 2Z'^2 E_1^{(0)} + (Z' - 2)e^2 \int_{\mathbf{r}_1} \int_{\mathbf{r}_2} \bar{\psi}_1 \left(\frac{1}{r_1} + \frac{1}{r_2}\right)\psi_1 \, d\mathbf{r}_1 \, d\mathbf{r}_2$$

$$+ e^2 \int_{\mathbf{r}_1} \int_{\mathbf{r}_2} \bar{\psi}_1 \left(\frac{1}{r_{12}}\right)\psi_1 \, d\mathbf{r}_1 \, d\mathbf{r}_2 \qquad (8.2)$$

where

$$E_1^{(0)} = -\frac{e^2}{2a_0}.$$

The third term on the right-hand side of equation (8.2) was evaluated in section 5.3. It is

$$e^2 \int_{\mathbf{r}_1} \int_{\mathbf{r}_2} \bar{\psi}_1 \frac{1}{r_{12}} \psi_1 \, d\mathbf{r}_1 \, d\mathbf{r}_2 = -\tfrac{5}{4}Z'E_1^{(0)}.$$

The second term in the equation is

$$\int_{\mathbf{r}_1} \int_{\mathbf{r}_2} \bar{\psi}_1 \left(\frac{1}{r_1} + \frac{1}{r_2}\right)\psi_1 \, d\mathbf{r}_1 \, d\mathbf{r}_2 = 2\left(\frac{Z'^3}{\pi a_0^3}\right) \int_{\mathbf{r}} \exp\left(-2Z'r/a_0\right)r \, d\mathbf{r}$$

$$= 8\left(\frac{Z'}{a_0}\right)^3 \int_0^{\infty} \exp\left(-2Z'r/a_0\right)r \, dr$$

$$= 8\left(\frac{Z'}{a_0}\right)^3 \left(\frac{a_0}{2Z'}\right)^2 = -\frac{4Z'E_1^{(0)}}{e^2}$$

so that

$$(\psi_1, \mathscr{H}\psi_1) \equiv E(Z') = [2Z'^2 - \tfrac{5}{4}Z' - 4Z'(Z' - 2)]E_1^{(0)}.$$

We know from the variational principle that

$$(\psi_1, \mathscr{H}\psi_1) \geqslant E_1$$

where E_1 is the correct groundstate energy. In the present scheme, therefore, we shall achieve the most accurate groundstate energy by minimizing $E(Z')$ with respect to the variable parameter Z'. The condition to be imposed is the usual one:

$$\frac{dE}{dZ'} = (4Z' - \tfrac{5}{4} - 8Z' + 8)E_1^{(0)} \equiv 0.$$

(We note that

$$\frac{d^2E}{dZ'^2} = -4E_1^{(0)} \qquad (E_1^{(0)} < 0)$$

which shows that $E(Z')$ is indeed minimized by our procedure.) The parameter Z' is easily found to be

$$Z' = \tfrac{27}{16}$$

so that

$$(\psi_1, \mathscr{H}\psi_1) = -5.70 \text{ Rydbergs}.$$

This result compares very well with the experimental value of -5.80744 Rydbergs and is better than the estimate we obtained previously by first-order perturbation theory.

We are to conclude from this analysis that a good wavefunction for the groundstate of helium is of the form

$$\psi_1(\mathbf{r}_1, \mathbf{r}_2) = \left[\frac{(1.6875)^3}{\pi a_0{}^3}\right] \exp\left[-\tfrac{27}{16}(r_1 + r_2)/a_0\right]. \tag{8.3}$$

(Indeed, the most probable electron-nucleus separation is found to be 0.31 Å in terms of the radial distribution function based upon equation (8.3). The accepted value is 0.30 Å.[1]) But how are we to interpret Z' physically? One way is to note that the parameter may be written

$$Z' = (2 - \tfrac{5}{16}).$$

The first figure is the nuclear charge and would be the coefficient of the coulomb potentials $-e^2/r_1$ and $-e^2/r_2$ in the Hamiltonian operator if helium were a hydrogenic atom. The second term may be thought of as a direct result of there being two electrons orbiting about the nucleus, such that one screens the other (through its negative charge) from the full effect of the attractive coulomb potential. The nucleus, in effect, possesses only the charge $(2 - (5/16))|e| = (\tfrac{27}{16})|e|$ as regards its interaction with the electrons. The potential function

$$U(r) = -\frac{27}{16}\frac{e^2}{r},$$

then, may be considered to represent in a certain sense *the average potential for the electron-nucleus interaction in helium.* Potential functions of the general form of $U(r)$, that is,

$$U(r) = -(Z - S)\frac{e^2}{r}$$

where $Z|e|$ is the bare nuclear charge, are called *screening-constant potentials.* The quantity S, of course, is the "screening constant," and represents, in an average way, the effect of all other electrons on a given one, in a many-electron atom.

The general method to be employed for getting at the structures of multi-electron atoms is well illustrated by our variational calculation on helium.

[1] J. C. Slater, *Quantum Theory of Atomic Structure*, McGraw-Hill Book Co., New York, 1960, Vol. I, p. 210.

A wavefunction of a reasonable form is assumed and used to compute the groundstate energy. The latter quantity, in turn, leads to an improved wavefunction. Evidently, we could continue this process by taking the new wavefunction through another calculation, which presumably would yield a more accurate groundstate energy and, again, a better wavefunction. The key to the procedure, speaking physically, is the idea that an electron in a complex atom may be regarded as moving in an *average* potential field engendered by its coparticles. If this or some other equivalent assumption were not made, no way of describing atomic structure could be had; for, it is the potential function which makes the quantum many-body problem incapable of an exact solution.

It should be remarked before going on that the variational procedure outlined above has been extended, with very good results, in calculations on the groundstate of helium. The initial work was done by E. A. Hylleraas in 1929, wherein wavefunctions of the form

$$\psi(r_1, r_2) = \exp(-Z's) \sum_{l=0} \sum_{m=0} \sum_{n=0} c_{lmn} s^l u^m t^n$$

were used, with the notation

$$s = r_1 + r_2$$
$$u = r_{12}$$
$$t = r_2 - r_1,$$

the c_{lmn} being variable parameters. Hylleraas obtained a best value of -5.80748 Rydberg by using a polynomial with 14 terms as the coefficient of $\exp(-Z's)$; a more recent attempt,[2] with $l = 0, \frac{1}{2}, 1$, etc., and $m, n = 0, 1, 2$, etc., employs a polynomial of 189 terms to obtain an extrapolated value of

$$E(1.75) = -5.8074487542 \pm 0.0000000002 \text{ Rydberg.}$$

The computations necessary for getting this accurate a value for the groundstate energy are done on a digital computer. The value for $E(Z')$ given above may be regarded as more reliable than the empirical value.

8.2. THE HARTREE-FOCK EQUATION

It is quite important to see clearly that our investigation of complex atomic structure involves, in an essential way, the notion of an average potential field. Because of this requirement, we shall not derive the most general equation of motion for an electron in a complex atom right away, but instead shall develop an expression uncomplicated by the use of spin variables. Later

---

[2] C. Schwartz, Ground state of the helium atom, *Phys. Rev.* **128**: 1146–1148 (1962).

on these variables will be introduced to obtain an equation of motion whose form fully reflects the content of the exclusion postulate.

The Hamiltonian operator for an atom (or ion) composed of N electrons is

$$\mathcal{H} = -\frac{\hbar^2}{2\mu_e} \sum_{i=1}^{N} \nabla_{\mathbf{r}_1}^2 - Ze^2 \sum_{i=1}^{N} \frac{1}{r_i} + \frac{e^2}{2} \sum_{\substack{i,j=1 \\ (i \neq j)}} \frac{1}{r_{ij}}$$

where the third term is a sum over all the number pairs i, j such that $i \neq j$ and each pair is effectively counted but once. It will turn out to be very convenient to recast $\mathcal{H}$ into *atomic units*. Recall that, in that system, a_0 (with the appropriate reduced electron mass) is the unit of length; it is consistent to put μ_e as the unit of mass and $|e|$ as the unit of charge also, so that the unit of energy is e^2/a_0. Numerically, we have the conversion relations

atomic unit of length $\doteq 5.29 \times 10^{-9}$ cm $= 1$ Bohr hydrogen radius

atomic unit of energy $\doteq 27.196$ eV $= 2$ Rydbergs.

The many-electron Hamiltonian operator, in atomic units, is then

$$\mathcal{H} = -\frac{1}{2} \sum_{i=1}^{N} \nabla_{\mathbf{r}_i}^2 - \sum_{i=1}^{N} \frac{Z}{r_i} + \frac{1}{2} \sum_{\substack{i,j \\ (i \neq j)}} \frac{1}{r_{ij}}. \tag{8.4}$$

We shall *assume* that the operator $\mathcal{H}$ acts upon the product of normalized one-electron wavefunctions (or *atomic orbitals*)

$$\psi(\mathbf{r}^N) = \prod_{i=1}^{N} \frac{P_{n_i l_i}(r_i) \Upsilon_{l_i}^{m_i}(\vartheta_i, \varphi_i)}{r_i} \equiv \prod_{i=1}^{N} |i\rangle \tag{8.5}$$

where $\mathbf{r}^N$ refers to the set of variables $\{\mathbf{r}_1, \mathbf{r}_2, \ldots, \mathbf{r}_N\}$,

$$\Upsilon_l^m(\vartheta, \varphi) = (-1)^{\frac{1}{2}(m + |m|)} \left\{ \frac{(2l + 1)(l - |m|)!}{4\pi(l + |m|)!} \right\}^{\frac{1}{2}} P_l^m(\cos \vartheta) \exp(im\vartheta)$$

and $P_{nl}(r)$ is an as yet unknown, real-valued function of the radial coordinate in spherical polar space. Physically speaking, equation (8.5) reflects the fact that the potential functions in $\mathcal{H}$ are independent of the angular coordinates, that they form a *central field*, in the sense of Newtonian mechanics. Moreover, we are saying that each electron in the atom considered moves more or less independently of its neighbors, the only effect of the latter appearing when the radial portion of the wavefunction is computed. This is a fundamental hypothesis in our procedure.

The mean value of $\mathcal{H}$ is, by equations (8.4) and (8.5),

$$\langle \mathcal{H} \rangle = \sum_{i=1}^{N} \langle i | f_i | i \rangle + \frac{1}{2} \sum_{\substack{i,j \\ (i \neq j)}} \langle ij | g_{ij} | ij \rangle \tag{8.6}$$

where

$$|ij\rangle \equiv |i\rangle|j\rangle$$

$$f_i = -\left(\tfrac{1}{2}\nabla_{\mathbf{r}_i}^2 + \frac{Z}{r_i}\right)$$

$$g_{ij} = \frac{1}{r_{ij}}$$

and advantage has been taken of the normality of the $|i\rangle$. The quantity $\langle i|f_i|i\rangle$ is called a *one-center integral*. It may be written

$$\langle i|f_i|i\rangle = -\langle i|\left(\tfrac{1}{2}\nabla_{\mathbf{r}_i}^2 + \frac{Z}{r_i}\right)|i\rangle$$

$$= -\frac{1}{2}\int_0^\infty P_{n_i l_i}(r_i)\left(\frac{d^2}{dr_i^2} - \frac{l_i(l_i+1)}{r_i^2} + \frac{2Z}{r_i}\right)P_{n_i l_i}(r_i)\,dr_i$$

$$\times \int_0^\pi \int_0^{2\pi} |\Upsilon_{l_i}^{m_i}(\vartheta_i, \varphi_i)|^2 \sin\vartheta_i\,d\vartheta_i\,d\varphi_i$$

$$= \frac{1}{2}\int_0^\infty P_{n_i l_i}(r_i)\left[-\frac{d^2}{dr_i^2} + \frac{l_i(l_i+1)}{r_i^2} - \frac{2Z}{r_i}\right]P_{n_i l_i}(r_i)\,dr_i \qquad (8.7)$$

where we have noted that

$$\nabla_{\mathbf{r}_i}^2|i\rangle = \left(\frac{1}{r_i^2}\right)\frac{d}{dr_i}\left[r_i^2 \frac{d(P_{n_i l_i}/r_i)}{dr_i}\right]\Upsilon_{l_i}^{m_i}(\vartheta_i, \varphi_i)$$

$$+ \left[\frac{1}{r_i^2 \sin\vartheta_i}\frac{\partial}{\partial\vartheta_i}\left(\sin\vartheta_i \frac{\partial}{\partial\vartheta_i}\Upsilon_{l_i}^{m_i}\right) + \frac{1}{r_i^2 \sin^2\vartheta_i}\frac{\partial^2\Upsilon_{l_i}^{m_i}}{\partial\varphi_i^2}\right]\left(\frac{P_{n_i l_i}}{r_i}\right)$$

$$= \frac{1}{r_i}\left[\left(\frac{d^2 P_{n_i l_i}}{dr_i^2}\right)\Upsilon_{l_i}^{m_i} - \frac{l_i(l_i+1)}{r_i^2}\Upsilon_{l_i}^{m_i}P_{n_i l_i}(r_i)\right]$$

and have employed the orthonormality of the spherical harmonics. The term $\langle ij|g_{ij}|ij\rangle$ is called a *two-center integral*. It is reduced as follows:

$$\langle ij|g_{ij}|ij\rangle = \int_{\mathbf{r}_i} \frac{P_{n_i l_i}}{r_i}\overline{\Upsilon_{l_i}^{m_i}}\int_{\mathbf{r}_j}\frac{P_{n_j l_j}}{r_j}\overline{\Upsilon_{l_j}^{m_j}}\left(\frac{1}{r_{ij}}\right)\frac{P_{n_i l_i}}{r_i}\Upsilon_{l_i}^{m_i}\frac{P_{n_j l_j}}{r_j}\Upsilon_{l_j}^{m_j}\,d\mathbf{r}_i\,d\mathbf{r}_j$$

$$= 2\pi \int_{\mathbf{r}_i} P_{n_i l_i}\overline{\Upsilon_{l_i}^{m_i}}\int_{r_i}^\infty |P_{n_j l_j}\Upsilon_{l_j}^{m_j}|^2 P_{n_i l_i}\Upsilon_{l_i}^{m_i}\frac{dr_j\,dr_i}{r_j\,r_i^3}\int_{r_j-r_i}^{r_j+r_i}dr_{ij}\ (r_j > r_i),$$
$$\text{etc.,}$$

$$= 4\pi \int_{\mathbf{r}_i} |P_{n_i l_i}\Upsilon_{l_i}^{m_i}|^2\int_{r_i}^\infty |P_{n_j l_j}\Upsilon_{l_j}^{m_j}|^2\frac{dr_j\,dr_i}{r_j\,r_i^2}\qquad (r_j > r_i),\ \text{etc.,}$$

$$= 4\pi \int_0^\infty |P_{n_i l_i}|^2\int_{r_i}^\infty |P_{n_j l_j}\Upsilon_{l_j}^{m_j}|^2\frac{dr_j}{r_j}\,dr_i\qquad (r_j > r_i),\ \text{etc.,}$$

$$= 4\pi \int_0^\infty |P_{n_i l_i}|^2\int_0^\infty |P_{n_j l_j}\Upsilon_{l_j}^{m_i}|^2\frac{dr_j}{\rho_r}\,dr_i$$

223

where

$$\rho_r = \begin{cases} r_j & r_j > r_i \\ r_i & r_i > r_j \end{cases}$$

and we have made use of the procedure described in section 5.3 for reducing two-center integrals. We notice that $\langle ij | g_{ij} | ij \rangle$ is angle-dependent, being a function of $|Y_{l_j}{}^{m_j}|^2$. This property is undesirable in applications and so will be removed by performing a spherical average of the two-center integral. We shall use

$$(\langle ij | g_{ij} | ij \rangle)_{av} \equiv \frac{\int_{\Omega_j} \langle ij | g_{ij} | ij \rangle \, d\Omega_j}{\int_{\Omega_j} d\Omega_j}$$

$$= \int_0^\infty \int_0^\infty |P_{n_i l_i}|^2 |P_{n_j l_j}|^2 \frac{dr_j}{\rho_r} \, dr_i \qquad (8.8)$$

where

$$d\Omega_j = \sin \vartheta_j \, d\vartheta_j \, d\varphi_j$$

is an element of solid angle. This averaging of the two-center integral is clearly an approximation on our part; we expect that it will be a reasonable thing to do only when the atom under consideration possesses spherical symmetry.

The mean value of the many-electron Hamiltonian operator, under the reductions just carried through, has the form

$$\langle \mathcal{H}_{av} \rangle = \frac{1}{2} \left[\sum_{i=1}^N \langle i | \left(-\frac{d^2}{dr_i{}^2} + \frac{l_i(l_i+1)}{r_i{}^2} - \frac{2Z}{r_i} \right) | i \rangle \right.$$

$$\left. + \sum_{\substack{i,j \\ (i \neq j)}} \left(\langle ij | \frac{1}{\rho_r} | ij \rangle \right)_{av} \right]. \qquad (8.9)$$

According to the variational principle, $\langle \mathcal{H}_{av} \rangle$ must be larger in magnitude or equal to the energy of the ith electron in the atom under consideration. We should like, then, to minimize $\langle \mathcal{H}_{av} \rangle$ by varying the $P_{n_i l_i}(r_i)$. This procedure, it turns out, leads to an equation of motion for the latter functions. Let us see how that comes about.

If $\langle \mathcal{H}_{av} \rangle$ is to be made a minimum by varying the $P_{n_j l_j}(r_i)$, it is reasonable to write

$$\delta \langle \mathcal{H}_{av}^{(i)} \rangle \propto \delta(P_{n_i l_i}, P_{n_i l_i})$$

or

$$\delta \langle \mathcal{H}_{av}^{(i)} \rangle - \epsilon_{n_i l_i} \delta(P_{n_i l_i}, P_{n_i l_i}) = 0 \qquad (8.10)$$

224

where

$$\langle \mathscr{H}_{\text{av}}^{(i)} \rangle \equiv \langle i|f|i \rangle + \sum_{\substack{j=1 \\ (j \neq i)}}^{N} \langle ij|g_{ij}|ij \rangle \qquad (i = 1, \ldots, N)$$

and $\epsilon_{n_i l_i}$ is an undetermined constant of proportionality (in atomic units). By carrying out the variations indicated in equation (8.10), we find

$$\delta \langle \mathscr{H}_{\text{av}}^{(i)} \rangle = 2 \int_0^\infty \delta P_{n_i l_i} \left[-\frac{1}{2} \frac{d^2}{dr_i^2} + \frac{l_i(l_i - 1)}{2r_i^2} - \frac{Z}{r_i} \right.$$

$$\left. + \sum_{\substack{j=1 \\ (j \neq i)}}^{N} \int_0^\infty |P_{n_j l_j}|^2 \frac{dr_j}{\rho_r} \right] P_{n_i l_i}(r_i) \, dr_i$$

and

$$\delta(P_{n_i l_i}, P_{n_i l_i}) = 2 \int_0^\infty \delta P_{n_i l_i} P_{n_i l_i}(r_i) \, dr_i$$

since the $P_{n_i l_i}(r_i)$ are real-valued functions. With these results equation (8.10) becomes

$$\int_0^\infty \delta P_{n_i l_i} \left[-\frac{1}{2} \frac{d^2}{dr_i^2} + \frac{l_i(l_i + 1)}{2r_i^2} - \frac{Z}{r_i} \right.$$

$$\left. + \sum_{\substack{j=1 \\ (j \neq i)}}^{N} \int_0^\infty |P_{n_j l_j}|^2 \frac{dr_j}{\rho_r} - \epsilon_{n_i l_i} \right] P_{n_i l_i}(r_i) \, dr_i = 0.$$

Because the variation in $P_{n_i l_i}(r_i)$ is quite arbitrary, the foregoing equality will hold in general only if the integrals vanish identically. In other words, we must have it for every i that

$$\left[-\frac{1}{2} \frac{d^2}{dr_i^2} + \frac{l_i(l_i + 1)}{2r_i^2} - \frac{Z}{r_i} + \sum_{\substack{j=1 \\ (j \neq i)}}^{N} \int_0^\infty \frac{|P_{n_j l_j}|^2}{\rho_r} \, dr_j \right] P_{n_i l_i}(r_i)$$

$$= \epsilon_{n_i l_i} P_{n_i l_i}(r_i) \quad (8.11)$$

is satisfied identically. Equation (8.11) is an equation of motion for the $P_{n_i l_i}(r_i)$ and is known as the *Hartree equation*, after D. R. Hartree, who first wrote it down in 1928.

The solutions of the Hartree equation, unlike their hydrogenic counterparts, do not form an orthogonal set. To see this, we need only note that

$$(\mathscr{H}_{\text{av}}^{(i)} P_{n_i l_i}, P_{n_j l_j}) - (P_{n_i l_i}, \mathscr{H}_{\text{av}}^{(j)} P_{n_j l_j}) \neq 0 \qquad (i \neq j),$$

contrary with the usual case, because

$$\mathscr{H}_{\mathrm{av}}^{(i)} \equiv -\frac{1}{2}\frac{d^2}{dr_i^2} + \frac{l_i(l_i + 1)}{2r_i^2} - \frac{Z}{r_i} + \sum_{\substack{j=1 \\ (j \neq i)}}^{N} \int_0^\infty \frac{|P_{n_j l_j}|^2}{\rho_r}\, dr_j,$$

owing to the integral term, is different for every $P_{nl}(r)$. Therefore, the ordinary proof of orthogonality fails for these wavefunctions. This problem may be alleviated by making the $P_{nl}(r)$ orthonormal with the Gram-Schmidt process or, better, by including spin variables in the definition of the wavefunction $\psi(\mathbf{r}^N)$, as we shall see.

If we define

$$Z_i \equiv Z - \sum_{\substack{j=1 \\ (j \neq i)}}^{N} \int_0^\infty |P_{n_j l_j}|^2 \frac{r_i}{\rho_r}\, dr_j,$$

then the Hartree equations become

$$\left[-\frac{1}{2}\frac{d^2}{dr_i^2} + \frac{l_i(l_i + 1)}{2r_i^2} - \frac{Z_i}{r_i}\right] P_{n_i l_i}(r_i) = \epsilon_{n_i l_i} P_{n_i l_i}(r_i) \qquad (8.12)$$

which form clearly indicates the physical significance of the integral term: *it represents the spherically averaged electronic charge which interacts with the* ith *electron*. In this sense it is closely related to the screening constant discussed in section 8.1.

The eigenvalue $(-\epsilon_{n_i l_i})$ is the energy required for the complete removal of the ith electron from a neutral atom, by definition. It is for this reason that

$$\sum_{i=1}^{N} \epsilon_{n_i l_i}$$

is *not* the total electronic energy. Indeed, by direct calculation,

$$\sum_{i=1}^{N} \epsilon_{n_i l_i} = \frac{1}{2}\sum_{i=1}^{N} \langle i|\left(-\frac{d^2}{dr_i^2} + \frac{l_i(l_i + 1)}{r_i^2} - \frac{2Z}{r_i}\right)|i\rangle$$

$$+ \sum_{i=1}^{N}\sum_{j=1}^{N} \langle i|\int_0^\infty \frac{|P_{n_j l_j}|^2}{\rho_r}\, dr_j|i\rangle$$

$$= \langle\mathscr{H}_{\mathrm{av}}\rangle + \frac{1}{2}\sum_{\substack{i,j \\ (i \neq j)}} (\langle ij|\rho_r^{-1}|ij\rangle)_{\mathrm{av}}.$$

The mean value $\langle\mathscr{H}_{\mathrm{av}}\rangle$ is the total electronic energy of the atom. We see that it is lower in value (more negative) than the sum of the one-electron energies by exactly the total (repulsive) interaction energy of the electrons. This is to be expected, since $-\langle \mathcal{H}_{av} \rangle$ represents the total energy required to remove the N electrons *in succession*, a task which is more difficult than removing N electrons from neutral atoms.

Now let us turn to the general multielectron problem wherein we must consider N spinning fermions. The most important aspect of the problem is our choice of the wavefunction $\psi(\mathbf{r}^N, s^N)$. We shall show now that a good selection (not the only one) is

$$\psi(\mathbf{r}^N, s^N) = (N!)^{-\frac{1}{2}} \begin{vmatrix} u_1(1) & u_1(2) & \cdots & u_1(N) \\ u_2(1) & & & \\ \vdots & & & \vdots \\ u_N(1) & & \cdots & u_N(N) \end{vmatrix} \tag{8.13}$$

where the *spin-orbital* $u_i(k)$ is given by

$$u_i(k) = \varphi_i(\mathbf{r}_k)\alpha_i(s_k) \quad \text{or} \quad \varphi_i(\mathbf{r}_k)\beta(s_k)$$

the orbital $\varphi_i(\mathbf{r}_k)$ being orthogonal, for a given s_k, and normalized to unity. The wavefunction defined in equation (8.13) is consistent with the exclusion postulate; for, according to the theory of determinants, an interchange of any two columns in a determinant produces a change in its sign. The column interchange in equation (8.13) may be accomplished by exchanging any one k $(1 \leqslant k \leqslant N)$ for another. Moreover, if, for any $l \neq k$,

$$u_i(k) = u_i(l) \qquad (i = 1, 2, \ldots, N), \tag{8.14}$$

then the determinant $\psi(\mathbf{r}^N, s^N)$ vanishes identically,[3] in agreement with the restricted form of the exclusion postulate.

The wavefunction $\psi(\mathbf{r}^N, s^N)$ is normalized because the spin-orbitals are orthonormal. The integral

$$(N!)^{-1} \sum_{\{s_i\}} \int_{\mathbf{r}^N} \begin{vmatrix} \bar{u}_1(1) & \cdots & \bar{u}_1(N) \\ \vdots & & \vdots \\ \bar{u}_N(1) & \cdots & \bar{u}_N(N) \end{vmatrix} \begin{vmatrix} u_{1'}(1) & \cdots & u_{1'}(N) \\ \vdots & & \vdots \\ u_{N'}(1) & \cdots & u_{N'}(N) \end{vmatrix} d\mathbf{r}^N$$

contains terms which differ from

$$(N!)^{-1} \sum_{\{s_i\}} \int_{\mathbf{r}_N} \prod_{i=1}^{N} \bar{u}_i(i) \begin{vmatrix} u_{1'}(1) & \cdots & u_{1'}(N) \\ \vdots & & \vdots \\ u_{N'}(1) & \cdots & u_{N'}(N) \end{vmatrix} d\mathbf{r}^N$$

[3] To see that the determinant vanishes when equation (8.14) is satisfied, we need only realize that interchanging l for k should change the sign of the determinant, and thus the determinant itself. But putting l for k causes no change in the determinant if equation (8.14) is adhered to; the inconsistency is resolved if the determinant is identically zero.

only by interchanges of the i $(1 \leqslant i \leqslant N)$, and by the corresponding signs. The signs can be made positive, if they are not so, by reordering the $u_i(i)$; this operation, of course, does not alter the magnitude of the component integral. We conclude from this fact that every contribution to (ψ, ψ) must be of the same value, the component integrands differing from one another only by permutations of the i $(1 \leqslant i \leqslant N)$. Because there are $N!$ permutations possible for a set of N different numbers, the factor $(N!)^{-1}$ is cancelled when the normalization integral is carried out. If any of the $u_{i'}(i')$ $(1 \leqslant i' \leqslant N)$ is different from all of the $u_i(i)$, the term

$$\sum_{\{s_i\}} \int_{\mathbf{r}^N} \bar{u}_1(1)\bar{u}_2(2)\cdots\bar{u}_N(N)u_{1'}(1')u_{2'}(2')\cdots u_{N'}(N')\, d\mathbf{r}^N$$

will vanish. It follows that (ψ, ψ) will vanish unless

$$u_1(1)u_2(2)\cdots u_N(N) = u_{1'}(1')u_{2'}(2')\cdots u_{N'}(N')$$

in at least one case; if the latter is true, the scalar product is equal to one, since the $u_i(i)$ are orthonormal.

From what has been said it is not hard to see that the diagonal matrix elements of operators of the form

$$\sum_{k=1}^{N} f_k,$$

where f_k is a one-electron operator, will be sums of $\sum_{s_1} \int_{\mathbf{r}_1} \bar{u}_i f_1 u_i\, d\mathbf{r}_1$. The diagonal matrix elements of

$$\sum_{i,j} g_{ij}$$

will be terms like

$$\sum_{i,j} \left[\sum_{s_1,s_2}' \int_{\mathbf{r}_1} \int_{\mathbf{r}_2} [(\bar{u}_i(1)\bar{u}_j(2)g_{12}u_i(1)u_j(2)) - (\bar{u}_i(1)\bar{u}_j(2)g_{12}u_i(2)u_j(1))]\, d\mathbf{r}_1\, d\mathbf{r}_2 \right] \cdot$$

It follows that the diagonal elements of the multielectron Hamiltonian operator are

$$(\psi, \mathscr{H}\psi) = \langle \mathscr{H} \rangle = \sum_{i=1}^{N} \sum_{s_1}' \int_{\mathbf{r}_1} \bar{u}_i(1)f_1 u_i(1)\, d\mathbf{r}_1$$

$$+ \sum_{i,j} \sum_{s_1,s_2}' \int_{\mathbf{r}_1} \int_{\mathbf{r}_2} \bar{u}_i(1)\bar{u}_j(2)g_{12}[u_i(1)u_j(2) - \delta_{s_1 s_2} u_i(2)u_j(1)]\, d\mathbf{r}_1\, d\mathbf{r}_2$$

$$(8.15)$$

where we have noted that

$$\sideset{}{'}\sum_{s_1, s_2} (u_i(1)u_j(2)g_{12}u_i(2)u_j(1))$$

vanishes unless $s_1 = s_2$ because the spin wavefunctions are orthogonal. What remains is to vary the u_i in equation (8.15) in order to minimize $\langle \mathscr{H} \rangle$ and thus obtain a new equation of motion.

The condition that $\langle \mathscr{H} \rangle$ remain stationary may be expressed

$$\delta\langle \mathscr{H} \rangle - \sum_{i=1}^{N} \epsilon_{ii}\delta(u_i, u_i) - \sum_{\substack{i,j \\ (i \neq j)}} \delta_{s_1 s_2}[\epsilon_{ij}\delta(u_i, u_j) + \epsilon_{ji}\delta(u_j, u_i)] = 0 \quad (8.16)$$

where the ϵ_{ii} and ϵ_{ij} are proportionality constants. The first term in addition to $\delta\langle \mathscr{H} \rangle$ is to preserve the normalization of the $u_i(1)$, while the second and third terms insure the continued orthogonality of any pair of $u_i(1)$ associated with identical spin variables. We are free to impose whatever relationship we wish between the ϵ_{ij} and ϵ_{ji}; we shall put

$$\epsilon_{ij} = \bar{\epsilon}_{ji}$$

so that the third and fourth terms in (8.16) become complex conjugates of one another.

Now let us expand equation (8.16). We have

$$\delta\langle \mathscr{H} \rangle = \sum_{i=1}^{N} \sideset{}{'}\sum_{s_1} \int_{\mathbf{r}_1} \{\delta\bar{u}_i f_1 u_i + \bar{u}_i f_1 \, \delta u_i\} \, d\mathbf{r}_1$$

$$+ \sideset{}{'}\sum_{i,j} \sideset{}{'}\sum_{s_1, s_2} \int_{\mathbf{r}_1} \int_{\mathbf{r}_2} \{\delta\bar{u}_i \bar{u}_j g_{12}[u_i u_j - \delta_{s_1 s_2} u_i u_j]$$

$$+ \bar{u}_i \bar{u}_j g_{12}[\delta u_i u_j - \delta_{s_1 s_2} \, \delta u_i u_j]\} \, d\mathbf{r}_1 \, d\mathbf{r}_2$$

$$= \sum_{i=1}^{N} \sideset{}{'}\sum_{s_1} \int_{\mathbf{r}_1} \delta\bar{u}_i(1)\left\{f_1 u_i(1) + \sum_{j=1}^{N} \sideset{}{'}\sum_{s_2} \int_{\mathbf{r}_2} \bar{u}_j(2)g_{12}\right.$$

$$\left. \times [u_i(1)u_j(2) - \delta_{s_1 s_2} u_i(2)u_j(1)]\right\} d\mathbf{r}_1 \, d\mathbf{r}_2 + \text{complex conjugate}$$

so that equation (8.16) takes on the form

$$\sum_{i=1}^{N} \sideset{}{'}\sum_{s_1} \int_{\mathbf{r}_1} \delta\bar{u}_i(1)\left\{f_1 u_i(1) + \int_{\mathbf{r}_2} \sideset{}{'}\sum_{s_2} \left[\sum_{j=1}^{N} \bar{u}_j(2)g_{12}(u_i(1)u_j(2) - \delta_{s_1 s_2} u_i(2)u_j(1))\right]\right.$$

$$\left. - \epsilon_{ii}u_i(1) - \sum_{\substack{j=1 \\ (j \neq i)}}^{N} \delta_{s_1^{(i)} s_1^{(j)}}\epsilon_{ij}u_j(1)\right\} d\mathbf{r}_1 \, d\mathbf{r}_2$$

$$+ \text{complex conjugate} = 0. \quad (8.17)$$

In order that equation (8.17) be satisfied, in general we must have it that the first series and its complex conjugate vanish individually, that each term in the sum over i be equal to zero, and, therefore, that

$$f_1 u_i(1) + \sum_{j=1}^{N} {\sum_{s_2}}' \int_{\mathbf{r}_2} \bar{u}_j(2) g_{12}[u_i(1)u_j(2) - \delta_{s_1 s_2} u_i(2)u_j(1)] \, d\mathbf{r}_2$$

$$= \sum_{j=1}^{N} \epsilon_{ij} \delta_{s_1^{(i)} s_1^{(j)}} u_j(1) \qquad (i = 1, 2, \ldots, N). \quad (8.18)$$

Equations (8.18) are the *Hartree-Fock equations*, which were first derived by V. Fock in 1930. They are different from the Hartree equation (8.11) in several important aspects. To begin with, the Hartree equation has but one energy eigenvalue for a given $u_i(1)$, while the Hartree-Fock equation contains a sum over the elements ϵ_{ij}. The difference in this case is illusory, however. Because each of the operators in equations (8.18) is linear, any linear combination of the $u_m(k)$ $(m = i, j; k = 1, 2)$ is also a solution of the equations. Therefore, we can choose without a loss of generality that linear combination which diagonalizes the Hermitian matrix comprising the ϵ_{ij}. If we denote the linear combinations by $\psi_i(1)$, $\psi_j(2)$, etc., then equations (8.18) may be written

$$f_1 \psi_i(1) + \sum_{j=1}^{N} {\sum_{s_2}}' \int_{\mathbf{r}_2} \bar{\psi}_j(2) g_{12}[\psi_i(1)\psi_j(2) - \delta_{s_1 s_2}\psi_i(2)\psi_j(1)] \, d\mathbf{r}_2$$

$$= \epsilon_i \psi_i(1) \qquad (i = 1, \ldots, N) \quad (8.19)$$

where we have put $\epsilon_{ii} \equiv \epsilon_i$ for simplicity. The eigenvalue $(-\epsilon_i)$ now represents the energy necessary to completely remove the ith electron from the atom *in succession*. This result is different from the corresponding interpretation of the eigenvalue in the Hartree equation.

The wavefunctions $\psi_i(1)$ are mutually orthogonal, in contrast with the $P_{n_i l_i}(r_i)$ in equation (8.11). To see this we need only note that

$$\mathcal{H}_H \psi_i(1) \equiv f_1 \psi_i(1) + \sum_{j=1}^{N} {\sum_{s_2}}' \int_{\mathbf{r}_2} \bar{\psi}_j(2) g_{12}[\psi_i(1)\psi_j(2) - \delta_{s_1 s_2}\psi_i(2)\psi_j(1)] \, d\mathbf{r}_2$$

is the same for all the $\psi_i(1)$, so that

$$(\mathcal{H}_H \psi_m, \psi_n) - (\psi_m, \mathcal{H}_H \psi_n) = 0 \qquad (m, n = 1, 2, \ldots, N)$$

and the usual proof of orthogonality is valid. The $\psi_i(1)$ can, of course, be normalized and used to form a complete orthonormal set in the invariant subspace of $\mathcal{H}_H$.

In equations (8.19) spherical averaging has not been performed. This makes the Hartree-Fock equation more exact than the Hartree equation, but still presents a disadvantage when solving for the ψ_i. If the atom under consideration contains only complete groups the problem is removed, however, since this case it is possible to prove that integration over $\mathbf{r}_2$ results in an equation with no dependence upon angles.[4] When the atom described by (8.19) contains an incomplete group, the solution for ψ_i can be made less difficult by spherical averaging in a special way[4] that is a bit more elaborate than what is necessary to describe here.

The physical interpretation of equations (8.19) is similar to that of equation (8.11). The one-electron operator f_1 accounts for the kinetic energy and nuclear interaction of the electron in the state $\psi_i(1)$. The term involving the two-electron operator g_{12} represents the average potential energy of the ith electron as it interacts with the N electrons in the atom, less a correction term—the exchange integral. Now, the exchange term represents a charge of the same spin as is the electron represented by $\psi_i(1)$, which fact is evidenced by the presence of $\delta_{s_1 s_2}$. *Therefore, the purpose of the exchange term would appear to be for preventing electrons of the same spin as the one under consideration from approaching it too closely.* The effect of this exclusion of charge is a lowering of the Hartree-Fock average potential, relative to the Hartree potential. The lower potential, in turn, makes the Hartree-Fock eigenvalue lower (as it should be by virtue of its physical significance) and tends to increase the magnitude of the wavefunction $\psi_i(1)$ in the vicinity of the nucleus.

Doc 1
Doc 3

8.3. THE METHOD OF THE SELF-CONSISTENT FIELD

Now to the solutions of equations (8.19). As these expressions are the most suitable for describing atoms which possess completed groups, we shall restrict the discussions here accordingly. The Hartree-Fock wave functions are thus of the form

$$\psi_i(1) = \left[\frac{P_{n_i l_i}(r_1) \Upsilon_{l_i}{}^{m_i}(\vartheta_1, \varphi_1)}{r_1} \right] \binom{\alpha(s_1)}{\beta(s_1)}.$$

Suppose we choose the spin-up state to work with. Then we attack equations (8.19) as follows.

(i) $f_1 \psi_i(1) = \dfrac{1}{2} \left[-\dfrac{d^2}{dr_1{}^2} + \dfrac{l_i(l_i + 1)}{r_1{}^2} - \dfrac{2Z}{r_1} \right] \psi_i(1)$

[4] For many such details see J. C. Slater, *Quantum Theory of Atomic Structure*, McGraw-Hill Book Co., New York, 1960, Vol. II, Chapter 17.

(ii) The coulomb term is converted from a sum over electrons (j) to a sum over orbital quantum numbers (n_j, l_j, m_j). Then we carry out the integration over (ϑ_2, φ_2, s_2) explicitly to get

$$\sum_j {\sum_{s_2}}' \int_{\mathbf{r}_2} \bar{\psi}_j(2) g_{12} \psi_i(1) \psi_j(2)\, d\mathbf{r}_2 = 2 \sum_{\{n_j, l_j\}} (2l_j + 1) \int_0^\infty |P_{n_j l_j}|^2 \frac{dr_2}{\rho_r} \psi_i(1)$$

(iii) The sum over electrons in the exchange term is converted as in (ii) above. Then we use the addition theorem for spherical harmonics,[5]

$$\frac{(2l + 1)}{4\pi} P_l(\cos\gamma) = \sum_{m=-l}^{l} \bar{Y}_l^m(\vartheta_2, \varphi_2) Y_l^m(\vartheta_1, \varphi_1),$$

and the expansions

$$r_{12}^{-1} = \sum_{k=0}^{\infty} \frac{\rho_r'^k}{\rho_r^{k+1}} P_k(\cos\gamma),$$

$$P_{l_j}(\cos\gamma) P_k(\cos\gamma) = \sum_{l=0}^{\infty} C_{l_j l}^k P_l(\cos\gamma),$$

to reduce the term to

$$-\sum_j {\sum_{s_2}}' \int_{\mathbf{r}_2} \bar{\psi}_j(2) g_{12} \psi_i(2) \psi_j(1) \delta_{s_1 s_2}\, d\mathbf{r}_2$$

$$= -\sum_{\{n_j, l_j\}} \frac{2l_j + 1}{2l_i + 1} \sum_{k=0}^{l_i + l_j}{}' C_{l_i l_j}^k \int_0^\infty \frac{\rho_r'^k}{\rho_r^{k+1}} P_{n_i l_i} P_{n_j l_j}\, dr_2 \psi_j(1)$$

where

$$\rho_r' = \begin{cases} r_1 & r_2 > r_1 \\ r_2 & r_2 < r_1 \end{cases} \qquad \rho_r = \begin{cases} r_1 & r_2 < r_1 \\ r_2 & r_2 > r_1 \end{cases}$$

and the sum over orbitals includes only those which have spins up.

Equations (8.19) may now be written

$$\frac{1}{2}\left[-\frac{d^2}{dr_1^2} + \frac{l_i(l_i + 1)}{r_1^2} - \frac{2Z}{r_1}\right] P_{n_i l_i}(r_1) + 2 \sum_{\{n_j, l_j\}} (2l_j + 1)$$

$$\int_0^\infty |P_{n_j l_j}|^2 \frac{dr_2}{\rho_r} P_{n_i l_i}(r_1) - \sum_{\{n_j, l_j\}} \frac{2l_j + 1}{2l_i + 1} \sum_{k=0}^{l_i + l_j}{}' C_{l_i l_j}^k$$

$$\int_0^\infty \frac{\rho_r'^k}{\rho_r^{k+1}} P_{n_i l_i} P_{n_j l_j}\, dr_2 P_{n_j l_j}(r_1) = \epsilon_i P_{n_i l_i}(r_1). \quad (8.20)$$

[5] These three expansions are discussed in detail in reference 4. The addition theorem is presented in problem 7 of Chapter 4. The second expansion is just a multipole expansion; the third expansion is the principle of superposition at work!

Equations (8.20) are a set of a remarkably complicated integrodifferential equations for the $P_{n_il_i}(r_1)$! In general such an equation will be quite difficult to solve because its solutions are coupled through the integral terms. However, a straightforward method for attaching the equation was developed by D. R. Hartree; it is called the *method of the self-consistent* field. We shall outline it here, leaving its details for the interested to seek elsewhere.[6]

(a) First, equation (8.20) is rewritten in terms of a function of the $P_{nl}(r)$ called $Y_k(n_il_i, n_jl_j; r_i)$ and defined by

$$Y_k \equiv Z_k + \int_{r_1}^{\infty} \left(\frac{r_1}{r_2}\right)^{k+1} P_{n_il_i}(r_2) P_{n_jl_j}(r_2) \, dr_2 \qquad (8.21)$$

where

$$Z_k(n_il_i, n_jl_j; r_1) \equiv \int_0^{r_1} \left(\frac{r_1}{r_2}\right)^{k} P_{n_il_i}(r_2) P_{n_jl_j}(r_2) \, dr_2.$$

Equation (8.21) is used to replace the integrated parts of the second and third terms on the left-hand side of equation (8.20) with quantities which can be estimated numerically.

(b) Y_k is computed by multiplying equation (8.21) by $r_1^{-(k+1)}$ and differentiating it with respect to r_1. The result is

$$\frac{dY_k}{dr_1} = \frac{1}{r_1}[(k+1)Y_k - (2k+1)Z_k]. \qquad (8.22)$$

This equation is integrated approximately by estimating Z_k.

(c) Once the Y_k have been calculated, they are put into equation (8.20). The eigenvalue ϵ_i is then estimated and a numerical integration of the equation is begun. (The coefficients $C^k_{l_il_j}$ are assumed to be known. They have, in fact, been tabulated.[7])

(d) The numerical integration of equation (8.20) proceeds both inward and outward, subject to

$$P_{nl}(0) = P_{nl}(\infty) = 0 \qquad (P_{n'l'}, P_{nl}) = \delta_{n'n}\delta_{ll'}.$$

At some value of r_1 between those chosen for the starting points of the integration it is required that

$$\left(\frac{d}{dr_1} \ln P_{nl}(\text{out})\right)_{r_1 = r_0} = \left(\frac{d}{dr_1} \ln P_{nl}(\text{in})\right)_{r_1 = r_0} \qquad (8.23)$$

---

[6] See D. R. Hartree, *The Calculation of Atomic Structures*, John Wiley and Sons, New York, 1957.

[7] See J. C. Slater, *Quantum Theory of Atomic Structure*, McGraw-Hill Book Co., New York, 1960, Vol. II, Appendix 20 and p. 319.

where "out" and "in" refer to the integration path used to get $P_{nl}(r_1)$. Equation (8.23) is the condition of consistency for the wavefunction $P_{n_l l_i}(r_1)$.

(e) In general it is found that the solutions do not match at $r_1 = r_0$. One then creates a new $P_{nl}(r_1)$ as the weighted mean of the two estimated solutions. This new wavefunction is then used to recalculate Z_k. A fresh estimate of Y_k follows from integrating equation (8.22) once again.

(f) Steps (c) and (d) above are carried out for the second time to get new wavefunctions. If the logarithmic derivatives match at r_0 to within some prescribed tolerance, the "correct" wavefunction and eigenvalue have been secured. Otherwise, the process is repeated over and over until self-consistency is achieved.

It has probably been obvious that the method of the self-consistent field contains several steps in which intuition plays a major role. Intuition is certainly an important aspect of the procedure. For there is really no "best" way of making the wavefunction and eigenvalue estimates: just what values of these quantities to employ is a decision occasioned by experience and the degree of tolerance imposed upon the criterion for self-consistency. This fact, perhaps, is a bit unsatisfying; we might get a better feeling for the whole thing by carrying through at least part of a self-consistent field calculation for some system whose physical aspects are already familiar. In this way the attributes of the computation procedure, especially the role played by physical intuition, might be seen more clearly.

Let us choose for study once again the groundstate of the helium atom. The Hartree equation (8.12) for this system is of the form

$$\left(-\frac{1}{2}\frac{d^2}{dr_1^2} + \frac{Z_1}{r_1}\right)P_{10}(r_1) = \epsilon_{10}P_{10}(r_1) \tag{8.24}$$

where

$$Z_1 = 2 - \int_0^\infty |P_{10}(r_2)|^2 \frac{r_1}{\rho_r}\,dr_2$$

and $P_{10}(r_1)$ is considered to be the wavefunction for either one of the electrons. We are not really neglecting here any special correlations arising from the exclusion postulate since the groundstate is a singlet state. The first step in solving equation (8.24) is the estimation of the integral term. Ordinarily we would proceed, as suggested above, with a numerical estimate. In this case, however, we are in a better situation because of our experience with groundstate helium in section 8.1: a good beginning approximation for $P_{10}(r)$ should be the function

$$P_{10}^{(l)}(r) = 2^{-1/2}\left(\frac{27}{8}\right)^{3/2} r \exp(-27r/16). \tag{8.25}$$

The integral term is, accordingly,

$$\int_0^\infty |P_{10}^{(1)}(r_2)|^2 \frac{r_1}{\rho_r} \, dr_2 = \frac{1}{2} \left(\frac{27}{8}\right)^3 \int_0^{r_1} x^2 \exp(-27x/8) \, dx$$

$$+ \left(\frac{r_1}{2}\right)\left(\frac{27}{8}\right)^3 \int_{r_1}^\infty x \exp(-27x/8) \, dx$$

$$= 1 - \exp(-27r_1/8) - \tfrac{27}{16}r_1 \exp(-27r_1/8).$$

(8.26)

Using equations (8.25) and (8.26) we can write equation (8.24) in a form suitable for integration; that is,

$$\left(\frac{d^2}{dr_1^2} - 1.70\right)P_{10}(r_1)$$

$$= \left(\frac{2}{r_1}\right)\left[1 + \exp(-27r_1/8) + \frac{27}{16}r_1 \exp(-27r_1/8)\right]P_{10}(r_1) \quad (8.27)$$

where we have chosen ϵ_{10} to be equal to the difference between the total electronic energy of the atom as found in section 8.1 (-2.85 atomic units) and the total electronic energy of He$^+$ (-2.00 atomic units). This estimate is certainly in line with equation (8.25). Rather than go through the iteration procedure, which would entail twofold outward and inward integrations of equation (8.27), we shall rough out estimates of $P_{10}(\text{out})$ and $P_{10}(\text{in})$ by considering extremum forms of this equation. In particular, we shall see what happens when r_1 becomes very small or very large. When the approximation

$$\exp(-27r/8) = O\left(1 - \frac{27r}{8}\right)$$

is valid, equation (8.27) may be written, to first order in r_1,

$$\left(\frac{d^2}{dr_1^2} + \frac{4}{r_1} + 2\epsilon_{10}^{(1)}\right)P_{10}^{(1)}(r_1) = 0.$$

The solution of this equation is

$$P_{10}^{(1)}(r_1) \propto r_1 \exp(-2r_1)$$

with the eigenvalue $\epsilon_{10}^{(1)} = -2.00$ atomic units. This result is expected: $P_{10}(r_1)$ behaves like the hydrogenic wavefunction for the groundstate of a two-electron atom for small values of the radial coordinate. The magnitude of the eigenvalue is -2.00 atomic units (that for the comparable hydrogenic system), showing that the second electron does little to hinder the attractive force on the first electron when the latter is near the nucleus.

When r_1 is large enough to put

$$\exp(-27r_1/8) = O(0)$$

into equation (8.27), that expression becomes

$$\left(\frac{d^2}{dr_1{}^2} + \frac{2}{r_1} + 2\epsilon_{10}^{(2)}\right)P_{10}^{(2)}(r_1) = 0.$$

The solution here is

$$P_{10}^{(2)}(r_1) \propto r_1 e^{-r_1}$$

with the eigenvalue $\epsilon_{10}^{(2)} = -0.50$ atomic unit. We see that when the first electron is far from the nucleus, the latter is completely screened by the second electron. This action results in an effective nuclear charge of unity and a wavefunction and energy eigenvalue no different from that for the comparable hydrogenic state (-0.50 atomic unit).

We may conclude that the initial results of the method of the self-consistent field applied to helium are quite interpretable within the context of our previous experience. The physical intuition gained in our earlier work has evidently led us to a reasonably good starting estimate for $P_{10}(r)$. To be more precise, of course, we should compare the logarithmic derivatives of $P_{10}^{(1)}(r)$ and $P_{10}^{(2)}(r)$. This is not hard to do. We find, for any value of r,

$$\left|\frac{d}{dr}\ln P_{10}^{(1)} - \frac{d}{dr}\ln P_{10}^{(2)}\right| = 1.00$$

which result shows that self-consistency is not yet too close. It seems clear after a glance at $\epsilon_{10}^{(1)}$, $\epsilon_{10}^{(2)}$, and the empirical value of ϵ_{10} (-0.90372 atomic unit), that we should choose as a second trial wavefunction a mean of $P_{10}^{(1)}(r_1)$ and $P_{10}^{(2)}(r_1)$ weighted somewhat toward the latter function. A good working function might be

$$P_{10}^{(II)}(r_1) \propto 0.27P_{10}^{(1)}(r_1) + 0.73P_{10}^{(2)}(r_1).$$

We expect that $P_{10}^{(II)}(r_1)$ will give more self-consistent results than did its predecessor.

In Figure 48 are shown the radial wavefunctions $P_{nl}(r)$ for the 1s orbitals in helium and neon, as calculated by the method of the self-consistent field.[8] In doing the iterations, equations (8.2) were simplified by the use of an average exchange term. The average is performed by supposing the atomic electrons

[8] The data are from F. Herman and S. Skillman, *Atomic Structure Calculations*, Prentice-Hall, Englewood Cliffs, N.J., 1963.

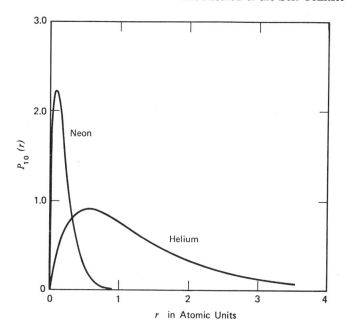

Figure 48. A plot of the groundstate Hartree-Fock wave-function for two atoms possessing closed-shell structures. These wavefunctions are normalized when multiplied by the factor $0.8853/Z^{1/3}$.

to form a free-electron gas, such that $\psi_i(r)$ may be replaced by the plane-wave state obtained in section 3.5. The result is that[9]

$$\sum_{j=1}^{N}\sideset{}{'}\sum_{s_2}\int_{\mathbf{r}_2}\bar\psi_j(2)g_{12}\psi_i(2)\psi_j(1)\,dr_2\delta_{s_1s_2}\rightarrow 6\left[\frac{3}{8\pi}\sum_{j=1}^{N}\bar\psi_j(1)\psi_j(1)\right]^{1/3}\psi_i(1)$$

which is the same function of r for all the $\psi_i(1)$. Although the physical assumption is rather drastic, the free-electron approximation does not produce eigenvalues which differ from the empirical ones very much[8] and so greatly expedites calculation with small loss of accuracy.

The figure demonstrates very clearly the effect of increasing nuclear charge on the radial wavefunction: the most probable value of the $1s$ orbital radius decreases uniformly with a growth in atomic number.

Dec. 6

[9] For details see J. C. Slater, *Quantum Theory of Atomic Structure*, McGraw-Hill Book Co., New York, 1960, Vol. II, Appendix 22.

Dec 8 } Free
10 } electron
etc.

PROBLEMS

1. Compute the groundstate energies of Li^+, Be^{2+}, B^{3+}, C^{4+}, N^{5+}, and O^{6+} by generalizing and then minimizing the expression for $E(Z')$ given in section 8.1. Tabulate your results and compare them to the appropriate experimental data as given in Chapter 5.

2. Apply the screening-constant method to the groundstate of a three-electron atom.

   (a) Neglect spin by assuming a variation function in the form

$$\psi_1(\mathbf{r}_1, \mathbf{r}_2, \mathbf{r}_3) = \left(\frac{Z'^3}{\pi a_0{}^3}\right)^{\!\! 3/2} \exp\left[-Z'(r_1 + r_2 + r_3)/a_0\right].$$

   Calculate the contributions to $E(Z')$ from the coulomb integrals and minimize the function with respect to Z'. What is the value of the screening constant?

   (b) Using the results of section 8.1 and the theory of the hydrogen atom, compute the electronic energies of the singly and doubly ionized three-electron atom. Then compute a general expression for the total electronic energy of this three-body system.

   (c) Calculate the groundstate energies of Li, Be^+, and B^{2+}. The empirical values are -14.9573, -28.6533, and -46.8597 Rydbergs, respectively.

3. Let us see if we can provide a more rigorous basis for the equation

$$\delta\langle\mathcal{H}_{av}^{(i)}\rangle - \epsilon_{n_i l_i}\delta(P_{n_i l_i}, P_{n_i l_i}) = 0$$

which was used to derive the Hartree equation.

   (a) Write down the conditions for $\langle\mathcal{H}_{av}^{(i)}\rangle$ and $(P_{n_i l_i}, P_{n_i l_i})$, considered as functions of $P_{n_i l_i}$, to be extrema.

   (b) Show that the Hartree equation is equivalent to

$$\left(\frac{\delta\langle\mathcal{H}_{av}^{(i)}\rangle}{\delta P_{n_i l_i}}\right)_{P_{n_i l_i} = P_{n_i l_i}^0} = \alpha_i\left(\frac{\delta(P_{n_i l_i}, P_{n_i l_i})}{\delta P_{n_i l_i}}\right)_{P_{n_i l_i} = P_{n_i l_i}^0}$$

   where $P_{n_i l_i}^0$ is the $P_{n_i l_i}$ making $\langle\mathcal{H}_{av}^{(i)}\rangle$ an extremum. What is the physical interpretation of $P_{n_i l_i}^0$ and the arbitrary constant α_i?

   (c) Show that the extremum condition on $(P_{n_i l_i}, P_{n_i l_i})$ insures that the $P_{n_i l_i}^0$ will be normalized wavefunctions if the $P_{n_i l_i}$ are so.

4. Consider the definition of Z_i given in section 8.2. We can write

$$Z_i = Z - \sum_{\substack{j=1 \\ (j \neq i)}}^{N} \int_0^{r_i} |P_{n_j l_j}|^2 \, dr_j - \sum_{\substack{j=1 \\ (j \neq i)}}^{N} \int_{r_i}^{\infty} |P_{n_j l_j}|^2 \frac{r_i}{r_j} \, dr_j$$

$$\equiv Z_i' - \sum_{\substack{j=1 \\ (i \neq 1)}}^{N} \int_{r_i}^{\infty} |P_{n_j l_j}|^2 \frac{r_i}{r_j} \, dr_j \qquad (i = 1, \ldots, N).$$

238

(a) Interpret the term Z_i' in light of the concept of screening by atomic electrons. (Recall that r_i is the radial coordinate for the ith electron.)

(b) Which term, Z_i or Z_i', should affect the wavefunction for the ith electron the most?

(c) Show that

$$\frac{dZ_i}{dr_i} = \frac{Z_i - Z_i'}{r_i} \qquad (i = 1, \ldots, N).$$

5. Derive a general expression for the constants $C_{ti_j}^k$ which were introduced in section 8.3.

6. Show the relationship between equation (8.22) and the result of the last part of problem 4.

7. Write down the Hartree-Fock equation for the $^3S_1$ state in helium.

8. In our last consideration of the helium atom we noticed that the wavefunction for one of the electrons is very nearly hydrogenic in character when the particle is far from the nucleus.

(a) Suppose an atom consists of a number of closed shells plus just one electron orbiting *outside* these shells. Make an application of our conclusions about helium to show that the wavefunction for the outer electron should be nearly hydrogenic. (Is the potential field acting upon the electron spherically symmetric?)

(b) Extend your conclusions to atoms composed of complete groups plus *two* outer electrons. What kind of wavefunction is applicable here?

(c) From what you have deduced, comment on the likely structure for the emission spectra of alkali and alkaline-earth atoms. Are your thoughts in agreement with experiment?

CHAPTER 9

Many-Electron Atoms

C. The Spectra of Complex Atoms

Symmetry is the mother of Invariance;
Invariance, the father of Conservation.

9.1. THE IDEA OF A GROUP

It has become apparent that complex atoms are amenable to the formalism of the quantum theory only after some fundamental simplifications have been introduced. The line spectrum of helium, for instance, is understood in its overall characteristics once a simple method is postulated for coupling the orbital and spin angular momenta of the two electrons. Heavier atoms, we expect, will not be so easy to describe. The same thing follows for atomic structure: electronic configurations in complex systems should not be completely described unless we go beyond the simplifying assumption that all the repulsive coulomb interactions may be averaged. On the other hand, it is also true that an excessively detailed calculation may have a detrimental effect upon our understanding of the general aspects of atomic spectra and structure. The problem of the many-electron atom in quantum theory, then, is a dichotomy. We wish to describe the motions of bound electrons in a nearly exact way while yet perceiving the essential properties of atoms that link these systems together under very general quantum-physical relations.

The "essential properties" to which we refer are not intuitively obvious; they have been isolated from the mass of empirical data only after a long and careful examination of the quantum behavior of many complex atoms. It has been deduced that the common attribute of atomic systems, whatever their complexity, is their symmetry with respect to certain transformations of the electronic position and spin coordinates. Indeed, *nearly all the selection rules for atomic emission spectra follow directly from considerations of symmetry.*

It is our purpose now to see how this result comes about. For this reason we must first deal, in an introductory way, with the idea of a *group*.

Suppose a set exists whose members share the following properties:

(I) Any two members of the set can be combined to form a third member of the set. We shall call this combination *group multiplication*, although we do not mean multiplication in the ordinary sense, but in an abstract sense.

(II) The group multiplication is associative:

$$(A \cdot B) \cdot C = A \cdot (B \cdot C),$$

where A, B, and C are elements in the set and $\cdot$ stands for group multiplication.

(III) One member of the set, I, the identity element, has the property

$$I \cdot A = A \cdot I = A,$$

where A is *any* member of the set. It is not hard to show that there is only one such element I in the set.

(IV) Each member of the set is associated with an inverse, such that

$$A^{-1} \cdot A = A \cdot A^{-1} = I,$$

where A^{-1} is the inverse of A.

A set comprising elements which satisfy the foregoing axioms is called a *group*. If the set does not contain an infinite number of members, it is said to be of *order h*, where h is the number of members. Two groups which are of the same order are said to be *isomorphic* if one can associate with each element of one group an element of the other, such that $(A_1 \cdot B_1)$ in the first group corresponds uniquely to $(A_2 \cdot B_2)$ in the second group.

A *subgroup* of a group is a set whose members are in the group and satisfy, under the same kind of "multiplication" as does the whole group, the axioms (I) through (IV). As an example of a subgroup we might consider the *cyclic subgroup* associated with every member of any finite group. This subgroup contains the elements which are "powers" of a given element,

$$I, A, A \cdot A, A \cdot A \cdot A, \ldots,$$

and so on. The members of a cyclic subgroup must be repeated after a finite number (say, $n \leqslant h$) of self-multiplications. In fact, the nth power of A is the identity element I since the product of any power of A with I is that power once again. The reappearance of I therefore must signal the end of a cycle of *non*-repeated group elements. Thus,

$$A \cdot A \cdot A \cdot \cdots \cdot A \equiv A^n = I,$$

and the integer n is called the *order of the group element A*. The cyclic subgroup associated with A is an example of a special kind of group, the *abelian* group. An abelian group is one for which the operation of group multiplication is commutative as well as associative.

Suppose a group contains the elements $I, A_1, A_2, \ldots, A_n$. The elements

$$I \cdot A_i \cdot I^{-1} = A_i, \qquad A_j \cdot A_i \cdot A_j^{-1}, \qquad (j = 1, 2, \ldots, n; j \neq i; n < +\infty)$$

are said to be *conjugate* to the element A_i $(1 \leqslant i \leqslant n)$ and, therefore, to form a *conjugate class*. It follows from this definition that each class of an abelian group contains just one element and that the identity element is in a class by itself. A subgroup which comprises only entire classes is called an *invariant subgroup*. The elements in the sequence

$$B \cdot I \cdot B^{-1} = I, \; B \cdot A_1 \cdot B^{-1}, \; B \cdot A_2 \cdot B^{-1}, \ldots, B \cdot A_m \cdot B^{-1},$$

where B is some member of the whole group and A_i is the ith member of an mth-order invariant subgroup, are themselves members of the invariant subgroup. This is so because each of the elements in the above sequence is conjugate to an element I or A_i of the invariant subgroup, which, by definition, contains only whole classes. (Remember that each A_i determines a conjugate class.) Moreover, because an element A_j of the invariant subgroup *is* equal to some element $B \cdot A_i \cdot B^{-1}$ in the foregoing sequence, all the elements of the subgroup must be contained in the sequence. It follows that the sequence is identical with the subgroup, except for the ordering of its elements, and that, in this sense, invariant subgroups are *self-conjugate*.

A group that contains an infinite number of elements, each of which may be associated with one or more parameters varying in a continuous manner, is called a *continuous group*. The domains of the parameters comprise the *group space* in the sense that there is a unique correspondence between the elements of a continuous group and the group space. Continuous group elements having parameters only slightly different from one another are termed *adjacent*. It is required that the products of adjacent group elements, as well as their inverses, also be adjacent. Those elements adjacent to the identity element in a continuous group are contained in the *infinitesimal group*. The infinitesimal group is abelian, since the products of any two of its elements, $A \cdot B$ and $B \cdot A$, must be adjacent to a greater degree than are the elements themselves. (If A and B are of order ϵ different from I, then their products contain terms which differ at most by order ϵ^2.)

As an example of a finite group, we may consider the *symmetric group* of order six. This abstract group is characterized by the multiplication table

	I	A	B	C	D	E
I	I	A	B	C	D	E
A	A	B	I	E	C	D
B	B	I	A	D	E	C
C	C	D	E	I	A	B
D	D	E	C	B	I	A
E	E	C	D	A	B	I

This array summarizes the multiplications of the group members in that the product of the ith member in the outside column and the jth member in the outside row is the element G_{ij} between the lines. The array is known as the *group table* and fully describes the abstract group. We see immediately from the group table that the group is not abelian, that

$$A^{-1} = B \qquad C^{-1} = C \qquad E^{-1} = E$$
$$B^{-1} = A \qquad D^{-1} = D$$

and that the subgroups are $\{I\}$, $\{I, A, B\}$, $\{I, C\}$, $\{I, D\}$, and $\{I, E\}$. The conjugate classes are $\{I\}$, $\{A, B\}$, $\{C, D, E\}$, since, for example,

$$I \cdot A \cdot I^{-1} = A, \qquad B \cdot A \cdot B^{-1} = A, \qquad C \cdot A \cdot C^{-1} = B,$$
$$D \cdot A \cdot D^{-1} = B, \qquad E \cdot A \cdot E^{-1} = B$$

are the elements conjugate to A, and

$$I \cdot C \cdot I^{-1} = C, \qquad A \cdot C \cdot A^{-1} = D, \qquad B \cdot C \cdot B^{-1} = E,$$
$$D \cdot C \cdot D^{-1} = E, \qquad E \cdot C \cdot E^{-1} = D$$

are those conjugate to C. We see from the listing of subgroups and conjugate classes that $\{I, A, B\}$ is the only non-trivial invariant subgroup.

Of great importance to quantum physics is the idea of a *realization* of a group. A realization comes about when to every member A_i $(i = 1, 2, \ldots)$ of a group there corresponds a transformation $T(A_i)$ such that

$$T(A_i \cdot A_j) = T(A_i)T(A_j) \qquad (i, j = 1, 2, \ldots). \tag{9.1}$$

From equation (9.1) we see that the identity transformation is associated with the group element I, since

$$T(I \cdot A_j) = T(I)T(A_j) \equiv T(A_j).$$

Moreover,

$$T(A^{-1} \cdot A) = T(I) \equiv T^{-1}(A)T(A) = T(A^{-1})T(A)$$

or

$$T(A^{-1}) = T^{-1}(A),$$

where T^{-1} is an inverse transformation, also follows from equation (9.1). A realization is called *faithful* whenever

$$T(A) = T(I)$$

implies *only* that

$$A = I.$$

For, if A_1 and A_2 are different group elements and the realization of the group is faithful,

$$T(A_1)T^{-1}(A_2) = T(A_1 \cdot A_2^{-1}) \neq T(I)$$

243

unless $A_1 = A_2$. It follows that $T(A_1)$ and $T(A_2)$ are *always* different if $A_1 \neq A_2$ and the realization is faithful.

As an example of a faithful realization of a finite group, consider the permutations of three distinguishable objects $\{123\}$. The transformations realize the symmetric group of order six and are expressed by

$$T(I) = \begin{Bmatrix} 1 & 2 & 3 \\ 1 & 2 & 3 \end{Bmatrix} \qquad T(C) = \begin{Bmatrix} 1 & 2 & 3 \\ 1 & 3 & 2 \end{Bmatrix}$$

$$T(A) = \begin{Bmatrix} 1 & 2 & 3 \\ 3 & 1 & 2 \end{Bmatrix} \qquad T(D) = \begin{Bmatrix} 1 & 2 & 3 \\ 3 & 2 & 1 \end{Bmatrix} \tag{9.2}$$

$$T(B) = \begin{Bmatrix} 1 & 2 & 3 \\ 2 & 3 & 1 \end{Bmatrix} \qquad T(E) = \begin{Bmatrix} 1 & 2 & 3 \\ 2 & 1 & 3 \end{Bmatrix}$$

where $\begin{Bmatrix} 1 & 2 & 3 \\ 3 & 1 & 2 \end{Bmatrix}$ means put "3" for "1," "1" for "2," and "2" for "3," and so on. To see that we do have a realization, we need to say what is the realization of group multiplication in this case. This is done by writing, for example,

$$T(A)T(B) = \begin{Bmatrix} 123 \\ 312 \end{Bmatrix}\begin{Bmatrix} 123 \\ 231 \end{Bmatrix} \equiv \begin{Bmatrix} 123 \\ 312 \end{Bmatrix}\begin{Bmatrix} 312 \\ 123 \end{Bmatrix} \equiv \begin{Bmatrix} 123 \\ 123 \end{Bmatrix} \equiv T(A \cdot B) = T(I).$$

To carry out the transformation $T(A \cdot B)$, we do $T(A)$ first, then $T(B)$: replace "1" by "3," then "3" by "1"; replace "2" by "1," then "1" by "2"; replace "3" by "2," then "2" by "3." As another example (verify it!)

$$T(B)T(C) = \begin{Bmatrix} 123 \\ 231 \end{Bmatrix}\begin{Bmatrix} 123 \\ 132 \end{Bmatrix} = \begin{Bmatrix} 123 \\ 231 \end{Bmatrix}\begin{Bmatrix} 231 \\ 321 \end{Bmatrix} = \begin{Bmatrix} 123 \\ 321 \end{Bmatrix} = T(B \cdot C) = T(D).$$

We also note that multiplication is associative,

$$(T(A)T(B))T(C) = \begin{Bmatrix} 123 \\ 123 \end{Bmatrix}\begin{Bmatrix} 123 \\ 132 \end{Bmatrix} = \begin{Bmatrix} 123 \\ 132 \end{Bmatrix} = \begin{Bmatrix} 123 \\ 312 \end{Bmatrix}\begin{Bmatrix} 123 \\ 321 \end{Bmatrix}$$
$$= T(A)(T(B)T(C)),$$

that

$$T(A^{-1}) = T(B) = T^{-1}(A), \qquad T(D^{-1}) = T(D) = T^{-1}(D),$$
$$T(B^{-1}) = T(A) = T^{-1}(B), \qquad T(E^{-1}) = T(E) = T^{-1}(E),$$
$$T(C^{-1}) = T(C) = T^{-1}(C), \qquad T(I^{-1}) = T(I) = T^{-1}(I),$$

and that the even permutations (those differing from the original ordering by an even number of interchanges) form a subgroup, as do $T(I)$ and $T(I)$ with each of the odd permutations. The even permutations are an invariant *abelian* subgroup, as a matter of fact, even though the entire group is not abelian. (Check this with the abstract group by using the group table.)

One realization of a continuous group is the rotation of coordinate axes about an origin fixed in three-dimensional space. The *rotation group*, as it is called, is a three-parameter group, the three parameters being the angles of rotation about the cartesian axes emanating from the origin, or their equivalents. The group is not abelian—unless it is an infinitesimal group—because two consecutive finite rotations carried out in different order lead to different orientations of the rotated axes. (If you do not believe this, try a rotation yourself using, say, the edges of a book as the cartesian axes.) However, the subgroups of the rotation group which are realized by rotations about an axis fixed in space do form abelian groups. This is because plane rotations are realizations of infinitesimal one-parameter groups and so consist of only the identity transformation and (possibly infinitely many) successive applications of an infinitesimal rotation.

9.2. REPRESENTATIONS OF GROUPS

The connection between the theory of groups and quantum physics has already begun to emerge. We have seen that groups may be realized by transformations; we also know that transformations play an important part in quantum theory. Thus we expect that linear operators, which transform the vectors of Hilbert space, will turn out to be closely related to groups. Indeed, they will, and the relationship is put in the clearest terms by the notion of a group representation. This we shall discuss right now.

A *representation* of a group is an association between one element of the group and an element of a set composed of square matrices with non-vanishing determinants,[1] such that

$$M(A \cdot B) = M(A) \times M(B) \qquad (9.3)$$

for all group elements A, B and all matrices $M(A)$, $M(B)$, where the $\times$ multiplication on the right-hand side is to be understood in the sense of matrix algebra. Equation (9.3) is sufficient to show that the matrix $M(I)$, to be associated with the identity element, is the unit matrix and that

$$M(A^{-1}) = M^{-1}(A). \qquad (9.4)$$

A representation is said to be *faithful* if it is isomorphic to the group represented. The *dimension of a representation* is equal to the number of rows (or columns) in its constituent matrices.

---

[1] The $M(A)$ must possess non-vanishing determinants in order that they have inverses (equation (9.4)). See H. Margenau and G. M. Murphy, *The Mathematics of Physics and Chemistry*, D. Van Nostrand Co., Princeton, 1956, Vol. I, Chapter 10.

The conjugation operation, defined earlier for group elements, leads to the idea of equivalent representation matrices. The matrices

$$M(I) \times M(A) \times M^{-1}(I), \qquad M(B) \times M(A) \times M^{-1}(B),$$
$$M(C) \times M(A) \times M^{-1}(C), \ldots$$

are said to be *equivalent* to the representation matrix $M(A)$. (The *general* definition of equivalence actually permits the matrices multiplying $M(A)$ to be any possessing inverses. Here we are looking at equivalence within a group.) The significance of equivalence can be made a little more obvious by defining

$$\chi^{(i)}(A) \equiv \sum_{n=1}^{l_i} [M^{(i)}(A)]_{nn}$$

where $[M^{(i)}(A)]_{nn}$ is a diagonal element of a matrix in the ith representation of the group containing A and l_i is the dimension of the representation. The quantity defined in the foregoing equation is called the *character* of the representation matrix $M^{(i)}(A)$. The character is endowed with a very important attribute: it is unchanged in value if a matrix equivalent to $M^{(i)}(A)$ is used in computing it. To see this we need only write down

$$\sum_{n=1}^{l} [M(B) \times M(A) \times M^{-1}(B)]_{nn} = \sum_{n=1}^{l} \sum_{i=1}^{l} [M(B)]_{ni} \sum_{k=1}^{l} [M(A)]_{ik}$$

$$[M^{-1}(B)]_{kn} = \sum_{i=1}^{l} \sum_{n=1}^{l} \sum_{k=1}^{l} [M(A)]_{ik} [M^{-1}(B)]_{kn} [M(B)]_{ni}$$

$$= \sum_{i=1}^{l} \sum_{k=1}^{l} [M(A)]_{ik} [M(I)]_{ki} = \sum_{i=1}^{l} \sum_{k=1}^{l} [M(A)]_{ik} \delta_{ki}$$

$$= \sum_{i=1}^{l} [M(A)]_{ii}.$$

Moreover, because of equation (9.3), we can show in precisely the same way that *group elements in the same conjugate class may be associated with the same character.*

If a representation matrix $M(A)$ can be expressed as a direct sum of two or more matrices $M^{(j)}(A)$, that is, if

$$M(A) = M^{(1)}(A) \boxplus M^{(2)}(A) \boxplus \cdots, \tag{9.5}$$

then it is said to belong to a *reducible* representation. On the other hand, if $M(A)$ is not in the form of a direct sum and is not equivalent (in the *general* sense) to a matrix in the form of a direct sum, then it belongs to an *irreducible* representation. It follows that any finite-dimensional member of a reducible representation can be put into the form of equation (9.5) with each $M^{(j)}(A)$ now being a member of an irreducible representation, the reduction taking place by an appropriate (*general*) equivalence transformation. When this has been done $M(A)$ has the appearance of a diagonal matrix, but with matrices rather than numbers along the main diagonal, as shown in Chapter 6. The rule for matrix multiplication then demonstrates something quite important: products of the reduced $M(A)$ are also direct sums of matrices in irreducible form. This means that any set of the $M^{(j)}(A)$ for fixed j is an irreducible representation of the group originally represented by the $\{M(A)\}$. Reduction of the $\{M(A)\}$, therefore, generates irreducible representations! The irreducible representations so generated need not be distinct, of course.

A number of other important statements can be made about representations and characters. We shall take a look at two of these now, but without the benefit of proofs. The proofs, of course, are essential to the mathematical structure (although not to our purpose in becoming acquainted with the theory of groups and their representations) and may be found in most texts discussing group representations.[2]

(a) *Any representation can be transformed into a unitary representation.* A matrix in the unitary representation is related to the one in the original representation by

$$U_M(A) \equiv d^{-\frac{1}{2}} \times U^{-1} \times M(A) \times U \times d^{\frac{1}{2}},$$

where

$$d = U^{-1} \times \sum_{\{A\}} M(A) \times M(A)^\dagger \times U,$$

U is the unitary matrix diagonalizing the Hermitian matrix

$$\sum_{\{A\}} M(A) \times M(A)^\dagger,$$

the sum being over all elements $\{A\}$ in the group, and $d^{\frac{1}{2}}$ is the diagonal matrix formed by taking the square root of each element of d.

(b) *Two inequivalent, irreducible, unitary representations of the same group are orthonormal in the sense that*

$$\sum_{\{A\}} \left(\frac{L_l}{h}\right)^{\frac{1}{2}} \overline{[U_M^{(l)}(A)]}_{ij} \left(\frac{L_m}{h}\right)^{\frac{1}{2}} [U_M^{(m)}(A)]_{st} = \delta_{lm}\delta_{is}\delta_{jt} \tag{9.6}$$

[2] See, for example, E. P. Wigner, *Group Theory and its Application to the Quantum Mechanics of Atomic Spectra*, Academic Press, New York, 1959, Chapter 9; or L. M. Falicov, *Group Theory and its Physical Applications*, University of Chicago Press, Chicago, 1966, Chapter II. (See also the books cited in the Bibliography for this part of the text.)

for all l, m, i, s, j, and t, where L is the dimension of the representation, h is the order of the group, and the sum extends over all members of the group. A special case of this statement appears when we put

$$j = i, t = s:$$

$$\sum_{\{A\}} \overline{[U_M^{(l)}(A)]_{ii}} [U_M^{(m)}(A)]_{ss} = \frac{h}{L_l} \delta_{lm}\delta_{is}.$$

If we sum both sides of the equation over i and s, noting the definition of the character, we get

$$\sum_{\{A\}} \overline{\chi^{(l)}(A)}\chi^{(m)}(A) = h\delta_{lm} \tag{9.7}$$

which shows that the characters of the irreducible representations are orthogonal. From equation (9.7) we may conclude that inequivalent irreducible representations of the same group cannot have the same characters. Since each character uniquely specifies a class of group elements, the sum in equation (9.7) can be transformed from a sum over elements to a sum over classes. Moreover, because

$$h = \sum_{c=1}^{k} f_c$$

where f_c is the number of group elements in the cth class, there being k classes, we can rewrite equation (9.7) as

$$\sum_{c=1}^{k} \frac{f_c}{h} \overline{\chi^{(l)}(C_c)}\chi^{(m)}(C_c) = \delta_{lm}. \tag{9.8}$$

Equation (9.8) and the remarks made when characters were introduced expose the importance of these numbers clearly. The characters of a given irreducible representation of a group are unique and correspond uniquely to the conjugate classes in the group, one character for each class. Moreover, the characters of inequivalent irreducible representations corresponding to a given class are unique as well as distinct from those corresponding to other classes.

For continuous groups the definitions and statements just given still hold, provided that the corresponding elements of matrices representing adjacent group members differ only by infinitesimal distances in the group space and that the substitution

$$\sum_{\{A\}} \rightarrow \frac{\int_D g(A) \, dp_1 \, dp_2 \cdots dp_n}{\int_D dp_1 \, dp_2 \cdots dp_n}$$

is made, where the p are the parameters associated with the group $\{A\}$, $g(A)$ is the density of points in the space of the group elements which are near the element A, and the integrals are taken over all of group space. The precise definition of the integral $\int_p g(A) \, d\mathbf{p}$ is a matter of some mathematical complexity that is best left for discussion in other circumstances.[2] In all of the applications we shall make its meaning will be assumed clear.

As an example of an irreducible representation, we might take another look at the symmetric group of order six. In the notation of equation (9.2), an irreducible representation of this group is

$$M(I) = \begin{pmatrix} 1 & 0 \\ 0 & 1 \end{pmatrix} \qquad M(C) = \frac{1}{2}\begin{pmatrix} -1 & \sqrt{3} \\ \sqrt{3} & 1 \end{pmatrix}$$

$$M(A) = \frac{1}{2}\begin{pmatrix} -1 & \sqrt{3} \\ -\sqrt{3} & -1 \end{pmatrix} \qquad M(D) = \frac{1}{2}\begin{pmatrix} -1 & -\sqrt{3} \\ -\sqrt{3} & 1 \end{pmatrix}$$

$$M(B) = \frac{1}{2}\begin{pmatrix} -1 & -\sqrt{3} \\ \sqrt{3} & -1 \end{pmatrix} \qquad M(E) = \begin{pmatrix} 1 & 0 \\ 0 & -1 \end{pmatrix}.$$

The characters of this representation are accordingly

$$\chi(I) = 2, \qquad \chi(A) = \chi(B) = -1, \qquad \chi(C) = \chi(D) = \chi(E) = 0$$

which shows immediately that A and B are in one conjugate class and C, D, and E are in another, as has already been demonstrated by direct calculation.

9.3. THE GROUP OF THE SCHRÖDINGER EQUATION

We are in a position now to enlist the help of group-theoretical ideas in making some very general statements about complex atoms. Let us begin by assuming that the Schrödinger eigenvalue problem contains all the pertinent information for any experimental study of the energy states in a many-electron system. It is that equation, then, whose symmetry properties we want to examine.

Suppose there exists a set of realizations $\{\mathbf{T(A)}\}$ which are transformations of the position and spin coordinates and, when carried out on the Schrödinger eigenvalue problem, *do not cause any change in its formal appearance*. These transformations, we expect, can be represented by linear operators mapping in Hilbert space, such that

$$O_A \psi(\mathbf{r}'^N, s'^N) \equiv \psi(\mathbf{r}^N, s^N) \tag{9.9}$$

where $\mathbf{r}'^N$, s'^N are the points in Euclidean space and spin space, respectively, reached by applying the transformation $T(A)$ to $\mathbf{r}^N$, s^N, and $\psi(\mathbf{r}^N, s^N)$ is a solution of the Schrödinger eigenvalue problem. The physical meaning of the transformations may be seen by noting that by hypothesis

$$\mathcal{H}\psi(\mathbf{r}^N, s^N) = \mathcal{H}O_A\psi(\mathbf{r}'^N, s'^N)$$

and

$$O_A\mathcal{H}\psi(\mathbf{r}'^N, s'^N) = O_A E\psi(\mathbf{r}'^N, s'^N) = E\psi(\mathbf{r}^N, s^N) = \mathcal{H}\psi(\mathbf{r}^N, s^N).$$

This means that

$$[\mathcal{H}, O_A] = O \tag{9.10}$$

for any operation O_A which leaves the form of the Schrödinger eigenvalue problem invariant. Equation (9.10) is just a way of saying that the energy of the atom is left unchanged by the transformation $T(A)$. More specifically, it says that $\mathcal{H}$ and O_A share a complete set of eigenstates and, therefore, that the symmetry properties of $\mathcal{H}$ are manifest in the wavefunctions describing the states of the atom.

Now, the set of realizations $\{T(A)\}$ must certainly contain reciprocals, since $(\mathbf{r}, s)$ are reachable from $(\mathbf{r}', s')$ if the points are physically equivalent, and the members of $\{T(A)\}$ must, when multiplied together, lead to coordinate configurations which are physically possible. [If $T(A_1)$ and $T(A_2)$ lead to physical configurations separately, $T(A_1)T(A_2)$ must lead to a possible configuration.] Moreover, an identity transformation must exist because each $(\mathbf{r}, s)$ can be transformed into itself. It follows that the set of realizations $\{T(A)\}$ which do not alter the Schrödinger eigenvalue problem constitute in themselves the realization of a group. We call this group *the group of the Schrödinger equation.*

Given that the $\{T(A)\}$ are realizations of a group, we must ask as to how the group is to be represented. In general, the operation O_A enacted upon a wavefunction $\psi_m^{(j)}(\mathbf{r}^N, s^N)$ will yield a solution of the Schrödinger eigenvalue problem composed of the L orthonormal wavefunctions spanning the mth invariant subspace of $\mathcal{H}$. Therefore, we must write

$$O_A\psi_m^{(j)}(\mathbf{r}^N, s^N) = \sum_{k=1}^{L} [M(A)]_{kj}\psi_m^{(k)}(\mathbf{r}^N, s^N) \tag{9.11}$$

where $\psi_m^{(k)}(\mathbf{r}^N, s^N)$ corresponds to the L-fold degenerate energy eigenvalue E_m. Similarly, for the transformation $T(B)$, we have

$$O_B\psi_m^{(k)}(\mathbf{r}^N, s^N) = \sum_{l=1}^{L} [M(B)]_{lk}\psi_m^{(l)}(\mathbf{r}^N, s^N) \tag{9.12}$$

where, as in equation (9.11), the coefficients of $\psi_m^{(l)}$ in the sum are constants to be determined by the operation. Because (9.11) and (9.12) are linear equations, it is clear that the coefficients may be construed as elements of the matrices $M(A)$ or $M(B)$, respectively. Moreover,

$$O_B O_A \psi_m^{(j)}(\mathbf{r}^N, s^N) = O_B \sum_{k=1}^{L} [M(A)]_{kj} \psi_m^{(k)}(\mathbf{r}^N, s^N)$$

$$= \sum_{k=1}^{L} [M(A)]_{kj} O_B \psi_m^{(k)}(\mathbf{r}^N, s^N)$$

$$= \sum_{k=1}^{L} [M(A)]_{kj} \sum_{l=1}^{L} [M(B)]_{lk} \psi_m^{(l)}(\mathbf{r}^N, s^N)$$

$$= \sum_{l=1}^{L} \sum_{k=1}^{L} [M(B)]_{lk}[M(A)]_{kj} \psi_m^{(l)}(\mathbf{r}^N, s^N)$$

and

$$O_B O_A \psi_m^{(j)}(\mathbf{r}^N, s^N) \equiv O_{B \cdot A} \psi_m^{(j)}(\mathbf{r}^N, s^N) = \sum_{l=1}^{L} [M(B \cdot A)]_{lj} \psi_m^{(l)}(\mathbf{r}^N, s^N)$$

so that

$$[M(B \cdot A)]_{lj} = \sum_{k=1}^{L} [M(B)]_{lk}[M(A)]_{kj}. \tag{9.13}$$

Equation (9.13) shows that we can associate a matrix $M(A)$ with the operation O_A and a matrix $M(B)$ with O_B such that

$$M(B \cdot A) = M(B) \times M(A).$$

In other words, the $M(A)$ and $M(B)$ are matrices in a representation of the group of the Schrödinger equation! The dimension of the representation is equal to the multiplicity of the eigenvalue corresponding to the wavefunctions $\psi_m^{(j)}(\mathbf{r}^N, s^N)$. It is to be expected, of course, that different representations will be associated with different eigenvalues.

We know from the foregoing section that every representation can be made unitary. In the present case, however, the conditions that O_A be a unitary operator and the $\psi_m^{(j)}$ be orthonormal simply guarantee unitarity. For,

$$(O_A \psi_m^{(j)}, O_A \psi_m^{(k)}) \equiv (\psi_m^{(j)}, \psi_m^{(k)}) = \delta_{jk}$$

$$= \left(\sum_{p=1}^{L} [M(A)_{pj} \psi_m^{(p)}, \sum_{l=1}^{L} [M(A)_{lk} \psi_m^{(l)} \right)$$

$$= \sum_{p=1}^{L} \sum_{l=1}^{L} \overline{[M(A)]}_{pj}[M(A)]_{lk} (\psi_m^{(p)}, \psi_m^{(l)})$$

$$= \sum_{l=1}^{L} \overline{[M(A)]}_{lj}[M(A)]_{lk}.$$

It follows that the statements made about unitary representations in section 9.2 are automatically valid for the $M(A)$.

Just what are the members of the group of the Schrödinger equation is a question to be answered from general physical considerations; that is, only after the Hamiltonian operator has been explicitly written out. Since we are at the present time interested in the behavior of complex atoms, we shall be concerned with a multielectron Hamiltonian operator. The general form of this operator will be different, insofar as it contains spin-dependent terms, depending upon how large is the atomic number of the atom being investigated. Thus far we have dealt with what may be called "effectively light" atoms, whose spectra reflect that the forces arising from spin engender but a small contribution to the total electronic energy. These will continue to be the subject of our inquiry. Therefore, we shall be examining a Hamiltonian operator of the form

$$\mathscr{H} = -\frac{\hbar^2}{2\mu_e} \sum_{i=1}^{N} \nabla_{\mathbf{r}_i}^2 - Ze^2 \sum_{i=1}^{N} \frac{1}{r_i} + \frac{e^2}{2} \sum_{\substack{i,j \\ (i \neq j)}} \frac{1}{r_{ij}} + K(\mathbf{L} \cdot \mathbf{S})$$

where K is a constant depending upon the coulomb fields in the atom. Now we must discover the operations which leave $\mathscr{H}$ unaltered.

Imagine a complex atom composed of N electrons orbiting about a central nucleus. Each of these electrons can be designated by a pair of coordinates, $\mathbf{r}_i$, s_i, which can take on all values consistent with the structures of Euclidean and spin spaces, respectively. But these designations are quite *arbitrary*: it cannot matter which electron we call the ith one, which the jth one, and so on, insofar as the total electronic energy is concerned. In terms of theory, we can say that

$$E_n = \langle \mathscr{H} \rangle_n = {\sum_{s_N}}' \cdots {\sum_{s_1}}' \int_{\mathbf{r}_N} \cdots \int_{\mathbf{r}_1} \bar{\psi}_n(\mathbf{r}^N, s^N) \mathscr{H} \psi_n(\mathbf{r}^N, s^N) \, d\mathbf{r}_1 \cdots d\mathbf{r}_N$$

does not depend on the order of the integrations or summations. It would seem, then, that the transformation

$$T(P) \equiv \left\{ \begin{matrix} \mathbf{r}_1 s_1 & \cdots & \mathbf{r}_N s_N \\ \vdots & & \vdots \\ \mathbf{r}_1' s_1' & \cdots & \mathbf{r}_N' s_N' \end{matrix} \right\}$$

which is a permutation of the coordinate labels of the N electrons, is a realization of one member of the group of the Schrödinger equation. The general operation corresponding to $T(P)$ is

$$O_P \psi_m^{(j)}(\mathbf{r}^N, s^N) = \sum_{k=1}^{L} [M(P)]_{kj} \psi_m^{(k)}(\mathbf{r}^N, s^N) \tag{9.14}$$

where $\psi_m^{(j)}(\mathbf{r}^N, s^N)$ is one of the L wavefunctions corresponding to the eigenvalue E_m belonging to $\mathcal{H}$ and $[M(P)]_{kj}$ is a matrix element in the representation of dimension L of the symmetric group (of order $N!$). Equation (9.14) reduces to a very simple form after an important observation about permutations is made: every permutation may be expressed as a product of *transpositions*, which are interchanges of two coordinate labels. For example, in the case of just three electrons,

$$\begin{Bmatrix} \mathbf{r}_1 s_1 & \mathbf{r}_2 s_2 & \mathbf{r}_3 s_3 \\ \mathbf{r}_2 s_2 & \mathbf{r}_3 s_3 & \mathbf{r}_1 s_1 \end{Bmatrix} = \begin{Bmatrix} \mathbf{r}_1 s_1 & \mathbf{r}_2 s_2 & \mathbf{r}_3 s_3 \\ \mathbf{r}_2 s_2 & \mathbf{r}_1 s_1 & \mathbf{r}_3 s_3 \end{Bmatrix} \begin{Bmatrix} \mathbf{r}_1 s_1 & \mathbf{r}_2 s_2 & \mathbf{r}_3 s_3 \\ \mathbf{r}_3 s_3 & \mathbf{r}_2 s_2 & \mathbf{r}_1 s_1 \end{Bmatrix}.$$

Moreover, an even permutation (such as the one just above) may always be written as a product of an even number of transpositions, while an odd permutation is always a product of an odd number of transpositions. It follows that, trivially, we may assign the one-dimensional matrix (1) to represent *every* transposition and to the identity permutation, so that equation (9.14) becomes

$$O_P \psi_m^{(j)}(\mathbf{r}^N, s^N) = \sum_{k=1}^{L} \delta_{kj} \psi_m^{(k)}(\mathbf{r}^N, s^N) = \psi_m^{(j)}(\mathbf{r}^N, s^N) \tag{9.15}$$

which is to say that we are making $M(P)$ the direct sum of L one-dimensional (and thus irreducible!) representation matrices. Otherwise, we might assign the matrix (1) to represent the identity and the elements of the invariant subgroup of the symmetric group (realized by the even permutations) and assign the matrix (-1) to the remainder of the group (realized by the odd permutations). Equation (9.14) is, in the latter case,

$$O_P \psi_m^{(j)}(\mathbf{r}^N, s^N) = (-1)^P \psi_m^{(j)}(\mathbf{r}^N, s^N) \tag{9.16}$$

where P is the number of transpositions produced by O_P. These two ways of portraying the representation matrix $M(P)$ are certainly arbitrary on our part. In general, of course, $M(P)$ could be the direct sum of matrices in irreducible form corresponding to representations of dimension as high as L. If wave quantum mechanics were based solely upon the Schrödinger postulate, we should have to consider these higher-dimensional irreducible representations as well. However, for electrons, *the exclusion postulate ascribes physical meaning only to the wavefunctions for which equation* (9.16) *is valid*; we therefore can neglect the irreducible representation portrayed in equation (9.15) and those of dimension greater than one.

Now let us take another look at our complex atom and ask the question: What happens to the total electronic energy if the coordinates designating the electrons are rotated in ordinary and spin space? The answer is that the energy will not change so long as the spin-dependent forces are small compared to the charge-dependent forces. Indeed, a rotation of the coordinates in space only reorients the atom as a whole and thus cannot change the coulomb potentials, which depend solely upon the *relative* positions of the electrons. Moreover, if a rotation in spin space occurs along with the one in ordinary space, the spin-orbit term in $\mathscr{H}$ is unaffected since the relative orientation of L and S is unaffected. We conclude, then, that rotations are a realization of another member of the group of the Schrödinger equation. These rotations may be broken down into two kinds of transformation: "pure" rotation, wherein the coordinates are simply revolved about a fixed point or axis, and inversion, wherein the position coordinates are replaced by their counterparts of opposite sign.

The general pure rotation is represented by

$$O_R \psi^{(k)}(\mathbf{r}^N, s^N) = \sum_{k'} [M(R)]_{k'k} \psi^{(k')}(\mathbf{r}^N, s^N). \tag{9.17}$$

The irreducible representations of the pure rotation group can be obtained in several ways.[3] It will suffice for our purposes to note only the following.

$$O_R \psi^{(k)}(\mathbf{r}^N, s^N) = \sum_{k'} [D^{(L)} \otimes D^{(S)}]_{k'k} \psi^{(k')}(\mathbf{r}^N, s^N) \tag{9.18}$$

$$M(R) \equiv D^{(L)} \otimes D^{(S)} \tag{9.19}$$

$$D^{(L)} \equiv D^{(l_1)} \otimes D^{(l_2)} \cdots \qquad D^{(S)} \equiv D^{(l_{s_1})} \otimes D^{(l_{s_2})} \cdots \tag{9.20}$$

$$[D^{(l)}(\phi, \theta, \psi)]_{m'm}$$

$$= \exp(-im'\phi) \sum_{t=0}^{\infty} \frac{[(l+m)! \, (l-m)! \, (l+m')! \, (l-m')!]^{1/2}}{(l+m-t)! \, (l-m'-t)! \, t! \, (t+m'-m)!}$$

$$\times (-1)^t \cos^{2l+m-m'-2t}(\theta/2) \, [-\sin(\theta/2)]^{2t+m'-m} \exp(-im\psi)$$

$$(-l \leqslant m', m \leqslant l) \quad (9.21)$$

where (ϕ, θ, ψ) are the Euler angles, defined in Figure 49, l, l_s, m, and m_s are orbital and spin angular momentum quantum numbers for a single electron, and L, S, and J are the total quantum numbers defined in Chapter 7. Equations (9.19) and (9.20) tell us that the multielectron wavefunctions in equation (9.17) are to be built up according to Russell-Saunders coupling and that this corresponds to creating the *direct product* of irreducible representation matrices. We note also that, according to equation (9.21), the identity rotations correspond to $l = 0$ or $l_s = 0$ and that the matrices $D^{(l)}$ are of dimension $2l + 1$ or $2l_s + 1$, respectively.

[3] See, for example, M. E. Rose, *Elementary Theory of Angular Momentum*, John Wiley and Sons, New York, 1957, Chapter IV.

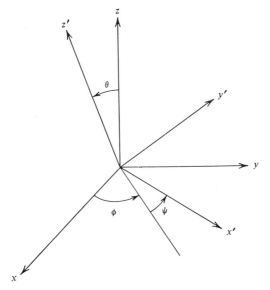

Figure 49. The Euler angles. These coordinates describe the orientation of a set of reference frames (x', y', z') in a set (x, y, z).

The general inversion operation is represented by

$$O_{Rf}\psi^{(j)}(\mathbf{r}^N, s^N) = \sum_{k=1}^{L} [M(Rf)]_{kj}\psi^{(k)}(\mathbf{r}^N, s^N). \tag{9.22}$$

The irreducible representations of the inversion group are obtained by noting that, according to the results of Hartree-Fock theory, the angular dependence of the $\psi^{(j)}(\mathbf{r}^N, s^N)$ is given by terms in

$$\Upsilon_l^m(\vartheta, \varphi)\begin{pmatrix}\alpha(s)\\\beta(s)\end{pmatrix}$$

and that inversion corresponds to $\vartheta \to \pi - \vartheta$, $\varphi \to \pi + \varphi$, $s \to -s$. There is no angular dependence of the spin wavefunctions other than the "up" or "down" denoted by the spin variable. It follows that an inversion in spin space cannot affect the magnitude of the spin wavefunction; what was "spin up" becomes "spin down" after inversion and so changes $\alpha(s)$ into $\beta(-s)$. From the definition of the spherical harmonics given in Chapter 4, it may be shown that

$$\Upsilon_l^m(\pi - \vartheta, \pi + \varphi) = (-1)^l\Upsilon_l^m(\vartheta, \varphi).$$

Therefore, equation (9.22) may be written

$$O_{Rf}\psi^{(j)}(\mathbf{r}^N, s^N) = P_{Rf}\psi^{(j)}(\mathbf{r}^N, s^N) \tag{9.23}$$

255

where the *parity* $P_{Rf} = +1$, if the sum of all the orbital quantum numbers l is an even number, and $P_{Rf} = -1$, if the sum is an odd number. We see that the representation matrix $M(Rf)$ may be expressed as a direct sum of *either* set of the one-dimensional representation matrices

(1) for the identity   (1) for the identity

(1) for inversion (-1) for inversion

In this case both representations correspond to wavefunctions (of "even" and "odd" parity, respectively) which are permitted within the framework of quantum mechanics. Indeed, we may state the important result that *the eigenfunctions of a reflection-invariant Hamiltonian operator may always be taken as either even or odd functions.*

There are no other transformations besides permutation, pure rotation, and inversion which leave the eigenvalue problem for $\mathscr{H}$ without change. These transformations, then, comprise the realization of the group of the Schrödinger equation. Moreover, because the operators representing the transformations commute with the Hamiltonian operator, the eigenfunctions of the operators representing the group of the Schrödinger equation are eigenfunctions of $\mathscr{H}$ and, indeed, the only eigenfunctions with the correct symmetry properties. These functions are uniquely designated by the fact that they are solutions of

$$O_P O_R O_{Rf} \psi(\mathbf{r}'''^N, s'''^N) = \psi(\mathbf{r}^N, s^N) \qquad (9.24)$$

where

$$O_{Rf} \psi(\mathbf{r}'''^N, s'''^N) = \psi(\mathbf{r}''^N, s''^N)$$

$$O_R \psi(\mathbf{r}''^N, s''^N) = \psi(\mathbf{r}'^N, s'^N)$$

$$O_P \psi(\mathbf{r}'^N, s'^N) = \psi(\mathbf{r}^N, s^N).$$

Equation (9.24) implies that the representation of the group of the Schrödinger equation is the direct product of the representations of the symmetric, pure rotation, and inversion groups:

$$M(P \otimes R \otimes Rf) \equiv M(P) \otimes M(R) \otimes M(Rf).$$

9.4. PERTURBATIONS AND IRREDUCIBLE GROUP REPRESENTATIONS

We can sharpen our ideas about the effect of symmetry upon the eigenfunctions of the Hamiltonian operator by considering their relationship with the irreducible representations of members of the group of the Schrödinger equation. These relationships, in turn, may be used to describe the result of adding a perturbation term of arbitrary magnitude to the Hamiltonian operator and thus to elucidate the cause of energy degeneracy. Let us see how this comes about.

Suppose O_A realizes some member of the group of the Schrödinger equation and acts upon the kth eigenfunction corresponding to an L_m-fold degenerate energy eigenvalue E_m. Then we may write

$$O_A \psi_m^{(k)} = \sum_{n=1}^{L_m} [U_M^{(m)}(A)]_{nk} \psi_m^{(n)} \qquad (9.25)$$

where $U_M^{(m)}(A)$ is a matrix in a unitary, irreducible representation of the group $\{A\}$. We shall say that $\psi_m^{(k)}$ "belongs to the kth row of the matrix $U_M^{(m)}(A)$" if it satisfies equation (9.25). The eigenfunctions $\psi_m^{(n)}$ ($1 \leqslant n \leqslant L_j$) are then termed "partner eigenfunctions" for $\psi_m^{(k)}$.

Now suppose there exist two wavefunctions belonging to different irreducible representations or to different rows of the same irreducible representation matrix:

$$O_A \psi_m^{(k)} = \sum_{l=1}^{L_m} [U_M^{(m)}(A)]_{lk} \psi_m^{(l)}$$

$$O_A \psi_{m'}^{(k')} = \sum_{l'=1}^{L_{m'}} [U_M^{(m')}(A)]_{l'k'} \psi_{m'}^{(l')}.$$

It is possible to make a very general statement concerning the $\psi_m^{(k)}$ and $\psi_{m'}^{(k')}$: the scalar product $(\psi_m^{(k)}, \psi_{m'}^{(k')})$ vanishes when $k \neq k'$ or $m \neq m'$ and has value independent of k when $k = k'$ and $m = m'$. To see this we can write

$$(\psi_m^{(k)}, \psi_{m'}^{(k')}) = (O_A \psi_m^{(k)}, O_A \psi_{m'}^{(k')})$$

$$= \sum_{l=1}^{L_m} \sum_{l'=1}^{L_{m'}} \overline{[U_M^{(m)}(A)]}_{lk} [U_M^{(m')}(A)]_{l'k'} (\psi_m^{(l)}, \psi_{m'}^{(l')}) \qquad (9.26)$$

where we have made use of the unitary character of the O_A. If both sides of equation (9.26) are summed (or integrated) over the $\{A\}$, and the right-hand side is summed over l' and l, we get, according to equation (9.6),

$$(\psi_m^{(k)}, \psi_{m'}^{(k')}) = \delta_{mm'} \delta_{kk'}. \qquad (9.27)$$

(Remember that the $\psi_m^{(k)}$ for a given value of m are assumed orthonormal so that the representation to which they belong will be unitary.) In light of the fact that different sets of partner functions correspond to different irreducible representations [equation (9.25) is unique!], equation (9.27) leads us to the important conclusion that *different sets of partner wavefunctions are orthogonal to one another*. With regard to the Hamiltonian operator, this fact indicates immediately that, since different energy eigenvalues correspond to different irreducible representations, the eigenvectors belonging to one eigenvalue are orthogonal to those belonging to another. Now, if two eigenvalues corresponding to different sets of partner eigenfunctions just happened to have the same numerical magnitude, we would be confronted with the peculiar situation known as *accidental degeneracy*. This is to say that a strictly numerical partitioning of eigenvalues would lead us to believe here that the corresponding eigenfunctions spanned a single invariant subspace of $\mathscr{H}$ when in fact they span two orthogonal invariant subspaces whose members have different symmetry properties as evinced by equation (9.25). In this case, then, the representation of the group of the Schrödinger equation will *never* be irreducible, but will always be the direct sum of two irreducible representations. Conversely, if an energy eigenvalue possesses but a single set of partner eigenfunctions, such that the associated representation is *always* irreducible, the eigenvalue is said to have *normal degeneracy*. The two kinds of degeneracy are sharply differentiated in a very physical way if their dispositions toward perturbations are examined.

Consider a perturbation Hamiltonian operator $\mathscr{H}'$ which is added to a Hamiltonian operator $\mathscr{H}$ and which commutes with all the operators representing the group of the Schrödinger equation constructed from $\mathscr{H}$. The alteration in the total energy that is caused by $\mathscr{H}'$ is expressed, in the first approximation, by the set of linear homogeneous equations

$$\sum_{k=1}^{L} a_k{}^m (H'_{m_{k'}m_k} - E^{(1)}\delta_{k'k}) = 0 \tag{9.28}$$

where

$$H_{m_{k'}m_k} = (\psi_m^{(k')}, \mathscr{H}'\psi_m^{(k)}) \tag{9.29}$$

and the $\psi_m^{(k)}$ are eigenfunctions of $\mathscr{H}$. If the L-fold degenerate eigenvalue

$$E_m = (\psi_m^{(k)}, \mathscr{H}\psi_m^{(k)}) \qquad (k = 1, 2, \ldots, L_m),$$

exhibits only *normal* degeneracy, it follows from the assumed invariance of $\mathscr{H}'$ that $\mathscr{H}'\psi_m^{(k)}$ belongs to the kth row of the irreducible representation matrix $U_M^{(m)}(A)$ if $\psi_m^{(k)}$ does. According to this conclusion, and to equation (9.27), equation (9.29) can be written

$$H'_{m_{k'}m_k} = E_m^{(1)}\delta_{k'k}.$$

The same result is obtained if the correction to E_m is calculated to all orders of approximation: for, if the corrections to the energy were ever to be unequal in value, this would mean that several *different* sets of eigenfunctions (each set corresponding to a distinct correction) would have been generated from the *single* set of $\psi_m^{(k)}$. As sets of eigenfunctions corresponding to different eigenvalues always belong to different irreducible representations, and thus to subspaces orthogonal to one another and to that in which their parent set resides, unequal values of the corrections would imply that vectors in one subspace could generate vectors in another, which is impossible. We may infer from this fact that *a perturbation Hamiltonian operator with the same symmetry properties as the unperturbed Hamiltonian operator can never cause a normally degenerate eigenvalue to split.*

On the other hand, if several of the E_m happen to coincide in value, then

$$E_m = (\psi_m^{(k)}, \mathscr{H}\psi_m^{(k)}) \qquad (k = 1, 2, \ldots, L_m),$$

may be written down for more than one value of m. This means that the representation corresponding to E_m is not irreducible, but may be written a direct sum of, say, n irreducible representations:

$$M(A) = M^{(1)}(A) \boxplus M^{(2)}(A) \boxplus \cdots \boxplus M^{(n)}(A),$$

and so on. If a perturbation $\mathscr{H}'$ that has the same invariance properties as does the operator to which E_m belongs is applied, then, in the first approximation, there can be different values of $E_m^{(1)}$ for each set of $\psi_m^{(k)}$ belonging to a different irreducible representation. Therefore, *a perturbation Hamiltonian operator with the same symmetry properties as the unperturbed Hamiltonian operator can split an accidentally degenerate eigenvalue into as many levels as is its accidental multiplicity.* The multiplicities of these levels are the same as the dimensions of the irreducible representations to which the levels correspond.

Let us consider two examples which illustrate the statements just made about perturbations. In Chapter 5 we discussed the Stark effect for a diatomic molecule approximated as the rigid rotator. There we found that the $(2l + 1)$-fold-degenerate energy levels of the rotator were split, for l and $|m| > 0$, into twofold-degenerate levels by the applied electric field. The reason for this residual twofold multiplicity, evidently, is that both the rotator Hamiltonian operator and the perturbation Hamiltonian operator

$$\mathscr{H}' = -\mu|\mathbf{E}|\cos\vartheta$$

are invariant under rotations in the xy-plane about the z-axis and that the unperturbed energy eigenvalues for the excited rotator possess a normal multiplicity of two arising from this rotational symmetry. On the other hand, in section 7.1 we found that the repulsive coulomb interaction split the levels for the excited states of a hydrogenic two-electron atom. This is because both the hydrogenic Hamiltonian operator and the perturbation Hamiltonian operator

$$\mathscr{H}' = \frac{e^2}{r_{12}}$$

259

are invariant under three-dimensional rotations and the excited-state energy levels of a hydrogenic atom

$$E_n = -\frac{Ze^2}{2a_0 n^2} \qquad (n = 2, 3, \ldots)$$

possess an accidental multiplicity[4] equal to n.

9.5. SELECTION RULES FOR COMPLEX SPECTRA

The selection rules for dipole radiation from an "effectively light" atom depend on the conditions under which the matrix elements of the dipole moment operator $\mathscr{M}$ vanish. These latter are proportional to sums of

$$M_{m'm}^{k'k} \propto (\psi_{m'}^{(k')}, x_n \psi_m^{(k)})$$

where $\psi_m^{(k)}$ is an eigenfunction belonging to the mth irreducible representation of some member of the group of the Schrödinger equation and x_n is the nth component of $\mathbf{r}$. We know that the matrix elements of the Hamiltonian operator whose eigenvalues correspond to the $\psi_m^{(k)}$ are, in part, determined by the symmetry properties of $\psi_{m'}^{(k')}$ and $\psi_m^{(k)}$: off-diagonal elements involving these functions are zero if $k' \neq k$ or $m' \neq m$. In just the same way we expect that the functions $\psi_{m'}^{(k')}$ and $\psi_m^{(k)}$ will determine whether or not the matrix element above vanishes. This expectation is indeed the case, and it gives rise to three kinds of selection rule:

   (a) The functions $\psi_{m'}^{(k')}$ and $\psi_m^{(k)}$ both correspond to irreducible representations of the pure *spin* rotation group because the dipole moment operator $\mathscr{M}$ is invariant under the rotation of spin coordinates. Therefore, the matrix elements of $\mathscr{M}$ will disappear unless $m' = m$; that is, unless both $\psi_{m'}^{(k')}$ and $\psi_m^{(k)}$ correspond to the same irreducible representation $D^{(S)}$, whose matrix elements are given by equations (9.20) and (9.21). But $D^{(S)}$, in turn, corresponds to just *one* value of the total spin angular momentum quantum number S. We conclude that *electronic transitions do not occur between energy levels of different spin multiplicity.* An example of this selection rule is provided by the relative absence of intercombination lines in the spectrum of helium: no transitions occur between singlet and triplet states. On the other hand, the selection rule cannot be expected to hold in general because the form of the dipole moment operator is dependent upon the absence of spin terms in the Hamiltonian operator. (See section 5.5.) Spin-orbit coupling permits a small probability for transitions between states of different total spin and so explains the very existence of triplet states.

---

[4] For a discussion of this fact see L. I. Schiff, *Quantum Mechanics*, McGraw Hill Book Co., New York, 1968, p. 94f.

(b) The operator $\mathcal{M}$ is not invariant under pure *position* rotations. For this reason the selection rule for the total angular momentum quantum number, J, will not be the same as for S, the total spin quantum number. Indeed, $\psi_{m'}^{(k')}$ must now belong to the same irreducible representation of the pure rotation group as does $x_n \psi_m^{(k)}$. The question is, then, what is this irreducible representation?

To begin with, x_n transforms under rotation, obviously, as a vector. This means that

$$O_R x_n = \sum_{n'=1}^{3} [D^{(1)}(R)]_{n'n} x_{n'}$$

where $D^{(1)}(R)$ is given by equation (9.21) with $l = 1$. [The choice of value for l is supported by calculating the matrix $D^{(1)}(\phi, \theta, \psi)$ explicitly to find that it is just the rotation matrix for a vector.] As regards $\psi_m^{(k)}$, we have

$$O_R \psi_m^{(k)} = \sum_{l=1}^{L_m} [D(R)]_{lk} \psi_m^{(l)}$$

where, by equation (9.18),

$$M(R) \equiv D(R) \equiv D^{(L)}(R) \otimes D^{(S)}(R).$$

The irreducible components of $D(R)$ may be found by using a method invented by E. P. Wigner,[3] which we shall illustrate for the special case of rotation about a given axis in ordinary space. If, in equation (9.21), we put $\theta = \psi = 0$, we have as the character of $D^{(l_1)} \otimes D^{(l_2)}$

$$\chi(\phi) = \sum_{m_{l_1} = -l_1}^{l_1} \sum_{m_{l_2} = -l_2}^{l_2} \exp(im_{l_1}\phi) \exp(-im_{l_2}\phi).$$

Now, as can be verified by explicit calculation, we can rewrite this character as

$$\chi(\phi) = \sum_{L=|l_2-l_1|}^{l_1+l_2} \sum_{M_L=-L}^{L} \exp(-iM_L\phi) \equiv \sum_{L=|l_2-l_1|}^{l_1+l_2} \chi^{(L)}(\phi)$$

where the character $\chi^{(L)}$ has precisely the form of a character of an irreducible representation. [Consider, for example, the character of the irreducible representation matrix $D^{(l)}(\phi, 0, 0)$.] Since each $\chi^{(L)}$ is different for different L-value, the foregoing expression is a decomposition of $\chi(\phi)$ into irreducible characters! Therefore, we can write as the representation matrix equivalent to $D^{(l_1)} \otimes D^{(l_2)}$:

$$\sum_{L} \boxplus D^{(L)} \equiv D^{(|l_2-l_1|)} \boxplus D^{(|l_2-l_1|+1)} \boxplus \cdots \boxplus D^{(l_2+l_1)}.$$

In an exactly analogous way, the representation matrix equivalent to $D^{(L)} \otimes D^{(S)}$ is

$$\sum_J \boxplus D^{(J)} \equiv D^{(|L-S|)} \boxplus D^{(|L-S|+1)} \boxplus \cdots \boxplus D^{(L+S)} \tag{9.30}$$

which shows that

$$J = |L - S|, |L - S| + 1, \ldots, L + S$$

in agreement with the definition of the total angular quantum number J as given in Chapter 7! Equation (9.30) together with equation (9.18) implies that

$$[D(R)]_{M_L' M_S'; M_L M_S} = \sum_J \sum_{M_J'} \sum_{M_J} \overline{C}_{J,M_J';M_L'M_S'}^{(LS)} [D^{(J)}]_{M_J'M_J} C_{J,M_J;M_L,M_S}^{(LS)}$$

where the sums over M_J are from $-J$ to J in integral steps and the $C_{J,M_J;M_L,M_S}^{(LS)}$ are elements of the matrix C satisfying

$$C^{-1} \times (D^{(L)} \otimes D^{(S)}) \times C = \sum_J \boxplus D^{(J)}.$$

Thus, the matrix C is a unitary matrix which is used in the reduction of $D(R)$. Its physical significance is made evident if we write the chain of identities

$$O_R \psi_J^{(M_J)} = \sum_{M_J'} [D^{(J)}]_{M_J'M_J} \psi_J^{(M_J')}$$

$$= \sum_{J'} \sum_{M_J'} [C^{-1} \times D(R) \times C]_{J'M_J';J,M_J} \psi_J^{(M_J')}$$

$$= \sum_{M_L',M_L} \sum_{M_S',M_S} \sum_{J',M_J'} \overline{C}_{J',M_J';M_L',M_S'}^{(LS)} [D(R)]_{M_L'M_S';M_LM_S}$$
$$\times C_{J,M_J;M_LM_S}^{(LS)} \psi_{J'}^{(M_J')}$$

$$\equiv \sum_{M_L',M_L} \sum_{M_S',M_S} [D(R)]_{M_L'M_S';M_LM_S} C_{J,M_J;M_LM_S}^{(LS)} \psi_L^{(M_L')} \psi_S^{(M_S')}$$

$$\equiv \sum_{M_L',M_L} \sum_{M_S',M_S} [D(R)]_{M_L'M_S';M_LM_S}$$
$$\times \langle L, S; M_L, M_S | L, S; J M_S \rangle | L, M_L' \rangle | S, M_S' \rangle \tag{9.31}$$

where in the third step we have interposed complete sets of angular momentum states $|L, S; M_L', M_S'\rangle$ and $|L, S; M_L, M_S\rangle$. Equation (9.31) as elaborated indicates that the matrix elements of C are none other than the vector coupling coefficients introduced in equation (7.17). That this is the case may be seen a little more readily by writing the latter equation in the form

$$\psi_J^{(M_J)} = \sum_{M_L=-L}^{L} \sum_{M_S=-S}^{S} C_{J,M_J;M_L,M_S}^{(LS)} \psi_L^{(M_L)}(\mathbf{r}^N) \psi_S^{(M_S)}(s^N) \qquad (9.32)$$

where $\psi_L^{(M_L)}$ is a multielectron wavefunction for a state of total orbital angular quantum number L and $\psi_S^{(M_S)}$ and $\psi_J^{(M_J)}$ are correspondingly defined. The vector-coupling coefficients are of great significance in the general theory of angular momentum as it is applied in the study of atomic, molecular, solid-state, and nuclear physics. Those interested in some of these applications in more detail might consult the books cited in references two and three in this chapter.

Returning to the dipole moment operator, we see that

$$O_R x_n \psi_J^{(M_J)} = \sum_{n'=1}^{3} \sum_{l=-J}^{J} [D^{(1)}]_{n'n} [D^{(J)}]_{lM_J} x_{n'} \psi_J^{(l)} \qquad (9.33)$$

which suggests that $\psi_{J'}^{(M'_J)}$ in $M_{J'J}^{M'_J M_J}$ must belong to one of the irreducible representations of

$$D^{(1)}(R) \otimes D^{(J)}(R).$$

According to what we discovered for the direct product $D(R)$, this means that the matrix element will vanish unless $\psi_{J'}^{(M'_J)}$ belongs to one of

$$D^{(J-1)} \ D^{(J)} \ D^{(J-1)} \qquad (J > 0),$$

the direct sum of these being equivalent to the direct product given above. We may infer from this that *electronic transitions do not occur between states of different total angular quantum number unless*

$$\Delta J = \pm 1, 0. \qquad (9.34)$$

If $J = 0$, then we have to do with

$$D^{(1)} \otimes D^{(0)} = D^{(1)}$$

and J' *must* be equal to one: *electronic transitions between states with $J = 0$ also do not occur.* In the case of a spinless one-electron atom, $J = L$, the orbital quantum number, and equation (9.34) becomes one of the familiar selection rules derived in Chapter 5. In the other extreme, when spin-dependent terms in the Hamiltonian operator are important, the quantum numbers L and S have no significance and J alone describes the state of an atom.

(c) The function $\mathscr{M}\psi_m^{(k)}$ will always correspond to a representation of the inversion group that is different from the representation to which $\psi_m^{(k)}$ belongs. This is because

$$O_{R_I}(-\mathbf{r}_i) = \mathbf{r}_i$$

by definition and the irreducible representations of the inversion group are one-dimensional. Therefore, $M_{m'm}^{k'k}$ will be zero unless $\psi_{m'}^{(k')}$ is of parity opposite to that of $\psi_m^{(k)}$. *Electronic dipole transitions occur only between states of opposite parity.* This selection rule, unlike its predecessors, is *exact.* The entire reason for this is that the spin-dependent part of the inversion operator never makes a change in the wavefunction. (The rule itself was formulated empirically before it was derived; it is sometimes called *Laporte's rule.*)

9.6. THE BUILDING-UP RULES

So far we have been able to deduce fairly general ways of finding out the internal structure of a given multielectron energy state. Through the use of selection rules and the expressions for Zeeman and spin-orbit splitting, the value of J, M_J, L, M_L, S, and M_S may be computed for any of the energy states permitted an "effectively light" atom. However, these devices do not lead to any conclusions about exactly which are the permitted energy states or what is their relative ordering as regards magnitude. The answers to those two questions are given by what may be called the *building-up rules.* We shall consider them now.

To begin with, we need not be concerned by the presence of spin-orbit interactions because these produce only a fine structure in the energy levels (already perturbed by coulomb repulsions) which can be accounted for easily by perturbation theory. Therefore, the complex atoms we examine will be supposed to possess electrons represented by the Hamiltonian operator

$$\mathscr{H} = -\frac{\hbar^2}{2\mu_e} \sum_{i=1}^{N} \nabla_{\mathbf{r}_i}^2 - Ze^2 \sum_{i=1}^{N} \frac{1}{r_i} + \frac{e^2}{2} \sum_{i=1}^{N} \sum_{j \neq i}^{N} \frac{1}{r_{ij}}.$$

Moreover, as we have done thus far, the wavefunctions upon which $\mathscr{H}$ operates will be taken to be determinants whose elements are products of position coordinate (with the radial part a solution of the Hartree-Fock equation) and spin wavefunctions. It follows from these assumptions that we may continue to speak reasonably of electronic configurations, even though such states have exact meanings only for hydrogenic atoms.

Suppose a helium atom exists in the (excited) configuration $(1s)(2s)$. Then $L = 0$ and $J = S$, according to the vector model. Because there are two electrons, we have $S = 0, 1$, corresponding to antiparallel and parallel spin directions, respectively. We conclude that a singlet state $^1S_0$ and a triplet state $^3S_1$ are possible for helium in the configuration $(1s)(2s)$. This is, of course, what is observed experimentally. With regard to the relative energies of these states, we must appeal to the analysis given in Chapter 7 as general-ized in the form of *Hund's rule*:

Energy levels of the greatest spin multiplicity lie lowest; of these, the lowest is that of the greatest total orbital angular momentum.

In the case of $(1s)(2s)$, then, the triplet state is of lower energy.

It should be understood that the foregoing discussion can be applied to any $(ns)(n + 1)s$ configuration outside completed groups in any atom; for, as was pointed out in section 8.2 *completed groups create a spherically sym-metric average potential energy*. It follows, then, that a singlet and a triplet state are also to be expected for beryllium in the $(1s)^2 2s 3s$ configuration, magnesium in the $(1s)^2(2s)^2(2p)^6 3s 4s$ configuration, and calcium in the $(1s)^2(2s)^2(2p)^6(3s)^2(3p)^6 4s 5s$ configuration. Each of these expectations is in agreement with experiment.

Now let us go a step farther and consider the configuration *closed shells* + $(ns)(n + 1)p$. Here we have $L = 1$ and $S = 0, 1$, so that $J = 0, 1, 2$. The states permitted this system are accordingly $^3P_0$, $^3P_1$, $^3P_2$, and $^1P_1$. Of these, the triplet states are of lower energy.

When two electrons possess the same quantum numbers n and l, they are said to be *equivalent*. Equivalent electrons pose a special problem in working out states because of the restrictions imposed by the exclusion postulate. For example, let us take a look at the groundstate configuration *closed shells* + $(2p)^2$ (the carbon atom). In this case $L = 2, 1, 0$; $S = 0, 1$; and $J = 0, 1, 2, 3$. If the electrons were not equivalent, the states permitted this system would be $^1S_0$, $^3S_1$, $^1P_1$, $^3P_2$, $^3P_1$, $^3P_0$, $^1D_2$, $^3D_3$, $^3D_2$, $^3D_1$. But not all of these states are permitted by the exclusion postulate. It turns out[5] that the only states permitted the $2p$ electrons are $^3P_0$, $^3P_1$, $^3P_2$, $^1S_0$, and $^1D_2$. As regards energy, the ^3P states are the lowest, followed by the $^1D_2$ and the $^1S_0$.

We see that the vector model and the exclusion postulate can provide all the observed states of two-electron configurations in "effectively light" atoms. Indeed, the method is also useful when three or more electrons must be considered. Take, as an example, the case of boron. We have *closed shell* + $(2s)^2(2p)$. Because the p-electron is of highest energy, we shall couple the angular momentum vectors of the s-electrons first. Thus we get $L = 0$, $S = 0, 1$, so that the energy states are $^1S_0$ and $^3S_1$. But the exclusion postulate eliminates $^3S_1$. The remaining singlet state may be called the *parent state*. This is to be coupled with the p-electron. From $^1S_0$ we obtain $L = 1$, $S = \frac{1}{2}$ and the allowed states $^2P_{1/2}$ and $^2P_{3/2}$.

[5] See, for example, B. W. Shore and D. H. Menzel, *Principles of Atomic Spectra*, John Wiley and Sons, New York, 1968, pp. 92–102.

We can summarize the building-up rules as:

(a) The atom under consideration must be "effectively light" in the sense that it contains only a few electrons, or, only a few electrons in addition to completed groups.

(b) The two electrons of lowest combined energy have their angular momentum vectors coupled first. All states differing only in the transposition of coordinate labels are called identical; all states corresponding to just one set of quantum numbers n, l, m, m_s are excluded. The remaining states are the parent energy states.

(c) The angular momenta of the electron of next highest energy are then coupled to the angular momenta of the parent states, following the procedure in (b). The same thing is done again until all the electrons have been taken into account.

(d) The final energy states are those permitted the electronic configuration under investigation. The relative order of their energies is deduced using Hund's rule.

In Table 9.1 are listed the permitted energy states for several electronic configurations. One property of the configurations of equivalent electrons is

Table 9.1. Allowed states in the Russell-Saunders approximation for one- and two-electron configurations.

Orbital	Allowed States
s	^2S
s^2	^1S
ss	^1S, ^3S
sp	^1P, ^3P
sd	^1D, ^3D
sf	^1F, ^3F
p (or p^5)	^2P
p^2 (or p^4)	^1S, ^3P, ^1D
pp	^1S, ^3S, ^1P, ^3P, ^1D, ^3D
pd	^1P, ^3P, ^1D, ^3D, ^1F, ^3F
pf	^1D, ^3D, ^1F, ^3F, ^1G, ^3G
d (or d^9)	^2D
d^2 (or d^8)	^1S, ^3P, ^1D, ^3F, ^1G
dd	^1S, ^3S, ^1P, ^3P, ^1D, ^3D, ^1F, ^3F, ^1G, ^3G
df	^1P, ^3P, ^1D, ^3D, ^1F, ^3F, ^1G, ^3G, ^1H, ^3H
f (or f^{13})	^2F
f^2 (or f^{12})	^1S, ^3P, ^1D, ^3F, ^1G, ^3H, ^1I
ff	^1S, ^3S, ^1P, ^3P, ..., ^1I, ^3I

apparent: *the energy states permitted the configuration* $(nl)^k$ *are the same as those permitted* $(nl)^{2(2l+1)-k}$. In a closed shell there are equal numbers of (m_l, m_s) and $(-m_l, -m_s)$ electrons, since all closed shells are spherically symmetrical. But, this, in turn, means that the coupling of the angular momenta of $(nl)^k$ to $(nl)^{2(2l+1)-k}$ must yield zero for the resultant J, L, and S; that is, the two configurations each possess the same values of L and S.

9.7. THE PERIODIC TABLE OF ELEMENTS

The building-up rules contain a very important idea concerning atomic spectra which leads to the general classification of chemical elements known as *the periodic table*. The idea is based upon the spherical symmetry of the coulomb field engendered by the nucleus and completed groups of electrons. It may be put: *the emission spectrum for an atom possessing M electrons outside closed shells is similar to that for an M-electron atom.* Empirical evidence for this statement is given in Figure 50, where the energy levels of hydrogen $(1s)$ and lithium $((1s)^2 2s)$, and helium $((1s)^2)$ and beryllium $((1s)^2 (2s)^2)$ are compared. The lower-lying levels of lithium and beryllium show a significant deviation from what is expected because electrons in the corresponding orbitals reap little of the effect of spherical symmetry in the closed shell. However, in general, we should find that the spectra of "effectively light" atoms appear much as the building-up rules would predict.

Let us suppose now that a classification of the chemical elements according to the structures of their atoms can be made using as the criterion for similarity the comparison of emission spectra. In the first group we would place the "effectively one-electron" configurations $(1s)$, $(1s)^2(2s)$, $(1s)^2(2s)^2$ $(2p)^6(3s)$, $(1s)^2(2s)^2(2p)^6(3s)^2(3p)^6(4s)$, and so on. These configurations are the groundstates of hydrogen, lithium, sodium, and potassium. [In the case of the last-named element we have supplied a $(4s)$ electron instead of a $(3d)$ electron because the emission spectrum of potassium demands this action.[6]] The next group would contain the "effectively two-electron" atoms helium, beryllium, magnesium, calcium, strontium, barium, and radium, so that we may write the table

H		He
Li	Be	
Na	Mg	
K	Ca	
Rb	Sr	
Cs	Ba	
Fr	Ra	

[6] See, for example, G. Herzberg, *Atomic Spectra and Atomic Structure*, Dover Publications, New York, 1944, p. 71f.

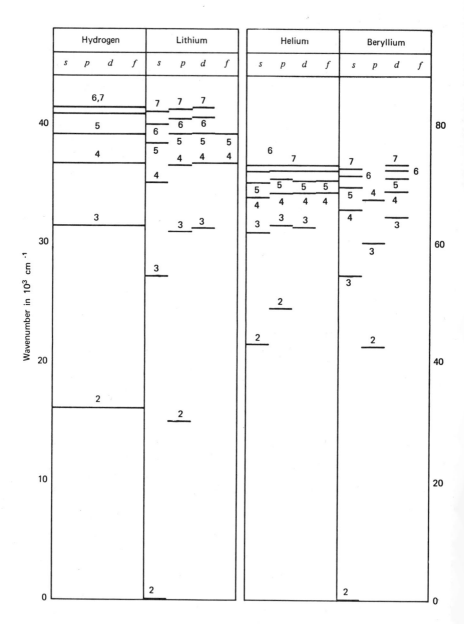

Figure 50. A comparison of the energy level schemes for "effectively one-electron" and "effectively two-electron" atoms. The levels are referred to the groundstate energies of lithium and beryllium, respectively, as zero.

The helium atom has been placed far off to the right for a reason to be made clear very shortly. The "effectively three-electron" atoms are not so easy to classify as were their predecessors. This is because of the competition for orbitals between the $4s$ and $3d$, $5s$ and $4d$, $6s$, $5d$, and $4f$, and $7s$, $6d$, and $4f$ electrons. The spectra of scandium, yttrium, lanthanum, and actinium have to be assigned the configuration *closed shells* $+ (nd)[(n + 1)s]^2$—which really makes them quite different from boron, aluminum, gallium, indium, and thallium, all of which are of the configuration *closed shells* $+ (ns)^2(np)$. In between calcium and gallium and strontium and indium, then, are ten elements whose configurations differ from scandium or yttrium by an increasing number of $(3d)$ or $(4d)$ electrons. These are known as *transition elements* because they represent a transition between the "effectively two-electron" configuration *closed shells* $+ (ns)^2$ and the *normal* "effectively three-electron" configuration *closed shells* $+ (ns)^2(np)$. The transition, of course, is by way of d electrons. Between lanthanum and thallium and after actinium there appears a transition within a transition. First, there is a transition from lanthanum to hafnium by way of $4f$ electrons and one from actinium (thorium to lawrencium) by way of $5f$ electrons. The fourteen elements in each transition are called *rare earths*. After the rare earths, there are nine elements, beginning with hafnium, which are transition elements by virtue of increasing numbers of $5d$ electrons. All these transitions may be summarized in the table:

H						He
Li	Be				B	Ne
Na	Mg				Al Groups Ar 4,5,6,7	
K	Ca	Sc + other transition elements	Cu	Zn	Ga	Kr
Rb	Sr	Y + other transition elements	Ag	Cd	In	Xe
Cs	Ba	La + other transition elements	Au	Hg	T	Rn
Fr	Ra	Ac Th + other rare earths				
		Ce + other rare earths				

Before going on, it is worthwhile to mention that the competition between nd and $(n + 1)s$ electrons gets very keen just at the end of each string of transition elements. As a result, copper, silver, and gold each have the groundstate configuration *closed shells* $+ (nd)^{10}[(n + 1)s]$ instead of the *closed shells* $+ (nd)^9[(n + 1)s]^2$ expected because of their places in the transition sequence. This means that they are "effectively one-electron" atoms and so are partners of the first group of elements in the left-most column of our table. Similarly, zinc, cadmium, and mercury are of the configuration *closed shells* $+ (nd)^{10}[(n + 1)s]^2$ and are accordingly "effectively two-electron" atoms.

The fourth, fifth, sixth, and seventh groups do not show any anomalies. The eighth group, on the other hand, has a very special property. Because of the way we have had to fill atomic orbitals, the atoms in the eighth group each must have the configuration *closed shells* $+ (ns)^2(np)^6$. *They are not "effectively eight-electron" atoms at all, but consist entirely of closed shells.* In this sense they are said to represent end-members in *periods* which comprise all the chemical elements between those in the first group and those in the eighth. Because helium contains only closed shells, it is placed at the end of the first period, as the first member of the eighth group. In this instance its spectrum characteristics are given a second place to its overall configuration. The second period goes from lithium to neon; the third, from sodium to argon; the fourth, from potassium through the transition elements to krypton, and so on. If we wish, advantage can be taken of the "period" notion to create a better notation for configurations. Any atom in the nth period can be expressed as [end-member of $(n - 1)$th period] (orbitals outside the $[(n - 1)p]$ electrons). For example, lithium may be expressed in this notation as [He]$(2s)$; magnesium, as [Ne]$(3s)^2$; scandium, as [Ar]$(3d)(4s)^2$, and iodine, as [Kr]$(4d)^{10}(5s)^2(5p)^5$.

The importance of our classification of the chemical elements would not be nearly so great (especially to chemists!) if it did not have some relevance to chemical behavior. The connection between chemistry and atomic structure is not obvious, however, but is based on an hypothesis, made by G. N. Lewis in 1916, which states that *the chemical bond joining two atoms may be thought of as a sharing of a pair of electrons.* It follows that the behavior of a given element will depend to a large extent on how easily its outer electrons can be disturbed. We can evaluate that property of an atom, of course, by taking a look at its energy levels. In general, it is expected that an electron outside closed shells will be of much larger energy than one inside a closed shell because the former moves in a more highly screened coulomb field than does the latter. Therefore, electrons outside closed shells should be easily perturbed, or even removed, in the presence of electrons from another atom. On the other hand, those elements at the ends of periods should be quite resistant to such perturbations. These expectations may be summarized by saying that *elements whose electronic configurations comprise complete groups will not form chemical bonds easily, while elements whose configurations comprise incomplete groups will be chemically reactive.* It is well known that this rule has a general validity. One corollary to the rule follows from the idea which we used to found the periodic table: *chemical elements in the same group will exhibit similar properties.* In other words, all the "effectively one-electron" species should behave alike, as should all the "effectively two-electron" species, and so on. This prediction is borne out by experience primarily for elements whose configurations are free from s-d-or s-d-f electron competition; that is, elements outside the transition and rare earth series.

In Figure 51 are shown plots of the first ionization potential and the heat of vaporization *versus* atomic number for several periods, exclusive of the transition and rare earth elements. It is quite clear from the figure that spectroscopic periodicity and chemical periodicity are equivalent concepts.

PROBLEMS

1. Demonstrate that the transformations of an equilateral triangle

$T(I)$ = no change in orientation

$T(A)$ = rotation through 120° from left to right

$T(B)$ = rotation through 240° from left to right

$T(C)$ = reflection about an altitude

$T(D)$ = reflection about an altitude

$T(E)$ = reflection about an altitude

are the realization of a group. Is the group abelian?

2. Show that the symmetry group of the equilateral triangle, introduced above, is isomorphic to the symmetric group of order six.

3. Consider the set of all real, positive integers $0, 1, \ldots, N - 2, N - 1$.

(a) Show that, if group multiplication is defined to be "addition (mod N),"

$$A + B = (A + B) \qquad \text{if } (A + B) < N,$$
$$A + B = (A + B) - N \quad \text{if } (A + B) \geqslant N, \qquad (0 \leqslant A, B < N)$$

these integers realize a group.

(b) Show that the group is abelian.

4. Make up a multiplication table for the group realized by the rotations of a square about its center point such that the original orientation of the square is preserved.

(a) What is the order of the group?

(b) Which are the subgroups? Are any invariant?

5. Write down the group table for the group of order two. Show that this group is isomorphic to the inversion group.

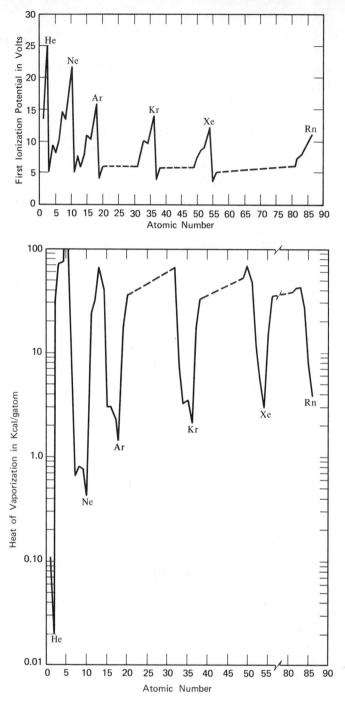

Figure 51. An illustration of structural and chemical periodicity.

6. Prove that every conjugate class of an abelian group contains only one element of the group.

7. Prove that, except for the conjugate class of the identity element, no class is a subgroup.

8. Prove that all elements in a conjugate class have the same order.

9. Suppose that A and B are members of an infinitesimal group represented by

$$M(A) = M(I) + \epsilon M_A$$
$$M(B) = M(I) + \epsilon M_B$$

where $M(I)$ is the unit matrix, M_A and M_B are matrices, and ϵ is a positive number of very small magnitude. Show that A and B commute to within terms of order ϵ^2.

10. Show that the symmetric group of order six may be assigned the irreducible representation

$$\begin{aligned}
M(I) &= (1) & M(C) &= (-1) \\
M(A) &= (1) & M(D) &= (-1) \\
M(B) &= (1) & M(E) &= (-1)
\end{aligned}$$

11. What is the relationship between the characters of a representation and the trace of a matrix (as defined in problem 2 of Chapter 6)?

12. Let us see if we can construct a part of the proof of the orthonormality of irreducible unitary representations. Consider a matrix of the form

$$M \equiv \sum_{\{A\}} U_M^{(l)}(A) \times P \times U_M^{(m)-1}(A)$$

where P is a matrix possessing L_l rows and L_m columns. (Assume $l \neq m$.)
  (a) Prove that

$$M = \sum_{\{A\}} U_M^{(l)}(A \cdot B) \times P \times U_M^{(m)-1}(A \cdot B).$$

  (b) Prove that

$$U_M^{(l)}(B) \times M = M \times U_M^{(m)}(B) \quad (l \neq m)$$

using the alternative forms of M in the order presented.

  (c) A theorem known as *Schur's Lemma* states that any matrix which satisfies the equality in (b) is a null matrix. Use this result to obtain an appropriate form of equation (9.6). (You must choose a form for the matrix P.)

13. Verify the orthogonality of the characters of the irreducible representations of the symmetric group of order six given in section 9.2 and Problem 10.

14. Show that

$$\psi_m^{(j)}(\mathbf{r}'^N, s'^N) = \sum_{k=1}^{L} [M(A^{-1})]_{kj}\psi_m^{(k)}(\mathbf{r}^N, s^N)$$

if

$$O_A\psi_m^{(j)}(\mathbf{r}'^N, s'^N) = \psi_m^{(j)}(\mathbf{r}^N, s^N).$$

15. Consider the irreducible representation of the pure rotation group in terms of the Euler angles. Show that
   (a) $D^{(0)}(\phi, \theta, \psi) = (I)$
   (b)

$$D^{(1)}(\phi, \theta, \psi) = \begin{pmatrix} e^{-i\phi}\dfrac{1+\cos\theta}{2}e^{-i\psi} & \dfrac{-e^{-i\phi}\sin\theta}{\sqrt{2}} & e^{-i\phi}\dfrac{1-\cos\theta}{2}e^{i\psi} \\[3mm] \dfrac{\sin\theta}{\sqrt{2}}e^{-i\psi} & \cos\theta & \dfrac{-\sin\theta}{\sqrt{2}}e^{i\psi} \\[3mm] e^{i\phi}\dfrac{1-\cos\theta}{2}e^{-i\psi} & e^{i\phi}\dfrac{\sin\theta}{\sqrt{2}} & e^{i\phi}\dfrac{1+\cos\theta}{2}e^{i\psi} \end{pmatrix}$$

16. Show that the parity of a multielectron system represented by a determinantal wavefunction which is a solution of the Hartree-Fock equation is given by

$$P_{Rf} = \prod_{i=1}^{N} (-1)^{l_i}$$

where l_i is the orbital angular momentum quantum number for the ith electron.

17. After equation (9.24) it was stated that the representation of the group of the Schrödinger equation is the direct product of the representations of the group members. Show that this is the case by directly computing, for example,

$$O_R O_{Rf}\psi(\mathbf{r}'''^N, s'''^N) = \psi(\mathbf{r}'^N, s'^N)$$

where

$$O_{Rf}\psi(\mathbf{r}'''^N, s'''^N) = \psi(\mathbf{r}''^N, s''^N)$$
$$O_R\psi(\mathbf{r}''^N, s''^N) = \psi(\mathbf{r}'^N, s'^N)$$

in terms of the elements of representation matrices.

18. Using group-theoretical arguments, prove that the perturbation Hamiltonian operator

$$\mathscr{H}' = \frac{1}{2}\sum_{i=1}^{N}\sum_{j\neq i}^{N}\frac{e^2}{r_{ij}}$$

will not alter the parity of the wavefunction for a multielectron hydrogenic atom.

19. *Quadrupole* radiation from an atom is determined by matrix elements of the form

$$(\psi_{m'}^{(k')},\ (x_i \cdot x_l)\psi_m^{(k)}).$$

   (a) Using the fact that $(x_i \cdot x_l)$ belongs to $D^{(2)}(R)$, deduce the angular momentum selection rules.
   (b) Deduce the parity selection rule.

20. The groundstates permitted the singly-ionized carbon atom ($Z = 6$) are $^2P_{\frac{1}{2}}$ and $^2P_{\frac{3}{2}}$, according to experiment.
   (a) Does this result follow from the building-up rules?
   (b) What are the relative energies associated with these states?

21. What should copper, silver, and gold have in common with hydrogen, lithium, and sodium, according to the building-up rules?

22. Prove the statement: The groundstates of atoms possessing only complete groups are always $^1S_0$ states.

CHAPTER 10

Molecular Spectra

Round about the center, in and out, round about.

10.1. ROTATIONAL AND VIBRATIONAL SPECTRA

A molecule under bombardment by photons may be expected to be set in motion. In general, the motion will be a complicated set of translations upon which rotations and vibrations have been superposed; nonetheless, we expect these motions to be comprehensible within a scheme deducible from the postulates of quantum mechanics. Our experience with atoms, moreover, leads us not to foresee a rotating, oscillating molecule which passes unrestricted through all imaginable states but instead to visualize this system as subject to whatever is the structure of the appropriate Schrödinger eigenvalue problem. This indeed does turn out to be the case, as may be seen from Figure 52, where a portion of the absorption spectrum for hydrogen chloride vapor is shown. We note from the figure that although a noticeable gap does exist between adjacent absorptions, as expected, the latter do not correspond to

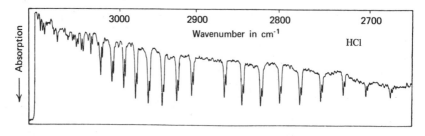

Figure 52. A part of the absorption spectrum for hydrogen chloride gas at 20 torr pressure. [After E. K. Plyler, *et al.*, Vibration-rotation structure in absorption bands for the calibration of spectrometers from 2 to 16 microns, *J. Res. Nat. Bureau of Standards*, 64A, No. 1 (1960).]

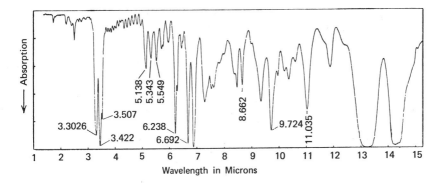

Figure 53. The infrared absorption spectrum for a 50-micron thickness of polystyrene film. [After E. K. Plyler, *et al.*, Reference wavelengths for calibrating prism spectrometers, *J. Res. Nat. Bureau of Standards*, 58A, No. 4 (1958).]

single wavelengths, as was so for atomic spectra. The reason for this lies more with the accuracy of the spectral measurement and the effect of environment (temperature and pressure) than with the quantum physical properties of the molecule.

If the translations are excluded from consideration, molecular spectra may be construed as the results of pure rotations, rotation-vibration motions, and pure vibrations. The rotations are engendered by microwave radiation, the rotation-vibration combinations by infrared radiation, and the vibrations by infrared radiation. Figure 52 is actually an example of a rotation-vibration spectrum; in Figure 53 we have part of a vibration spectrum—that of polystyrene.

10.2. THE BORN-OPPENHEIMER THEOREM

Before beginning a discussion of the quantum physics of molecular spectra, we might well ask a little about the nature of the chemical bond itself. We have seen how it is that atoms hold themselves together; now we should like to know the way atomic structure and Lewis' hypothesis, mentioned in Chapter 9, are combined to bring about molecules.

Immediately we are confronted with the rather significant complexity of the molecule. How can we visualize a "bond" between atoms in a molecule when the latter is described by a many-body Schrödinger eigenvalue problem involving the motions of many interacting nuclei and electrons? Indeed, can we expect the atoms to maintain their structural integrity when combined? Fortunately, a very great simplification, known as the Born-Oppenheimer

Theorem, may be made in our problem. The content of this simplification is that it is rigorously possible to suppress the electronic interactions while considering the motions of the nuclei in a molecule, and the converse. By this means the chemical bond shows up only as a parameter in the theory of molecular motion, and may be regarded even as a purely empirical entity. Conversely, when considering the theory of the chemical bond, nuclear motion may be ignored. Speaking physically, such a partitioning of the electron motions from those of the nuclei would appear to be the logical result of the much greater rapidity of the former. While the electrons are making many, many revolutions about the nuclei, the latter, because of their rather large masses, are carrying out only a few vibrations and only a very few rotations about the center of mass of the molecule. For example, in the case of the hydrogen molecule, we might say

$$\text{electronic energy} = O\left(\frac{-e^2}{a_0}\right),$$

in the groundstate, there being two electrons involved. However, if the two nuclei are displaced by $(a_0/2)$ from their equilibrium positions, we have

$$\text{vibrational energy} = O(M\omega^2 a_0^2),$$

where M is the mass of a nucleus and ω is the frequency of vibration. Such a displacement pulls the electrons the distance a_0 apart, so that

$$\text{vibrational energy} = O\left(\frac{e^2}{a_0}\right)$$

and

$$\omega = O\left(\frac{e}{(Ma_0^3)^{1/2}}\right) = O\left[\left(\frac{\mu_e}{M}\right)^{1/2}\frac{\mu_e e^4}{\hbar^3}\right]$$

upon noting the definition of the Bohr radius. On the other hand,

$$|\text{electronic energy}| = O\left(\frac{\mu_e e^4}{\hbar^2}\right)$$

which is to say

$$\left|\frac{\text{vibrational energy}}{\text{electronic energy}}\right| = O\left[\left(\frac{\mu_e}{M}\right)^{1/2}\right] = O(10^{-2}).$$

For the rotational energy we should find that the ratio is about a hundred-fold smaller.

The problem now is to provide a more rigorous basis for the conclusion just drawn on physical grounds. To do this we shall prove a theorem first given by M. Born and J. R. Oppenheimer in 1927. Our method of proof differs from theirs and from the more rigorous kinds based on the so-called Adiabatic

Theorem,[1] but possesses the advantages of conciseness and an obvious physical content.

The Born-Oppenheimer Theorem. In the first approximation, a solution of the Schrödinger eigenvalue problem constructed from the Hamiltonian operator

$$\mathcal{H} = -\sum_{i=1}^{N} \frac{\hbar^2}{2m_e} \nabla_i^2 - \sum_{k=1}^{M} \frac{\hbar^2}{2M_k} \nabla_k^2 + \frac{1}{2} \sum_{i=1}^{N} \sum_{\substack{j=1 \\ (j \neq i)}}^{N} \frac{e^2}{r_{ij}}$$

$$- \sum_{i=1}^{N} \sum_{k=1}^{M} \frac{Z_k e^2}{|\mathbf{r}_i - \mathbf{R}_k|} + \frac{1}{2} \sum_{k=1}^{M} \sum_{\substack{l=1 \\ (l \neq k)}}^{M} \frac{Z_k Z_l e^2}{R_{kl}}$$

for a molecule comprising N electrons and M nuclei, where $\mathbf{r}_i$ and $\mathbf{R}_k$ are the displacements of the ith electron and the kth nucleus from some reference point, respectively, is

$$\Psi_{n,m}(\mathbf{r}^N, \mathbf{R}^M) = \varphi_n(\mathbf{r}^N, \mathbf{R}^M)\psi_{n,m}(\mathbf{R}^M),$$

where n designates an electronic energy state, m designates a nuclear energy state, and $\varphi_n(\mathbf{r}^N, \mathbf{R}^M)$ and $\psi_{n,m}(\mathbf{R}^M)$ are the solutions of

$$-\sum_{i=1}^{N} \frac{\hbar^2}{2m_e} \nabla_i^2 \varphi_n + \left\{ \frac{1}{2} \sum_{i=1}^{N} \sum_{\substack{j=1 \\ (j \neq i)}}^{N} \frac{e^2}{r_{ij}} - \sum_{i=1}^{N} \sum_{k=1}^{M} \frac{Z_k e^2}{|\mathbf{r}_i - \mathbf{R}_k|} \right.$$

$$\left. + \frac{1}{2} \sum_{k=1}^{M} \sum_{\substack{l=1 \\ (l \neq k)}}^{M} \frac{Z_k Z_l e^2}{R_{kl}} \right\} \varphi_n(\mathbf{r}^N, \mathbf{R}^M) = U_n(\mathbf{R}^M)\varphi_n(\mathbf{r}^N, \mathbf{R}^M)$$

$$-\sum_{k=1}^{N} \frac{\hbar^2}{2M_k} \nabla_k^2 \psi_{n,m} + U_n(\mathbf{R}^M)\psi_{n,m}(\mathbf{R}^M) = E_{n,m}\psi_{n,m}(\mathbf{R}^M).$$

Proof. It should be noticed from the outset that the Hamiltonian operator $\mathcal{H}$ is not the complete Hamiltonian operator for the molecule since it does not contain spin-dependent terms. Because of this omission, we must acknowledge that we are proving a theorem strictly applicable only to "effectively light" atoms. This is pertinent because our method of proof will be taken over directly from the procedure used to derive the Hartree equation in Chapter 8.

---

[1] See, for example, A. Messiah, *Quantum Mechanics*, John Wiley and Sons, New York, 1962, Vol. II, pp. 781–792.

Molecular Spectra

Let us begin by computing the mean value of $\mathscr{H}$ in the state $\Psi_{n,m}(\mathbf{r}^N, \mathbf{R}^M)$. We have, in units of $((\Psi_{n,m}, \Psi_{n,m})_\mathbf{r})_\mathbf{R}$,

$$
\begin{aligned}
\langle \mathscr{H} \rangle_{n,m} &= ((\Psi_{n,m}, \mathscr{H}\Psi_{n,m})_\mathbf{r})_\mathbf{R} \\
&= -\sum_{i=1}^{N} \frac{\hbar^2}{2m_e} (\varphi_n, \nabla_i^2 \varphi_n)_\mathbf{r} (\psi_{n,m}, \psi_{n,m})_\mathbf{R} - \sum_{k=1}^{M} \frac{\hbar^2}{2M_k} \\
&\quad \times [((\varphi_n \psi_{n,m}, \nabla_k^2 \psi_{n,m} \varphi^n)_\mathbf{r})_\mathbf{R} + 2(\psi_{n,m}(\varphi_n, \nabla_k \varphi_n)_\mathbf{r} \cdot \nabla_k \psi_{n,m})_\mathbf{R} \\
&\qquad\qquad + (\psi_{n,m}(\varphi_n, \nabla_k^2 \varphi_n)_\mathbf{r} \psi_{n,m})_\mathbf{R}] \\
&\quad + \frac{1}{2} \sum_{\substack{i=1 \\ }}^{N} \sum_{\substack{j=1 \\ (j \neq i)}}^{N} \left((\psi_{n,m}\varphi_n, \frac{e^2}{r_{ij}} \varphi_n \psi_{n,m})_\mathbf{r} \right)_\mathbf{R} \\
&\quad - \sum_{i=1}^{N} \sum_{k=1}^{M} \left((\varphi_n \psi_{n,m}, \frac{Z_k e^2}{|\mathbf{r}_i - \mathbf{R}_k|} \varphi_n \psi_{n,m})_\mathbf{r} \right)_\mathbf{R} \\
&\quad + \frac{1}{2} \sum_{\substack{k=1 \\ }}^{M} \sum_{\substack{l=1 \\ (l \neq k)}}^{M} \left((\varphi_n \psi_{n,m}, \frac{Z_k Z_l e^2}{R_{kl}} \varphi_n \psi_{n,m})_\mathbf{r} \right)_\mathbf{R}
\end{aligned} \tag{10.1}
$$

We wish to minimize $\langle \mathscr{H} \rangle_{n,m}$ by varying the $\psi_{n,m}(\mathbf{R}^M)$. To this end we write

$$
\delta\langle \mathscr{H} \rangle_{n,m} - E_{n,m}\delta(\psi_{n,m}, \psi_{n,m}) = 0, \tag{10.2}
$$

where $E_{n,m}$ is a constant of proportionality. By carrying out the variations we find

$$
\begin{aligned}
\int_{\mathbf{R}^M} \delta\psi_{n,m} \Bigg\{ &-\sum_{i=1}^{N} \frac{\hbar^2}{2m_e} (\varphi_n, \nabla_i^2 \varphi_n)_\mathbf{r} \\
&- \sum_{k=1}^{M} \frac{\hbar^2}{2M_k} [(\varphi_n, \varphi_n)_\mathbf{r} \nabla_k^2 \psi_{n,m} + 2(\varphi_n, \nabla_k \varphi_n)_\mathbf{r} \cdot \nabla_k \psi_{n,m} + (\varphi_n, \nabla_k^2 \varphi_n)_\mathbf{r} \psi_{n,m}] \\
&+ \frac{1}{2} \sum_{\substack{i=1 \\ }}^{N} \sum_{\substack{j=1 \\ (j \neq i)}}^{N} \left(\varphi_n, \frac{e^2}{r_{ij}} \varphi_n \right)_\mathbf{r} \psi_{n,m} - \sum_{i=1}^{N} \sum_{k=1}^{M} \left(\varphi_n, \frac{Z_k e^2}{|\mathbf{r}_i - \mathbf{R}_k|} \varphi_n \right)_\mathbf{r} \psi_{n,m} \\
&+ \frac{1}{2} \sum_{\substack{k=1 \\ }}^{M} \sum_{\substack{l=1 \\ (l \neq k)}}^{M} \frac{Z_k Z_l e^2}{R_{kl}} (\varphi_n, \varphi_n)_\mathbf{r} \psi_{n,m} - E_{n,m}\psi_{n,m} \Bigg\} \, d\mathbf{R}^M = 0
\end{aligned} \tag{10.3}
$$

Equation (10.3) can be simplified to a large extent by making the *assumption* that, in the first approximation,

$$
\nabla_k \varphi_n = 0 \qquad (k = 1, \ldots, M).
$$

This assumption must be verified in every case wherein the Born-Oppenheimer theorem is applied. In general, however, we shall find that the gradient of $\varphi_n(\mathbf{r}^N, \mathbf{R}^M)$ with respect to $\mathbf{R}_k$ is quite small.[1] We shall proceed as if this were the case. Now, equation (10.3) will hold in general only if the coefficient of $\delta\psi_{n,m}$ vanishes identically. Therefore, we must have it that

$$-\sum_{k=1}^{M} \frac{\hbar^2}{2M_k} \nabla_k^2 \psi_{n,m} + [U_n(\mathbf{R}^M) - E_{n,m}]\psi_{n,m}(\mathbf{R}^M) = 0, \qquad (10.4)$$

where we have defined

$$U_n(\mathbf{R}^M)(\varphi_n, \varphi_n)_\mathbf{r} \equiv -\sum_{i=1}^{N} \frac{\hbar^2}{2m_e} (\varphi_n, \nabla_i^2 \varphi_n)_\mathbf{r} + \frac{1}{2}\sum_{i=1}^{N}\sum_{\substack{j=1 \\ (j \neq i)}}^{N} \left(\varphi_n, \frac{e^2}{r_{ij}} \varphi_n\right)_\mathbf{r}$$

$$-\sum_{i=1}^{N}\sum_{k=1}^{M} \left(\varphi_n, \frac{Z_k e^2}{|\mathbf{r}_i - \mathbf{R}_k|} \varphi_n\right)_\mathbf{r}$$

$$+\frac{1}{2}\sum_{k=1}^{M}\sum_{\substack{l=1 \\ (l \neq k)}}^{M} \frac{Z_k Z_l e^2}{R_{kl}} (\varphi_n, \varphi_n)_\mathbf{r}, \qquad (10.5)$$

and divided through by the constant $(\varphi_n, \varphi_n)_\mathbf{r}$. The formal structure of equation (10.5) leads us immediately to rewrite it as

$$-\sum_{i=1}^{N} \frac{\hbar^2}{2m_e} \nabla_i^2 \varphi_n + \left\{\frac{1}{2}\sum_{i=1}^{N}\sum_{\substack{j=1 \\ (j \neq i)}}^{N} \frac{e^2}{r_{ij}} - \sum_{i=1}^{N}\sum_{k=1}^{M} \frac{Z_k e^2}{|\mathbf{r}_i - \mathbf{R}^k|}\right.$$

$$\left.+\frac{1}{2}\sum_{k=1}^{M}\sum_{\substack{l=1 \\ (l \neq k)}}^{M} \frac{Z_k Z_l e^2}{R_{kl}}\right\}\varphi_n = U_n(\mathbf{R}^M)\varphi_n(\mathbf{r}^N, \mathbf{R}^M), \qquad (10.6)$$

which act completes the proof of the theorem. (Remember that $(\varphi_n, \varphi_n)_\mathbf{r}$ is an integration over $\mathbf{r}^N$ only.)

The function $U_n(\mathbf{R}^M)$ is by definition the mean value of the total electronic energy in the space of the electronic coordinates plus the sum of the coulomb energies for each interacting pair of nuclei. Accordingly, the function $\varphi_n(\mathbf{r}^N, \mathbf{R}^M)$ may be construed as an eigenfunction corresponding to $U_n(\mathbf{R}^M)$ (in the space of the $\mathbf{r}^N$) and termed an *electronic wavefunction*. The electronic wavefunction presumably may be found by solving equation (10.6), for a given set of $\mathbf{R}^M$, with the method of the self-consistent field. When this is done for all possible nuclear configurations, $U_n(\mathbf{R}^M)$ may be introduced into equation (10.4) which, in turn, yields the $\psi_{n,m}(\mathbf{R}^M)$. Equation (10.4) has the formal appearance of a Schrödinger eigenvalue problem if $U_n(\mathbf{R}^M)$ is designated as the internuclear potential function and $E_{n,m}$ is interpreted as the mean value of the total nuclear energy for a given electronic configuration. The $\psi_{n,m}(\mathbf{R}^M)$ are then called *nuclear wavefunctions*. It is clear that the electronic configuration of the molecule (that is, the chemical bond) affects the nuclear wavefunction only through the "parameter" $U_n(\mathbf{R}^M)$, as desired for our treatment of molecular spectra. It is also apparent that, as with any Schrödinger eigenvalue problem, $U_n(\mathbf{R}^M)$ need not be deduced theoretically, but instead could be derived from experiment or physical intuition.

10.3. THE CHEMICAL BOND

Let us look now at the quantum mechanical content of Lewis' hypothesis. Rather than deal with the most general possibility for a chemical bond, we shall restrict our analysis to a consideration of the simplest combination of two atoms—the hydrogen molecule. In doing so, we shall discover that nearly all the essential elements of bond theory are exposed.

Equation (10.6) applied to the hydrogen molecule has the form

$$-\tfrac{1}{2}\nabla_1^2\varphi_n - \tfrac{1}{2}\nabla_2^2\varphi_n + \left(\frac{1}{r_{12}} - \frac{1}{r_{A1}} - \frac{1}{r_{B1}} - \frac{1}{r_{A2}} - \frac{1}{r_{B2}} + \frac{1}{R_{AB}}\right)\varphi_n$$
$$= U_n(R_{AB})\varphi_n, \quad (10.7)$$

where we have written A and B to designate the protons,

$$r_{A1} = |\mathbf{r}_1 - \mathbf{R}_A|,$$

and so on, and we have cast the equation in atomic units. It is clear that equation (10.7) cannot be solved exactly. There are several ways we might get an approximate solution; the one we shall pursue has the merit of being physically intuitive. Supose we define

$$\mathcal{H}_i \equiv -\tfrac{1}{2}\nabla_i^2 - \frac{1}{r_{Ai}} - \frac{1}{r_{Bi}} + \frac{1}{R_{AB}},$$

so that equation (10.7) may be rewritten

$$\mathcal{H}_1\varphi_n + \mathcal{H}_2\varphi_n + \left(\frac{1}{r_{12}} - \frac{1}{R_{AB}}\right)\varphi_n = U_n(R_{AB})\varphi_n. \quad (10.8)$$

The $\mathcal{H}_i$ are Hamiltonian operators for a single electron in the field of two protons to which have been added repulsive interproton potential functions. Therefore,

$$\mathcal{H}_i\psi(\mathbf{r}_i, R_{AB}) = E_i(R_{AB})\psi(\mathbf{r}_i, R_{AB}) \quad (10.9)$$

is the electronic part of the Schrödinger eigenvalue problem, in the Born-Oppenheimer approximation, for H_2^+, the hydrogen molecular ion. If we regard the third term in equation (10.8) as a perturbation, then a starting point for our theory of the hydrogen molecule will be a description of the hydrogen molecular ion, through equation (10.9). It follows that the zeroth-order wavefunction for the hydrogen molecule will be

$$\varphi_n^{(0)}(\mathbf{r}_1, \mathbf{r}_2, R_{AB}) = \psi_n(\mathbf{r}_1, R_{AB})\psi_n(\mathbf{r}_2, R_{AB}).$$

In this way we are saying that the hydrogen molecule is a blend of two one-electron problems plus a perturbation, just as we said the helium atom was in Chapter 5. If it worked so well once, why not try it again?

The eigenvalue problem (10.9), amazingly enough, is separable if we transform to prolate spheroidal coordinates. These new coordinates are defined by [2]

$$\xi \equiv \frac{(r_A + r_B)}{R_{AB}} \qquad (1 \leqslant \xi < \infty)$$

$$\eta \equiv \frac{(r_A - r_B)}{R_{AB}} \qquad (-1 \leqslant \eta \leqslant 1)$$

with the additional coordinate ϕ, the angle of rotation about the molecular axis. (We have suppressed the subscript i for convenience.) Figure 54 shows these coordinates. The relation between prolate spheroidal and cartesian coordinates is, in the notation of section 4.2,

$$Q_\xi = \frac{R_{AB}}{2}\left(\frac{\xi^2 - \eta^2}{\xi^2 - 1}\right)^{1/2} \qquad Q_\eta = \frac{R_{AB}}{2}\left(\frac{\xi^2 - \eta^2}{1 - \eta^2}\right)^{1/2}$$

$$Q_\phi = \frac{R_{AB}}{2}\left((\xi^2 - 1)(1 - \eta^2)\right)^{1/2}.$$

Applying these expressions to equation (4.10), we find that equation (10.9) above has the explicit form

$$\frac{\partial}{\partial \xi}\left[(\xi^2 - 1)\frac{\partial \psi}{\partial \xi}\right] + \frac{\partial}{\partial \eta}\left[(1 - \eta^2)\frac{\partial \psi}{\partial \eta}\right] + \left[\frac{1}{(\xi^2 - 1)} + \frac{1}{(1 - \eta^2)}\right]\frac{\partial^2 \psi}{\partial \phi^2}$$

$$+ \frac{R_{AB}}{2}\left[R_{AB}U(R_{AB})(\xi^2 - \eta^2) + 4\xi\right]\psi(\xi, \eta, \phi) = 0, \quad (10.10)$$

where

$$U(R_{AB}) = E(R_{AB}) - \frac{1}{R_{AB}}.$$

[2] See, for example, H. Margenau and G. M. Murphy, *The Mathematics of Physics and Chemistry*, D. Van Nostrand Co., Princeton, 1956, Vol. I, pp. 180–183; 385–387.

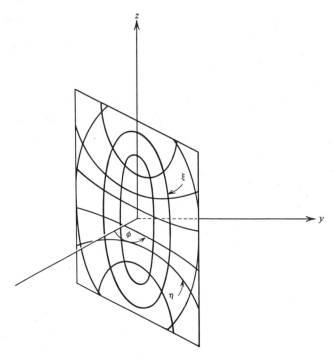

Figure 54. Prolate spheroidal coordinates. Shown here is a section through the coordinate system which makes an angle ϕ, in the xy-plane, about the z-axis. The surfaces ξ = constant are prolate spheroids (ellipses in section) while the surfaces η = constant are hyperboloids of two sheets (hyperbolas in section). A point is specified by its location on a section at the intersection of an ellipse and an hyperbola.

Equation (10.10) is separable into the three equations

$$\frac{d^2\Phi}{d\phi^2} = -m^2\Phi(\phi),$$

$$\frac{d}{d\eta}\left[(1-\eta^2)\frac{dM}{d\eta}\right] + \left(-\beta + p^2\eta^2 - \frac{m^2}{1-\eta^2}\right)M(\eta) = 0,$$

$$\frac{d}{d\xi}\left[(\xi^2-1)\frac{dL}{d\xi}\right] + \left(\beta + 2R_{AB} - p^2\xi^2 - \frac{m^2}{\xi^2-1}\right)L(\xi) = 0,$$

where

$$p^2 = \frac{-R_{AB}^2 U(R_{AB})}{2}$$

284

and β is a separation constant. The first of these equations is well known to us and corresponds to the eigenvalues $m = 0, \pm 1, \pm 2$, etc. The states of the system corresponding to these eigenvalues are labeled with Greek letters. We use σ for $m = 0$, π for $m = \pm 1$, δ for $m = \pm 2$, and so on. Unlike the case of the hydrogen atom, this designation is more than bookkeeping; for, the eigenvalue $U(R_{AB})$ turns out to depend upon m. We note, of course, that the primary physical significance of m is that it indicates that the component of the angular momentum about the molecular axis is a constant of the motion.

The solutions of the second and third equations are not particularly straightforward.[3] Both, in fact, are series solutions whose coefficients must be computed self-consistently with the calculation of the eigenvalues β and p^2. The solutions may be labeled according to the value of m and according to the behavior of $M(\eta)$ and $\Phi(\phi)$ under inversion. Upon reflection of the coordinates about an axis bisecting the molecular axis, $\phi \to \phi + \pi$ and $\eta \to -\eta$. Thus $\Phi(\phi) \to -\Phi(\phi)$ if m is odd and is unchanged otherwise; $M(\eta)$ is either an odd or even function of η. We recognize that these statements are results of the symmetry of the molecular ion. All we are reaffirming is that the inversion operation is a realization of the group of the Schrödinger equation (10.10) and, therefore, that the eigenfunctions must be classified as to which irreducible representation of the inversion group they belong. Eigenstates of $\mathscr{H}_i$ which are even functions of the coordinates ξ, η, and ϕ are denoted by a subscript g; odd functions are denoted by u (g for the German *gerade*, meaning "even"; u for *ungerade*). Thus, we have the states σ_g, σ_u, π_u, π_g, and so on.

The value of $E(R_{AB})$ is shown in Figure 55 for the lowest even and odd states. We note that the even state possesses a minimum at about two atomic units, and that both states correspond to an energy of about -0.5 atomic unit when the nuclei are far apart. The latter behavior is expected since the separated molecular ion is nothing more than a proton and a hydrogen atom (whose groundstate electronic energy is -0.5 atomic unit). The fact that the even state possesses a minimum indicates that it corresponds to a stable bound state of the molecular ion. Because the odd state does not show this behavior, we conclude that it is essentially a repulsive or *antibonding* state. Being odd, this state corresponds to zero probability of finding the electron *exactly* between the two protons and a small probability otherwise. On the other hand, the even state corresponds to a fairly significant probability of finding the electron between the protons. (See Figure 56.) It is in this sense that we have a quantum mechanical verification of Lewis' hypothesis: stable

[3] For a discussion, see J. C. Slater, *Quantum Theory of Molecules and Solids*, McGraw-Hill Book Co., New York, 1963, Vol. I, App. 1.

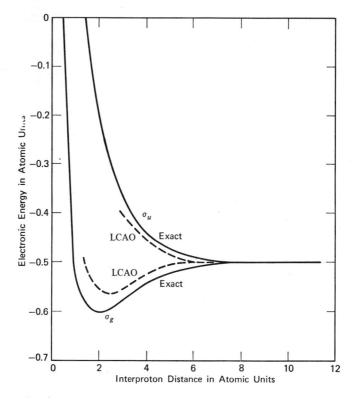

Figure 55. A plot of the electronic energy against interproton distance for the hydrogen molecular ion in the σ_g and σ_u states.

molecules possess significant amounts of electronic charge between their bonded nuclei.

For large values of R_{AB}, the even and odd states can be shown to have the asymptotic forms

$$\sigma_g: \psi \sim \exp\left(-R_{AB}\xi/2\right) \cosh\left(R_{AB}\eta/2\right) \propto e^{-r_A} + e^{-r_B},$$

$$\sigma_u: \psi \sim \exp\left(-R_{AB}/2\right) \sinh\left(R_{AB}\eta/2\right) \propto e^{-r_A} - e^{-r_B}.$$

Aside from normalization constants, the wavefunctions given above are just symmetric and antisymmetric combinations of groundstate hydrogenic wavefunctions. These combinations are *exact* for large R_{AB}, but are rather inaccurate at separations of the protons small enough to permit an appreciable overlap of the atomic wavefunctions. Nonetheless, they are simple enough and have such an intuitive appeal ("atoms in molecules") that a great deal of effort has gone into making them acceptable trial or zeroth-order wavefunc-

286

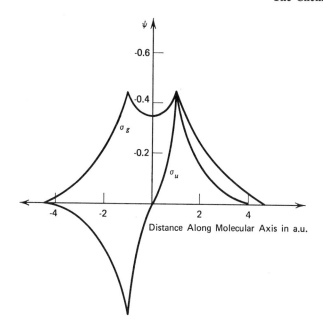

Figure 56. A sketch of the H_2^+ wavefunctions for the σ_g and σ_u states. The origin corresponds to a point midway between the protons, which are here taken to be at the equilibrium separation (2 a. u.).

tions for use in chemical bond theory. In that context, the atomic wavefunctions are termed *atomic orbitals* and the molecular wavefunctions are termed *molecular orbitals*. The practice of using *l*inear *c*ombinations of *a*tomic *o*rbitals as a starting point is called, obviously enough, the LCAO approximation.

We might take a look at the LCAO approximation for H_2^+ in order to get a better feeling for what it entails. To begin with, we shall normalize our MO wavefunctions:

$$\langle \sigma_g | \sigma_g \rangle = {}_A\langle 1s | 1s \rangle_A + 2{}_A\langle 1s | 1s \rangle_B + {}_B\langle 1s | 1s \rangle_B \equiv 2 + 2S,$$
$$\langle \sigma_u | \sigma_u \rangle = {}_A\langle 1s | 1s \rangle_A - 2{}_A\langle 1s | 1s \rangle_B + {}_B\langle 1s | 1s \rangle_B \equiv 2 - 2S,$$

where

$$|1s\rangle_A \Rightarrow \psi_{100}(r_A),$$

and so on, and

$$S = {}_A\langle 1s | 1s \rangle_B$$

287

is called the *overlap integral*. Therefore, our MO wavefunctions are

$$|\sigma_g\rangle = \frac{1}{\sqrt{2}} \frac{(|1s\rangle_A + |1s\rangle_B)}{(1 + S)^{1/2}},$$

$$|\sigma_u\rangle = \frac{1}{\sqrt{2}} \frac{(|1s\rangle_A - |1s\rangle_B)}{(1 - S)^{1/2}}.$$

With these wavefunctions we find

$$\langle\sigma_g|\mathscr{H}|\sigma_g\rangle = \frac{H_{AA} + H_{AB}}{1 + S} \equiv E^+(R_{AB}),$$

$$\langle\sigma_u|\mathscr{H}|\sigma_u\rangle = \frac{H_{AA} - H_{AB}}{1 - S} \equiv E^-(R_{AB}),$$

where

$$H_{AA} = {}_A\langle 1s|\mathscr{H}|1s\rangle_A = {}_B\langle 1s|\mathscr{H}|1s\rangle_B,$$
$$H_{AB} = {}_A\langle 1s|\mathscr{H}|1s\rangle_B = {}_B\langle 1s|\mathscr{H}|1s\rangle_A,$$

the latter equality holding because the $|1s\rangle$ states are real-valued functions. Explicitly, we find for the matrix elements and the overlap integral[2]

$$H_{AA} = -\tfrac{1}{2} + \exp(-2R_{AB})\left(1 + \frac{1}{R_{AB}}\right),$$

$$H_{AB} = \left(-\frac{1}{2} + \frac{1}{R_{AB}}\right)S - \exp(-R_{AB})(1 + R_{AB}),$$

$$S = \left(1 + R_{AB} + \frac{R_{AB}^2}{3}\right)\exp(-R_{AB}).$$

When these are inserted back into the expressions for E^+ and E^-, we get the dashed curves shown in Figure 55. We note that the former energy function has a minimum, which means that the simple LCAO method is in complete qualitative agreement with the exact calculation using spheroidal wavefunctions.

The general success of the LCAO calculation for H_2^+ has prompted a more careful designation of its electronic states. The common practice is to label states according to (a) the atomic orbitals from which they are formed, (b) the component of the angular momentum about the molecular axis, (c) the irreducible representations of the inversion group, and (d) the bonding or antibonding character of the molecular orbital. Thus, we have σ_u^*1s and $\sigma_g 1s$ for the lowest antibonding and bonding states, $\sigma_g 2s$ and σ_u^*2s for the bonding and antibonding states made from symmetric and antisymmetric combinations of $|2s\rangle$ states, and so on.

Now, back to the hydrogen molecule! We shall take advantage of the physically appealing LCAO approximation and write the zeroth-order wavefunction for the groundstate as

$$|\varphi_1^{(0)}\rangle = |\sigma_g 1s\rangle_1 |\sigma_g 1s\rangle_2 \qquad (n = 1)$$

where the subscripts refer to electrons "one" and "two," respectively. The groundstate energy, $U_1(R_{AB})$, is accordingly.

$$U_1(R_{AB}) = 2E^+(R_{AB}) + \langle \varphi_1^{(0)}|r_{12}^{-1}|\varphi_1^{(0)}\rangle + R_{AB}^{-1}.$$

Upon evaluating the two-center integral,[4] we find that $U_1(R_{AB})$ has a minimum at 1.6 a.u. corresponding to an energy of -1.10 a.u. The experimental values are 1.4 a.u. and -1.17 a.u., respectively. This disagreement is really not very much, considering the simplicity of our approximation scheme.

There are two straightforward ways to improve the zeroth-order wavefunction. One is to write each $|1s\rangle$ state in $|\sigma_g 1s\rangle$ in terms of a screening constant Z', as discussed earlier with regard to the helium atom. If this is done and Z' is chosen so as to minimize the energy, one finds the minimum to correspond to the energy -1.13 a.u., in better agreement with experiment. Of course, this is expected, since we are adding a variable parameter to the wavefunction. Another, perhaps more physical alternative is to examine the LCAO wavefunction to see why it might not be the best one, speaking physically. If $|\varphi_1^{(0)}\rangle$ is written out explicitly, we have

$$|\varphi_1^{(0)}\rangle = \frac{1}{2 + 2S}(|1s\rangle_{A1}|1s\rangle_{A2} + |1s\rangle_{A1}|1s\rangle_{B2} + |1s\rangle_{B1}|1s\rangle_{A2} + |1s\rangle_{B1}|1s\rangle_{B2}).$$

The first and last products represent the probability that both electrons are on one nucleus or the other. These may be thought of as representing an "ionic" structure for H_2, in the sense that the first term corresponds to the structure $H_A^- H_B^+$, while the last term corresponds to $H_A^+ H_B^-$. The middle products, on the other hand, represent the chances of finding structures with one electron per nucleus, what the chemist would call a "covalent" structure for the molecule. Experience indicates that the latter structure is the more likely for H_2. Evidently, the fault with our $|\varphi_1^{(0)}\rangle$ is that *it gives equal weight to ionic and covalent structures for the molecule*. We should be able to improve our calculation of the energy, then, by writing a new zeroth-order wavefunction as

$$|\varphi_1^{(0)}\rangle = \frac{1}{2 + 2S}[|1s\rangle_{A1}|1s\rangle_{B2} + |1s\rangle_{B1}|1s\rangle_{A2}$$
$$+ \alpha(|1s\rangle_{A1}|1s\rangle_{A2} + |1s\rangle_{B1}|1s\rangle_{B2})],$$

---

[4] *Ibid.*, Chapter 4.

where α is a variable parameter representing the "ionic character" of the bond and can vary between zero and one. If α is put equal to zero, we are saying that the bond is completely covalent. Calculations of the energy under this assumption are known as the *valence bond method*; the result of the *VB* method, in the present case, is a minimum energy of -1.116 a.u., only a slight improvement over the *MO* method. If α is varied to make the energy a minimum, we find -1.119 a.u. for $\alpha = 0.256$. In this case we say, somewhat facetiously, that "ionic-covalent resonance" occurs in the hydrogen molecule.

It should be evident that we now possess an understanding of the chemical bond equivalent to our understanding of the atom in section 7.1. Just as in that case, we can deepen our knowledge by introducing spin variables, self-consistent field calculations, and group theoretical descriptions of molecular symmetry. All of this would, of course, take us a bit far afield from our present concern, so we shall not do it. On the other hand, those interested might consult reference 3 in detail for excellent discussions of chemical bonding.

10.4. ROTATIONAL SPECTRA OF DIATOMIC MOLECULES

Consider a molecule composed of two atoms of mass M_1 and M_2, respectively. Equation (10.4) may be written for this case as

$$\frac{\hbar^2}{2M_1} \nabla_1^2 \psi_{n,m} + \frac{\hbar^2}{2M_2} \nabla_2^2 \psi_{n,m} + [E_{n,m} - U_n(\mathbf{R}_1, \mathbf{R}_2)]\psi_{n,m}(\mathbf{R}_1, \mathbf{R}_2) = 0,$$

or, upon introducing center-of-mass and relative coordinates, as in Chapter 4,

$$\frac{\hbar^2}{2M} \nabla_{\mathbf{R}}^2 \psi_{n,m} + \frac{\hbar^2}{2\mu} \nabla_{\mathbf{r}}^2 \psi_{n,m} + [E_{n,m} - U_n(\mathbf{r})]\psi_{n,m}(\mathbf{R}, \mathbf{r}) = 0, \quad (10.11)$$

where

$$\mathbf{R} = \frac{(M_1 \mathbf{R}_1 + M_2 \mathbf{R}_2)}{M}$$

$$\mathbf{r} = \mathbf{R}_2 - \mathbf{R}_1$$

$$M = M_1 + M_2$$

$$\mu = \frac{M_1 M_2}{M}.$$

Equation (10.11) readily separates into the two equations

$$\frac{\hbar^2}{2M} \nabla_{\mathbf{R}}^2 \varphi_{n,p} = E_p \varphi_{n,p}(\mathbf{R}) \quad (10.12)$$

$$\frac{\hbar^2}{2\mu} \nabla_{\mathbf{r}}^2 \psi_{n,N,L,M} + [E_{n,N,L} - U_n(\mathbf{r})]\psi_{n,N,L,M}(\mathbf{r}) = 0, \qquad (10.13)$$

where

$$E_{n,N,L} \equiv E_{n,m} - E_p.$$

The solutions of equation (10.12) were discussed in section 3.5; they are, of course, eigenvectors corresponding to continuous eigenvalues. Equation (10.13)—the one which interests us now—may be separated in spherical polar space if we assume that $U_n(\mathbf{r})$ is a central field potential and take advantage of the fact that

$$\left(\frac{\partial \psi_{n,N,L,M}}{\partial r}\right)_{r=r_e} = 0,$$

where r_e is the equilibrium internuclear separation. The result is

$$\frac{1}{\sin \vartheta} \frac{\partial}{\partial \vartheta} \left(\sin \vartheta \frac{\partial \Upsilon_L^M}{\partial \vartheta}\right) + \frac{1}{\sin^2 \vartheta} \frac{\partial^2 \Upsilon_L^M}{\partial \varphi^2} + \frac{2\mu r_e^2}{\hbar^2} E_L \Upsilon_L^M(\vartheta, \varphi) = 0, \quad (10.14)$$

$$\frac{1}{r^2} \frac{d}{dr} \left(r^2 \frac{dR}{dr}\right) + \frac{2\mu}{\hbar^2} \left[E_{n,N,L} - E_L\left(\frac{r_e}{r}\right)^2 - U_n(r)\right] R(r) = 0, \quad (10.15)$$

where

$$E_L \equiv E_{n,N,L} - U_n(r_e).$$

The solution of equation (10.14) is the spherical harmonic of degree $L - M$:

$$\Upsilon_L^M(\vartheta, \varphi) = \left[\frac{(2L + 1)(L - |M|)!}{4\pi(L + |M|)!}\right]^{1/2} P_L^M(\cos \vartheta) \exp(iM\varphi), \quad (10.16)$$

which is to say that

$$E_L = \frac{\hbar^2}{2I} L(L + 1) \qquad (L = 0, 1, 2, \ldots), \qquad (10.17)$$

where

$$I \equiv \mu r_e^2,$$

and is called the moment of inertia of the molecule. We notice that each rotational energy state is of multiplicity $(2L + 1)$.

The selection rules for dipole radiation from a rotating diatomic molecule follow from equation (10.16) and the arguments given in section 5.5 as

$$\Delta L = \pm 1$$
$$\Delta M = \pm 1, 0.$$

This means that

$$\frac{1}{\lambda_L} = \frac{E_L - E_{L-1}}{hc} = \frac{\hbar}{2\pi Ic} L \qquad (L = 1, 2, \ldots) \qquad (10.18)$$

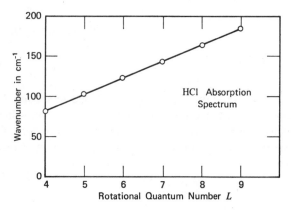

Figure 57. A plot of wavenumber against rotational quantum number for the far infrared spectrum of hydrogen chloride gas.

for the allowed transitions. Equation (10.18) offers a means for deducing the moment of inertia of a diatomic molecule: all that we need do is measure the rotational spectrum, assign rotational quantum numbers to the observed wavelengths, and compute I from the slope of a plot of λ_L^{-1} versus L. This has been done in Figure 57 for the hydrogen chloride molecule.

10.5. ROTATIONAL SPECTRA OF POLYATOMIC MOLECULES

The rotational motion of a polyatomic molecule, like that of the simpler diatomic species, may be studied the most easily by assuming that the nuclei are at their equilibrium separations. When that situation exists, we need concern ourselves only with the orientation of the molecule as a whole in some set of reference frames. In general, we can choose the reference frames to be at the center of mass of the molecule and so can write down the angular momentum (which is directly related to the kinetic energy) of the system in terms of the angular velocity $\boldsymbol{\omega}$ as[5]

$$
\begin{aligned}
L_x &= I_{xx}\omega_x + I_{xy}\omega_y + I_{xz}\omega_z \\
L_y &= I_{yx}\omega_x + I_{yy}\omega_y + I_{yz}\omega_z \\
L_z &= I_{zx}\omega_x + I_{zy}\omega_y + I_{zz}\omega_z
\end{aligned}
\tag{10.19}
$$

[5] See, for example, H. Goldstein, *Classical Mechanics*, Addison-Wesley Publ. Co., Reading, Mass., 1959, Chapter 5.

where

$$I_{qq} = \sum_{j=1}^{M} m_j(r_j^2 - q_j^2)$$

$$\left.\vphantom{\begin{matrix}I\\I\end{matrix}}\right\} \quad q, q' = x, y, z$$

$$I_{qq'} = -\sum_{j=1}^{M} m_j q_j q_j'$$

and r_j is the distance between the center of mass of the molecule and that of the jth nucleus. It is clear that equations (10.19) may be written in the matrix form

$$\begin{pmatrix} L_x \\ L_y \\ L_z \end{pmatrix} = \begin{pmatrix} I_{xx} & I_{xy} & I_{xz} \\ I_{yx} & I_{yy} & I_{yz} \\ I_{zx} & I_{zy} & I_{zz} \end{pmatrix} \begin{pmatrix} \omega_x \\ \omega_y \\ \omega_z \end{pmatrix} \tag{10.20}$$

and that we may construe the ω_q ($q = x, y, z$) as the components of a vector $\boldsymbol{\omega}$ which is transformed into the vector $\mathbf{L}$ by the (operator) matrix (I). We see that (I) is not generally in diagonal form when expressed in terms of the cartesian basis vectors $\hat{\mathbf{x}}$, $\hat{\mathbf{y}}$, $\hat{\mathbf{z}}$; however, there *is* one set of basis vectors $\hat{\mathbf{x}}'$, $\hat{\mathbf{y}}'$, $\hat{\mathbf{z}}'$ for which we may write equation (10.20) as[5]

$$\begin{pmatrix} L_{x'} \\ L_{y'} \\ L_{z'} \end{pmatrix} = \begin{pmatrix} I_{x'x'} & 0 & 0 \\ 0 & I_{y'y'} & 0 \\ 0 & 0 & I_{z'z'} \end{pmatrix} \begin{pmatrix} \omega_{x'} \\ \omega_{y'} \\ \omega_{z'} \end{pmatrix}. \tag{10.21}$$

These new basis vectors lie in subspaces known as *principal axes*. The means by which (I) is put into diagonal form, described in section 6.2, is called, in this instance, a *principal axis transformation*. The eigenvalues $I_{q'q'}$ ($q' = x'$, y', z') of the operator (I) are accordingly termed the *principal moments of inertia* and denoted I_A, I_B, and I_C, respectively.

Once the principal moments of inertia are known, the Hamiltonian function for the rotational motion of a polyatomic molecule may be expressed[5]

$$H = \tfrac{1}{2}\mathbf{L}\cdot\boldsymbol{\omega} = \tfrac{1}{2}(I_A\omega_x'^2 + I_B\omega_y'^2 + I_C\omega_z'^2)$$

$$= \frac{1}{2}\left[I_A\left(\sin\theta\sin\psi\frac{d\phi}{dt} + \cos\psi\frac{d\theta}{dt}\right)^2 + I_B\left(\sin\theta\cos\psi\frac{d\phi}{dt} - \sin\psi\frac{d\theta}{dt}\right)^2 \right.$$

$$\left. + I_C\left(\cos\theta\frac{d\phi}{dt} + \frac{d\psi}{dt}\right)^2 \right],$$

where the last expression of H is in terms of the Euler angles, defined in

Figure 49. The deduction of the Hamiltonian operator from H is quite a complicated procedure because the vectors $\hat{\phi}$, $\hat{\theta}$, and $\hat{\psi}$ are not orthogonal. We shall not go into the details of the computation of the Laplacian operator for this reason,[6] but shall only cite the result[7]

$$
\begin{aligned}
\mathscr{H} = \frac{\hbar^2}{4} \Bigg\{ & \left[-\left(\frac{1}{I_A} + \frac{1}{I_B}\right) + \left(\frac{1}{I_B} - \frac{1}{I_A}\right) \cos 2\psi \right] \frac{1}{\sin\theta} \frac{\partial}{\partial\theta} \left(\sin\theta \frac{\partial}{\partial\theta} \right) \\
& - \frac{1}{\sin^2\theta} \left[\left(\frac{1}{I_A} + \frac{1}{I_B}\right) - \left(\frac{1}{I_A} - \frac{1}{I_B}\right) \cos 2\psi \right] \left(\frac{\partial^2}{\partial\phi^2} - 2\cos\theta \frac{\partial^2}{\partial\phi\,\partial\psi} + \frac{\partial^2}{\partial\psi^2} \right) \\
& - \left[\frac{2}{I_C} - \left(\frac{1}{I_A} + \frac{1}{I_B}\right) + \left(\frac{1}{I_A} - \frac{1}{I_B}\right) \cos 2\psi \right] \frac{\partial^2}{\partial\psi^2} \\
& - 2\left(\frac{1}{I_A} - \frac{1}{I_B}\right) \frac{\sin 2\psi}{\sin\psi} \left(\frac{\partial^2}{\partial\theta\,\partial\psi} - \cos\theta \frac{\partial^2}{\partial\theta\,\partial\psi} \right) \\
& + 2\left(\frac{1}{I_A} - \frac{1}{I_B}\right) \cos 2\psi \cot\theta \frac{\partial}{\partial\theta} - \left(\frac{1}{I_A} - \frac{1}{I_B}\right)(1 + 2\cot^2\theta) \sin 2\psi \frac{\partial}{\partial\psi} \\
& + 2\left(\frac{1}{I_A} - \frac{1}{I_B}\right) \sin 2\psi \cot\theta \csc\theta \frac{\partial}{\partial\phi} \Bigg\}.
\end{aligned}
$$

We may distinguish three special cases of $\mathscr{H}$:

(a) The molecule is a *spherical top*. In this instance $I_A = I_B = I_C \equiv I$ and we have

$$
\mathscr{H} = -\frac{\hbar^2}{2I} \left[\frac{1}{\sin\theta} \frac{\partial}{\partial\theta} \left(\sin\theta \frac{\partial}{\partial\theta} \right) + \frac{1}{\sin^2\theta} \frac{\partial^2}{\partial\varphi^2} \right], \tag{10.22}
$$

where we have defined

$$
\frac{\partial^2}{\partial\varphi^2} \equiv \frac{\partial^2}{\partial\psi^2} + \frac{\partial^2}{\partial\phi^2} - 2\cos\theta \frac{\partial^2}{\partial\psi\,\partial\phi}.
$$

Equation (10.22) is of the same form as the expression given in the foregoing section for the Hamiltonian operator describing the rotation of a diatomic molecule. Evidently the spherical top is equivalent to any molecule wherein the nuclei are placed equidistant about the center of mass.

[6] The method for getting the Laplacian is given by H. Margenau and G. M. Murphy, *The Mathematics of Physics and Chemistry*, D. Van Nostrand, Co., Princeton, 1956, Vol. I, pp. 192–197.
[7] S. C. Wang, On the asymmetrical top in quantum mechanics, *Phys. Rev.* **34**: 243–52 (1929).

(b) The molecule is a *symmetrical top*. This requires the condition $I_A = I_B \neq I_C$. The Hamiltonian operator is then

$$\mathscr{H} = -\frac{\hbar^2}{2} \left\{ \frac{2}{I_A} \frac{1}{\sin \theta} \frac{\partial}{\partial \theta} \left(\sin \theta \frac{\partial}{\partial \theta} \right) + \frac{2}{I_A \sin^2 \theta} \left[\frac{\partial^2}{\partial \psi^2} - 2 \cos \theta \frac{\partial^2}{\partial \phi \, \partial \psi} + \frac{\partial^2}{\partial \phi^2} \right] \right.$$

$$\left. + \left(\frac{2}{I_C} - \frac{2}{I_A} \right) \frac{\partial^2}{\partial \psi^2} \right\}$$

$$= -\frac{\hbar^2}{2I_A} \left[\frac{1}{\sin \theta} \frac{\partial}{\partial \theta} \left(\sin \theta \frac{\partial}{\partial \theta} \right) + \frac{1}{\sin^2 \theta} \frac{\partial^2}{\partial \phi^2} + \left(\frac{I_A}{I_C} + \cot^2 \theta \right) \frac{\partial^2}{\partial \psi^2} \right.$$

$$\left. - \frac{2 \cos \theta}{\sin^2 \theta} \frac{\partial^2}{\partial \phi \, \partial \psi} \right]. \quad (10.23)$$

The Schrödinger eigenvalue problem constructed from $\mathscr{H}$ follows as

$$-\frac{\hbar^2}{2I_A} \left[\frac{1}{\sin \theta} \frac{\partial}{\partial \theta} \left(\sin \theta \frac{\partial \Psi}{\partial \theta} \right) + \frac{1}{\sin^2 \theta} \frac{\partial^2 \Psi}{\partial \phi^2} \right.$$

$$\left. + \left(\frac{I_A}{I_C} + \cot^2 \theta \right) \frac{\partial^2 \Psi}{\partial \psi^2} - \frac{2 \cos \theta}{\sin^2 \theta} \frac{\partial^2 \Psi}{\partial \phi \, \partial \psi} \right] = E \Psi(\phi, \theta, \psi). \quad (10.24)$$

If we put

$$\Psi(\phi, \theta, \psi) \equiv F(\theta) \exp \left[i(M\phi + K\psi) \right] \quad (10.25)$$

into equation (10.24), we get the differential equation

$$-\frac{\hbar^2}{2I_A} \left\{ \frac{1}{\sin \theta} \frac{d}{d\theta} \left(\sin \theta \frac{dF}{d\theta} \right) - \left[\frac{M^2}{\sin \theta} + \left(\frac{I_A}{I_C} + \cot^2 \theta \right) K^2 \right. \right.$$

$$\left. \left. - \frac{2 \cos \theta}{\sin^2 \theta} KM \right] \right\} F(\theta) = E \, F(\theta). \quad (10.26)$$

It is clear that $\Psi(\phi, \theta, \psi)$ as defined in equation (10.25) is a solution of equation (10.24). The quantum numbers M and K are restricted to positive and negative integral values and zero so that $\Psi(\phi, \theta, \psi)$ will be single-valued. Now let us put

$$z = \tfrac{1}{2}(1 - \cos \theta)$$

$$F(\theta) = T(z)$$

$$\alpha = \frac{2I_A E}{\hbar^2} - \frac{I_A}{I_C} K^2$$

in order to bring equation (10.26) into the form

$$\frac{d}{dz} \left[z(1 - z) \frac{dT}{dz} \right] + \left\{ \alpha - \frac{[M + K(2z - 1)]^2}{4z(1 - z)} \right\} T(z) = 0. \quad (10.27)$$

Molecular Spectra

This equation has two regular singular points—at $z = 0$ and $z = 1$. If we proceed as we did in section 4.3, wherein the angular separate of the Schrödinger eigenvalue problem for hydrogen was considered, we find that the solution of equation (10.27) should be of the form

$$T(z) = z^{\frac{1}{2}|K-M|}(1 - z)^{\frac{1}{2}|K+M|}F(z), \qquad (10.28)$$

where $F(z)$ is a power series in z. Upon substituting equation (10.28) into equation (10.27), we get

$$z(1 - z)\frac{d^2F}{dz^2} + [(1 + p)z - q]\frac{dF}{dz} - n(p + n)F(z) = 0, \quad (10.29)$$

where

$$p = 1 + |K - M| + |K + M|$$
$$q = 1 + |K - M|$$
$$n(p + n) = \alpha + K^2 - \{[\tfrac{1}{2}(p - 1) + 1]^2 - 1\}.$$

The solutions of equation (10.29) are the *Jacobi polynomials*, defined by[8]

$$F_n(p, q; z) = 1 + \sum_{m=1}^{n}(-1)^m \frac{n!}{m!\,(n - m)!}\frac{(p + n + m - 1)!\,(q - 1)!}{(p + n - 1)!\,(q + m - 1)!}z^m.$$
$$(10.30)$$

We see from equation (10.30) that q must be greater than zero (it is so, by definition) and that n must be a positive integer. Moreover, if we define the quantum number L by

$$L \equiv n + \tfrac{1}{2}(p - 1) = n + \tfrac{1}{2}|K - M| + \tfrac{1}{2}|K + M|,$$

then the equation defining $n(p + n)$ in terms of α, K, and p may be written

$$E \equiv E_{L,K} = \frac{\hbar^2}{2I_A}\left[L(L + 1) + \left(\frac{I_A}{I_C} - 1\right)K^2\right]$$
$$(L = 0, 1, 2, \ldots; |K| \leqslant L). \qquad (10.31)$$

These are the energy levels permitted the symmetric top. We note that they are of multiplicity $2(2L + 1)$ unless $K = 0$, when they are of multiplicity $(2L + 1)$. The factor of two comes from the fact that $E_{L,K}$ is independent of the sign of K.

Noting the definitions of the Euler angles and referring to our earlier study of the hydrogen atom, we may interpret $\sqrt{L(L + 1)}\hbar$ $(L = 0, 1, \ldots)$ as

[8] See H. Margenau and G. M. Murphy, *The Mathematics of Physics and Chemistry*, D. Van Nostrand Co., Princeton, 1956, Vol. I, pp. 72–74.

the total angular momentum of the molecule, $K\hbar$ $(-L \leqslant K \leqslant L)$, as the component of the total angular momentum about the symmetry axis of the molecule, and $M\hbar$ $(-L \leqslant M \leqslant L)$, as the angular momentum of the molecule along the "z-axis" of the center-of-mass coordinate system in which the Euler angles are defined.

The selection rules for dipole radiation from a symmetric top molecule are

$$\Delta L = 0, \pm 1, \Delta K = 0, K \neq 0,$$

$$\Delta L = \pm 1, \quad \Delta K = 0, K = 0.$$

It follows that the wavelength for a given spectral line is to be computed from

$$\frac{1}{\lambda_L} = \frac{E_{L,K} - E_{L-1,K}}{hc} = \frac{\hbar}{4\pi c I_A} L \quad (L = 1, 2, \ldots). \quad (10.32)$$

One of the moments of inertia of a symmetric top can therefore be determined with the same procedure suggested for diatomic molecules. [The other moment of inertia can be computed from the Stark effect spectrum or by developing a more accurate expression than (10.32) to account for deviations of the nuclei from their equilibrium positions.]

(c) The molecule is an *asymmetric top*. Here we have the most general case, wherein $I_A < I_B < I_C$. The usual procedure is to write the solution of the Schrödinger eigenvalue problem constructed from the asymmetric top Hamiltonian operator in the form

$$\Psi'_{LM}(\phi, \theta, \psi) = \sum_{K=-L}^{L} C_K \Psi_{LKM}(\phi, \theta, \psi), \quad (10.33)$$

where

$$\Psi_{LKM}(\phi, \theta, \psi) = \left\{ \frac{(2L+1)(L + \frac{1}{2}|K - M| + \frac{1}{2}|K + M|)!}{8\pi^2(L - \frac{1}{2}|K - M| - \frac{1}{2}|K + M|)! \, (|K - M|)!} \right\}^{1/2}$$
$$\times \, [\tfrac{1}{2}(1 - \cos\theta)]^{\frac{1}{2}|K - M|}[\tfrac{1}{2}(1 + \cos\theta)]^{\frac{1}{2}|K + M|} F_n[\tfrac{1}{2}(1 - \cos\theta)]$$
$$\times \, \exp\left[i(M\phi + K\psi)\right]$$

is the normalized symmetric top wavefunction. Equation (10.33) leads, through the usual generalized perturbation method, to the determinantal equation

$$\det |H_{lk} - \delta_{lk} E| = 0 \quad (-L \leqslant l, k \leqslant L), \quad (10.34)$$

where H_{lk} is the matrix element of the Hamiltonian operator in the basis composed of the $\Psi_{LKM}(\phi, \theta, \psi)$. Procedures for reducing the determinantal equation (10.34) have been devised,[7] and fairly extensive tables of the energy levels have been compiled.[9] Some of the levels, for small values of L, are

$$L = 0 \qquad E_{0M} = 0 \qquad\qquad (-L \leqslant M \leqslant L)$$

$$L = 1 \qquad E_{1M} = \frac{\hbar^2}{2}\left(\frac{1}{I_A} + \frac{1}{I_B}\right)$$

$$= \frac{\hbar^2}{2}\left(\frac{1}{I_A} + \frac{1}{I_C}\right) \qquad (-L \leqslant M \leqslant L)$$

$$= \frac{\hbar^2}{2}\left(\frac{1}{I_B} + \frac{1}{I_C}\right).$$

There are in general $(2L + 1)$ asymmetric top wavefunctions for a given value of L. Each of these corresponds to a different value of the energy but is not associated with a unique quantum number.

10.6. VIBRATIONAL SPECTRA OF DIATOMIC MOLECULES

Now let us return to equation (10.15), which describes the relative motion of the nuclei in a diatomic molecule. The expression may be rewritten for convenience as

$$\frac{d^2 S}{dr^2} + \left[\frac{2\mu}{\hbar^2}(E_{n,N,L} - U_n(r)) - \frac{L(L+1)}{r^2}\right]S(r) = 0 \qquad (10.35)$$

where

$$S(r) = rR(r)$$

and the expression for E_L has been introduced. Equation (10.35) can be solved only if $U_n(r)$ is known. As mentioned earlier, this knowledge depends either on a complete solution of the electronic eigenvalue problem for the molecule or a good intuitive selection of the internuclear potential function. The latter alternative is by far the easier in most cases and is the one we shall choose here. One possibility which has proved useful in the past is the potential function

$$U_n(r) = D \exp\left[-2a(r - r_e)\right] - 2D \exp\left[-a(r - r_e)\right] \qquad (10.36)$$

that was proposed by P. M. Morse in 1929.[10] In equation (10.36), D, a, and

---

[9] For a listing of references, see G. M. Barrow, *Introduction to Molecular Spectroscopy*, McGraw-Hill Book Co., New York, 1962, p. 112. For a good discussion, see G. Herzberg, *Infrared and Raman Spectra of Polyatomic Molecules*, D. Van Nostrand Co., Princeton, 1945, §4, Chapter I.

[10] P. M. Morse, Diatomic molecules according to the wave mechanics. II. Vibrational levels, *Phys. Rev.* **34**: 57–64 (1929).

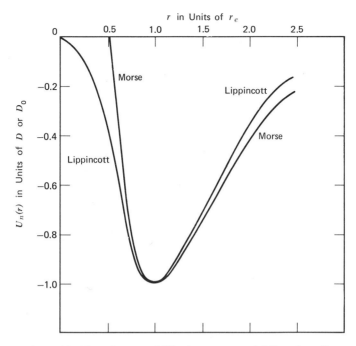

Figure 58. The Morse and Lippincott potential functions for the interaction of two hydrogen atoms. (The Lippincott potential has been renormalized so that it may be compared directly with the Morse potential.)

r_e are constants. The first of these is the value of $-U_n(r_e)$, the second is essentially the scale factor which determines the frequency of oscillation for the molecule (as we shall see), while the third is the equilibrium internuclear separation. A plot of the Morse potential is shown in Figure 58.

When equation (10.36) is put into equation (10.35) and the substitution

$$g = \frac{D}{\frac{1}{2}\hbar\omega_e}$$

is made, we find

$$\frac{d^2 S}{dr^2} + (ag)^2\left[E - \exp\left[-2a(r - r_e)\right]\right.$$

$$\left. + 2\exp\left[-a(r - r_e)\right] - E_L'\left(\frac{r_e}{r}\right)^2\right]S(r) = 0 \quad (10.37)$$

where

$$E = \frac{E_{n,N,L}}{D}$$

$$E'_L = \frac{L(L+1)\hbar^2}{2ID}$$

and

$$\omega_e = \left[\frac{(d^2U_n/dr^2)_{r=r_e}}{\mu}\right]^{1/2} = a\left(\frac{2D}{\mu}\right)^{1/2}$$

upon differentiating equation (10.36) twice. Now we put

$$S(r) = \exp\left[-g \exp\left[-a(r-r_e)\right]\right]y(r)$$

into equation (10.37) and get

$$\frac{d^2y}{dr^2} + 2ag \exp\left[-a(r-r_e)\right]\frac{dy}{dr}$$

$$+ a^2\left[g^2E + g(2g-1)\exp\left[-a(r-r_e)\right] - g^2E'_L\left(\frac{r_e}{r}\right)^2\right]y(r) = 0.$$

The equation is transformed once more by writing

$$x = 2g \exp\left[-a(r-r_e)\right]:$$

$$x^2\frac{d^2y}{dx^2} + x(1-x)\frac{dy}{dx} + \left[g^2E + (g-\tfrac{1}{2})x\right.$$

$$\left. - g^2E_L(ar_e)^2\left(ar_e - \ln\left(\frac{x}{2g}\right)\right)^{-2}\right]y(x) = 0.$$

The general solution of this equation would be quite difficult to get. Fortunately, we do not have to get it; for, we can expect with good reason that the separation of the nuclei will not usually be much different from that at equilibrium, which fact provides us with a method for simplifying the equation. Specifically, we can put

$$\left(ar_e - \ln\left(\frac{x}{2g}\right)\right)^{-2} = (ar_e)^{-2}\left[1 + \frac{2}{ar_e}\left(\frac{x}{2g} - 1\right) - \cdots\right]$$

into the foregoing equation to get the reduced expression

$$x^2\frac{d^2y}{dx^2} + x(1-x)\frac{dy}{dx} + \left\{g^2\left[E - E'_L\left(1 - \frac{2}{ar_e}\right)\right]\right.$$

$$\left. + \left(g - \frac{1}{2} - \frac{gE'_L}{ar_e}\right)x\right\}y(x) = 0. \quad (10.38)$$

Finally, we make the substitution

$$y(x) = x^{g\{E_L'[1 - (2/ar_e)] - E\}^{1/2}} F(x)$$

to bring equation (10.38) into the desired form:

$$x\frac{d^2F}{dx^2} + \left\{2g\left[E_L'\left(1 - \frac{2}{ar_e}\right) - E\right]^{1/2} + 1 - x\right\}\frac{dF}{dx}$$

$$- \left\{g\left[E_L'\left(1 - \frac{2}{ar_e}\right) - E\right]^{1/2} - g + \frac{1}{2} + \frac{gE_L'}{ar_e}\right\}F(x) = 0 \quad (10.39)$$

The solution of equation (10.39) is the confluent hypergeometric function[11]

$${}_1F_1\left(a; 2a + 2g\left(1 - \frac{E_L'}{ar_e}\right); x\right)$$

$$\equiv \sum_{n=0}^{\infty} \frac{\Gamma(a + n)\Gamma(2a + 2g(1 - (E_L'/ar_e)))}{\Gamma(a)\Gamma(2a + 2g(1 - (E_L'/ar_e)) + n)} \frac{x^n}{n!} \quad (10.40)$$

where

$$a = g\left[E_L'\left(1 - \frac{2}{ar_e}\right) - E\right]^{1/2} + \tfrac{1}{2} - g + \frac{gE_L'}{ar_e}.$$

If we impose the condition that a be a negative integer or zero, $F(x)$ will be a polynomial and E will be determined uniquely. We shall do this, although our motivation is *not* that $F(x)$ grows exponentially as x increases, since

$$0 \leqslant x \leqslant 2g \exp(ar_e)$$

in this case; we require a polynomial solution in order to determine E. Thus we find, upon setting a equal to $-N$ and squaring,

$$E = -1 + (N + \tfrac{1}{2})\frac{2}{g} - \left(\frac{N + \tfrac{1}{2}}{g^2}\right)^2 + E_L'\left(1 - \frac{E_L'}{(ar_e)^2}\right) - \frac{2}{ar_e}(N + \tfrac{1}{2})E_L'$$

or

$$E_{N,L} = -D + (N + \tfrac{1}{2})\hbar\omega_e - \frac{(N + \tfrac{1}{2})^2\hbar^2\omega_e^2}{4D}$$

$$+ \frac{L(L + 1)\hbar^2}{2I}\left[1 - \frac{L(L + 1)\hbar^2}{2I(ar_e)^2D}\right] - \frac{(N + \tfrac{1}{2})L(L + 1)\hbar^3\omega_e}{2Iar_eD}$$

$$(N = 0, 1, \ldots; L = 0, 1, \ldots). \quad (10.41)$$

The terms in equation (10.41) are interpreted as follows. The first term after $-D$ gives the vibrational energy levels of the molecule in the harmonic

---

[11] See, for example, G. Arfken, *Mathematical Methods for Physicists*, Academic Press, New York, 1966, pp. 500–503 for more information.

oscillator approximation. [*Cf.* equation (6.38).] The second term is a correction to the vibrational states which arises from the asymmetric form of equation (10.36). The constant

$$\frac{\hbar\omega_e^2}{4D} \equiv x_e\omega_e$$

is, accordingly, given the name *anharmonicity constant*. The third term is the rotational energy level for the equilibrium internuclear separation plus a correction term for the stretching of the internuclear bond which comes about from rotation. This interpretation is perhaps made more obvious when it is noted that

$$(ar_e)^2 D = \tfrac{1}{2}I\omega_e^2$$

is the vibrational potential energy, in the harmonic oscillator approximation, when the nuclei have been stretched apart to twice their separation at equilibrium. The final term in equation (10.41) represents the coupling between the vibrational and rotational motion of the nuclei. For this reason the parameter

$$\frac{\hbar^3\omega_e}{2Iar_eD} \equiv \alpha_e$$

is called the *vibrational-rotational coupling constant*.

The vibrational-rotational energy levels given by equation (10.41) have been found to be in good agreement with experiment for several different diatomic molecules. However, for many others it does not lead to very accurate estimates of the quantities ω_e, D_0, ω_ex_e, r_e, and α_e. As an extreme example, we might take a look at the molecule Se_2. The experimental values of D_e (the *bond dissociation energy*), given by

$$D_e = D - \tfrac{1}{2}\hbar\omega_e,$$

and α_e are 81.9 kcal per mole and 0.00027 cm^{-1}, respectively, although the Morse expression (10.41) predicts 102.9 kcal per mole and 0.00041 cm^{-1}, respectively. To alleviate this problem, a number of "modified" Morse potentials have been proposed from time to time. The most promising of these appears to be the function suggested by E. R. Lippincott in 1955.[12] In this case we have (see Figure 58)

$$U_n(r) = D_0\{1 - \exp\left[-a(r - r_e)^2/2r\right]\}. \tag{10.42}$$

[12] E. R. Lippincott and R. Schroeder, General relation between potential energy and internuclear distance for diatomic and polyatomic molecules. I., *J. Chem. Phys.* **23**: 1131–1141 (1955).

As with the Morse potential, equation (10.42) must be expanded about its minimum (at $r = r_e$) in order that equation (10.35) be readily soluble. The method used by Lippincott was the expansion of the exponential in $U_n(r)$, followed by power series expansions of the terms in r^{-1} and r^{-2} to get

$$U_n(r) = \left(\frac{aD_0}{2r_e}\right)(r - r_e)^2 - \left(\frac{aD_0}{2r_e^2}\right)(r - r_e)^3$$

$$+ \frac{aD_0}{2r_e^2}\left(\frac{1}{r_e} - \frac{a}{4}\right)(r - r_e)^4 + O[(r - r_e)^5].$$

This expression is first truncated after the leading term, put into equation (10.35), and then, after the latter (harmonic oscillator) equation is solved, "added back" term by term through the usual perturbation technique. The result is, in terms of the parameters in equation (10.41),

$$\omega_e = \left(\frac{aD_0}{\mu r_e}\right)^{\frac{1}{2}}$$

$$x_e\omega_e = \frac{3}{8\pi}\left(\frac{\hbar\omega_e^2}{4D_0} + \frac{\hbar}{4I}\right).$$

Computations of D_e and the anharmonicity constant made from these equations are compared in Table 10.1 with the appropriate empirical data for several molecules. The values of D_e which may be found from the Morse expression are also listed for an additional comparison. It is not hard to see that the Lippincott potential is, in general, in accord with what is observed experimentally; the Morse potential, as mentioned earlier, appears to be a cruder approximation to the truth.

10.7. VIBRATIONAL SPECTRA OF POLYATOMIC MOLECULES: NORMAL COORDINATES

For a molecule composed of M nuclei, we may write equation (10.4) in the form

$$\frac{\hbar^2}{2}\sum_{k=1}^{3M}\frac{1}{M_k}\frac{\partial^2}{\partial x_k^2}\psi_{n,m} + [E_{n,m} - U_n(\mathbf{q}^M)]\psi_{n,m}(\mathbf{q}^M) = 0, \qquad (10.43)$$

where $\mathbf{q}^M$ refers to the $3M$ position coordinates of the nuclei *relative to their equilibrium positions* and

$$q_1 = x^1 - x_e^1$$
$$q_2 = y^1 - y_e^1$$
$$q_3 = z^1 - z_e^1$$

Molecular Spectra

Table 10.1. Calculated and experimental values of the bond dissociation energy and the anharmonicity constant for several diatomic molecules.[12]

Molecule	D_e (obsd.) kcal/mole	D_e (Lipp.) kcal/mole	D_e (Morse) kcal/mole	$x_e\omega_e$ (obsd.) cm^{-1}	$x_e\omega_e$ (Lipp.) cm^{-1}
As$_2$	91.3	90.4	116.9	1.12	1.19
CH	80.0	71.6	86.7	64.3	61.6
CO	210.5	195.0	247.0	13.5	12.7
Cs$_2$	10.5	10.3	15.6	—	—
H$_2$	103.2	102.0	111.0	118.0	117.5
HBr	86.4	85.7	107.0	45.2	47.3
HCl	102.0	95.7	119.0	52.0	50.7
HI	70.4	71.9	92.7	39.7	41.3
I$_2$	35.6	40.9	53.5	0.61	0.63
O$_2$	117.0	114.0	146.0	12.1	11.1
S$_2$	101.4	102.0	133.0	2.85	2.77
SO	118.5	114.0	146.0	6.12	5.96
Se$_2$	81.9	79.5	102.9	1.06	1.08
OH	100.5	93.9	116.0	82.8	72.1

and so on, where $(x_e{}^1, y_e{}^1, z_e{}^1)$ is the equilibrium position of the first nucleus. Equation (10.43) is in general very difficult to solve, but can be drastically simplified after the coordinates

$$x_k = \sqrt{M_k}\, q_k \qquad (k = 1, \ldots, 3M),$$

are introduced and $U_n(\mathbf{x}^M)$ is expanded in a Taylor series to second order:

$$U_n(\mathbf{x}^M) = U_n(0) + \sum_{k=1}^{3M} \left(\frac{\partial U_n}{\partial x_k}\right)_{x_k=0} x_k + \frac{1}{2}\sum_{k=1}^{3M}\sum_{j=1}^{3M}\left(\frac{\partial^2 U_n}{\partial x_k\, \partial x_j}\right)_{x_k=x_j=0} x_k x_j.$$

If we impose the equilibrium conditions

$$U_n(0) = 0, \qquad \left(\frac{\partial U_n}{\partial x_k}\right)_{x_k=0} = 0 \qquad (k = 1, \ldots, 3M),$$

our approximate $U_n(\mathbf{x}^M)$ takes on the form

$$U_n(\mathbf{x}^M) = \frac{1}{2}\sum_{k=1}^{3M}\sum_{j=1}^{3M} V_{kj} x_k x_j, \tag{10.44}$$

where the meaning of V_{kj} is the obvious one. Equation (10.43) now may be written

$$\frac{\hbar^2}{2} \sum_{k=1}^{3M} \frac{\partial^2}{\partial x_k^2} \psi_{n,m} + \left[E_{n,m} - \frac{1}{2} \sum_{k=1}^{3M} \sum_{j=1}^{3M} V_{kj} x_k x_j \right] \psi_{n,m}(\mathbf{x}^M) = 0. \quad (10.45)$$

This expression looks quite a lot like the Schrödinger eigenvalue problem for a system of $3M$ non-interacting harmonic oscillators and, indeed, would be just that if it were not for the appearance of the cross products $x_k x_j$ $(k, j = 1, \ldots, 3M)$ in the potential function. Let us see if we can get rid of them.

The V_{kj} $(k, j = 1, \ldots, 3M)$ might be imagined as the (real) elements of a matrix V which has been represented in a basis composed of the x_k $(k = 1, \ldots, 3M)$. Evidently V could be brought into diagonal form if a new basis were found as some linear combination of the members of the old one. In other words, we might look at the problem as one analogous to the diagonalization of the Hamiltonian operator, as outlined in section 6.2. There we sought a unitary operator which would act upon the old basis to create the proper new one. Here, because the basis has only real-valued constituents, we seek an *orthogonal operator* O which has the properties

$$O\tilde{O} = \tilde{O}O = I \quad (10.46)$$

$$[OVO^{-1}]_{lm} = \omega_l^2 \delta_{lm} \quad (l, m = 1, \ldots, 3M), \quad (10.47)$$

where I is the identity operator and the ω_l^2 are the eigenvalues of the diagonalized V. Equation (10.46) is the exact counterpart of the definition of a unitary operator.

Suppose the basis in which V is diagonal is composed of the coordinates Q_l $(l = 1, \ldots, 3M)$. Then we must have it that

$$Q_l = Ox_l$$

and, by equation (10.46),

$$x_l = \tilde{O}Q_l.$$

In matrix notation the latter equation is expressed

$$\begin{pmatrix} x_1 \\ \vdots \\ x_{3M} \end{pmatrix} = \begin{pmatrix} \tilde{O}_{11} & \cdots & \tilde{O}_{13}{}^M \\ \vdots & & \\ \tilde{O}_{3M1} & \cdots & \tilde{O}_{3M3M} \end{pmatrix} \begin{pmatrix} Q_1 \\ \vdots \\ Q_{3M} \end{pmatrix}$$

which, in turn, suggests

$$x_k = \sum_{j=1}^{3M} \tilde{O}_{kj} Q_j \quad (k = 1, \ldots, 3M). \quad (10.48)$$

Molecular Spectra

Equation (10.48), when introduced into equation (10.45), should produce the harmonic oscillator equation we are after. Upon noting that

$$\frac{\partial}{\partial x_k} = \sum_{l=1}^{3M} \frac{\partial Q_l}{\partial x_k} \frac{\partial}{\partial Q_l} = \sum_{l=1}^{3M} O_{lk} \frac{\partial}{\partial Q_l}$$

we may write equation (10.45) as

$$\frac{\hbar^2}{2} \sum_{k=1}^{3M} \sum_{j=1}^{3M} \sum_{l=1}^{3M} \tilde{O}_{kj} \tilde{O}_{kl} \frac{\partial^2 \psi_{n,m}}{\partial Q_j \partial Q_l}$$

$$+ \left[E_{n,m} - \frac{1}{2} \sum_{k=1}^{3M} \sum_{j=1}^{3M} \sum_{l=1}^{3M} \sum_{m=1}^{3M} V_{kj} \tilde{O}_{kl} \tilde{O}_{jm} Q_l Q_m \right] \psi_{n,m} = 0.$$

Now, by equation (10.46),

$$\sum_{k=1}^{3M} \sum_{j=1}^{3M} \sum_{l=1}^{3M} O_{kj} O_{kl} \frac{\partial^2 \psi_{n,m}}{\partial Q_j \partial Q_l} = \sum_{l=1}^{3M} \sum_{j=1}^{3M} \sum_{k=1}^{3M} O_{jk} \tilde{O}_{kl} \frac{\partial^2 \psi_{n,m}}{\partial Q_j \partial Q_l}$$

$$= \sum_{l=1}^{3M} \sum_{j=1}^{3M} \delta_{jl} \frac{\partial^2 \psi_{n,m}}{\partial Q_j \partial Q_l} = \sum_{l=1}^{3M} \frac{\partial^2 \psi_{n,m}}{\partial Q_l^2}.$$

Moreover, by equation (10.47),

$$\sum_{k=1}^{3M} \sum_{j=1}^{3M} \sum_{l=1}^{3M} \sum_{m=1}^{3M} V_{kj} \tilde{O}_{kl} \tilde{O}_{jm} Q_l Q_m = \sum_{l=1}^{3M} \sum_{m=1}^{3M} \sum_{j=1}^{3M} \sum_{k=1}^{3M} O_{lk} V_{kj} \tilde{O}_{jm} Q_l Q_m$$

$$= \sum_{l=1}^{3M} \sum_{m=1}^{3M} \omega_l^2 \delta_{lm} Q_l Q_m = \sum_{l=1}^{3M} \omega_l^2 Q_l^2$$

since, of course, $\tilde{O} = O^{-1}$. With these results we can write equation (10.45) in the form

$$\sum_{l=1}^{3M} \left(\frac{\hbar^2}{2} \frac{\partial^2 \psi_{n,m}}{\partial Q_l^2} + [E_{n,m} - \tfrac{1}{2}\omega_l^2 Q_l^2]\psi_{n,m}(\mathbf{Q}^M) \right) = 0. \tag{10.49}$$

The solution of this equation is

$$\psi_{n,m}(\mathbf{Q}^M) = \prod_{l=1}^{3M} \varphi_{N_l}(Q_l)$$

where

$$\varphi_{N_l}(Q_l) = (N_l!)^{-\frac{1}{2}}(2\hbar\omega_l)^{-N_l/2}\left(\frac{\omega_l}{\pi\hbar}\right)^{\frac{1}{4}}\left(-\hbar\frac{d}{dQ_l} + \omega_l Q_l\right)^{N_l} \exp\left[-((\omega_l/2\hbar)Q_l^2)\right],$$

as given in Chapter 6, and

$$E_{n,m} = \sum_{l=1}^{3M} (N_l + \tfrac{1}{2})\hbar\omega_l \qquad (N_l = 0, 1, \ldots; m = \{N_l\}).$$

The physical interpretation of the Q is that they represent certain linear combinations of the displacements of *all* the nuclei in the molecule. The vibration frequency ω_l, then, must be construed as that which each nucleus maintains as it carries out its contribution to the displacement Q. This vibration frequency is called a *normal vibration frequency* of the molecule; the Q are accordingly called *normal coordinates*.

What remains is to specify how the elements of the matrix (O) and the eigenvalues $\omega_l{}^2$ are to be computed. The procedure is quite the same as what was done in Chapter 6. We rewrite equation (10.47) as

$$\begin{pmatrix} O_{11} & \cdots & O_{13M} \\ \vdots & & \\ O_{3M1} & \cdots & O_{3M3M} \end{pmatrix} \begin{pmatrix} V_{11} & \cdots & V_{13M} \\ \vdots & & \vdots \\ V_{3M1} & \cdots & V_{3M3M} \end{pmatrix}$$

$$= \begin{pmatrix} \omega_1{}^2 & \cdots & 0 \\ & \vdots & \\ 0 & \cdots & \omega_{3M}{}^2 \end{pmatrix} \begin{pmatrix} O_{11} & \cdots & O_{13M} \\ \vdots & & \vdots \\ O_{3M1} & \cdots & O_{3M3M} \end{pmatrix}$$

which form implies the equation

$$\sum_{k=1}^{3M} O_{lk}(V_{kl} - \omega_l{}^2 \delta_{lk}) = 0 \qquad (l = 1, \ldots, 3M). \tag{10.50}$$

Equations (10.50) have no solutions for the O_{lk} but the trivial ones unless

$$\det |V_{kl} - \omega_l{}^2 \delta_{kl}| = 0 \qquad (k, l = 1, \ldots, 3M). \tag{10.51}$$

The $\omega_l{}^2$ are found by solving equation (10.51); the matrix elements of (O) are then computed from equations (10.50).

PROBLEMS

1. Compute the moment of inertia of BeO from the data given. (The wavenumbers are from an electronic transition-rotation spectrum.)

L	$\lambda_L{}^{-1}$ cm^{-1}
1	21,199.81
2	21,202.88
3	21,205.74
4	21,208.52
5	21,211.12

2. Below is a table of the quantity $\hbar/2Ic$ for HCl in different vibrational levels of its electronic groundstate.

(a) Calculate the apparent internuclear separation for each vibrational level.

(b) Calculate the mean value of the vibrational-rotational coupling constant from equation (10.41) by considering $\hbar^2/2I_N$ to be the coefficient of $L(L + 1)$. Ignore the centrifugal stretching term in doing the calculation.

(c) Calculate the equilibrium internuclear separation from

$$\frac{\hbar^2}{2I} = \frac{\hbar^2}{2I_0} + \tfrac{1}{2}\alpha_e.$$

N	$(\hbar/2I_Nc)$ cm^{-1}	
0	10.4400	
1	10.1366	The observed values for $\alpha_e/\hbar c$ and r_{HCl} are 0.302
2	9.8329	cm^{-1} and 1.275 Å, respectively
3	9.5343	
4	9.232	
5	8.933	

3. In the harmonic oscillator approximation, what should be the ratio of vibrational frequencies (in a given vibrational state) for two diatomic molecules of different reduced mass but the same stretching force constant? Verify your expression in terms of the following spectral data for HCl^{35} and HCl^{37}. The first line gives the change in quantum number, while the second gives $\lambda_{35}{}^{-1} - \lambda_{37}{}^{-1}$ for the transition.

Vibrational Transition: $0 \to 1$ $0 \to 2$ $0 \to 3$

Difference in wavenumber: 2.01 cm^{-1} 4.00 cm^{-1} 5.834 cm^{-1}

For HCl^{35} and HCl^{37}, $(\mu_{35}/\mu_{37})^{\frac{1}{2}} = 0.99924$. For HCl^{35}, λ^{-1} is 2651.98 cm^{-1}. for $0 \to 1$.

4. Calculate the moment of inertia and bond length for CO_2 from the following data. (Carbon dioxide is a spherical top polyatomic molecule.) The accepted values are $I = 71.1 \times 10^{-40}$ gm-cm^2 and $r_{CO} = 1.16$ Å.

L	$\lambda_L{}^{-1}$ cm^{-1}
5	8.93
7	11.63
9	14.84
11	18.14
13	21.53
15	24.60

5. The determinantal equation from which the normal vibration frequencies are calculated contains a determinant of order $3M$ when a molecule composed of M nuclei is considered. Show that this determinant can be reduced to one of order $3M - 6$ by involving the conservation of total linear and angular momentum when considering the eigenvalue problem equation (10.49), provided that the molecule under consideration is not linear.

6. The normal vibration wavenumbers for the water molecule are 3756, 3652, and 1595 cm^{-1}. Assuming the harmonic oscillator approximation is adequate, calculate the wave numbers for the following transitions

$$\text{(a) } E(0 \ 1 \ 0) \rightarrow E(0 \ 2 \ 0)$$

$$\text{(b) } E(0 \ 1 \ 1) \rightarrow E(1 \ 1 \ 1)$$

$$\text{(c) } E(0 \ 0 \ 0) \rightarrow E(2 \ 1 \ 1)$$

where

$$\frac{E(N_1 N_2 N_3)}{hc} = [(N_1 + \tfrac{1}{2})3652 + (N_2 + \tfrac{1}{2})1595 + (N_3 + \tfrac{1}{2})3756].$$

The observed values are 1556, 3475, and 12,151 cm^{-1}, respectively.

PART IV

HIGH-ENERGY PHENOMENA

CHAPTER 11

Scattering Theory

Nuclear structure à la billiards.

11.1. SCATTERING CROSS SECTIONS

One of the very important experiments which led us to our present notions about atomic structure was that performed by Franck and Hertz. Their experiment involved the collisions of free electrons with atoms, phenomena which are comprised in the general term "scattering process." Scattering processes are said to occur whenever incident particles strike a material substance (the "target") and are scattered by it. The scattering may involve more than one constituent of the target and may leave those constituents in excited states. It will always leave the incident particle in a new state.

The study of scattering processes is the best tool available for discerning the internal properties of atoms, nuclei, and fundamental particles. In a typical experiment involving nuclei, we might bombard the appropriate target with, say, protons of variable energy and determine afterward the fate of both the protons and the nuclei. The latter will usually be transformed in some way and so must be detected by chemical means. If the material containing the nuclei is thin enough to preclude more than one scattering event per incident proton, and light enough to make the internuclear distance large compared to the de Broglie wavelength of the protons, and if the density of the incident beam is low enough to all but prevent proton-proton interactions, the scattering process may be described by

$$J\sigma_t \mathcal{N} = N, \tag{11.1}$$

where J is the number of incident protons crossing unit area of the target in unit time, $\mathcal{N}$ is the number of target nuclei per unit area, N is the number of transformed nuclei (or the number of scattered protons) produced in unit time and σ_t is the scattering *cross section* for the given process. The scattering cross section may be viewed as a constant parameter characteristic of how well the scattering process proceeds, so long as the energy of the protons is fixed. For a given incident proton energy, σ_t is measured by measuring J, $\mathcal{N}$, and N. One way of doing this is to collect the protons in a Faraday box (*Cf*. section 2.6) so as to deduce their number from their total charge, measure the thickness of the target to get $\mathcal{N}$, and detect the transformed nuclei as suggested before. The scattering cross section as a function of energy might be obtained by placing rows of targets before the incident beam and considering the energy loss of the protons as well as the three quantities J, $\mathcal{N}$, and N. σ_t is ordinarily recorded in the convenient units of *barns*, where

$$1 \text{ barn} = 10^{-24} \text{ cm}^2.$$

For small cross sections, the unit millibarn is also used. Some typical experimental results are shown in Figure 59. If the scattering process involves neutrons as the probe particles, the scattered corpuscles cannot be detected in a Faraday box; the attenuation of the neutron beam, as determined with the appropriate counter,[1] is then the primary measurement. Accordingly, equation (11.1) is cast into the form

$$\frac{dI}{I} = \sigma_t \mathcal{N}, \tag{11.2}$$

where I is the incident beam intensity, $(I - dI)$ is the transmitted beam intensity, and the remaining symbols have their previous meanings. Some measured values of σ_t for neutrons incident with energy 14 MeV are as follows.[2]

Element	σ_t in barns
Al	1.92
B	1.16
Be	0.65
Fe	2.75
Hg	5.64

In this chapter we shall be concerned with elastic scattering, wherein there are no internal changes in the scattered particle or the target. This process is conveniently described in terms of the *differential scattering cross section*, defined by

$$dN_\Omega = J\mathcal{N}\sigma(\Omega)\,d\Omega, \tag{11.3}$$

where $d\Omega$ is an element of solid angle about the target. In equation (11.3)

---

[1] For a discussion of particle detectors see E. Segre, *Nuclei and Particles*, W. A. Benjamin Co., New York, 1965, Chapter III.

[2] The data are taken from J. M. Blatt and V. F. Weisskopf, *Theoretical Nuclear Physics*, John Wiley and Sons, New York, 1952, p. 482. The values of σ_t given contain contributions from elastic and inelastic scattering processes.

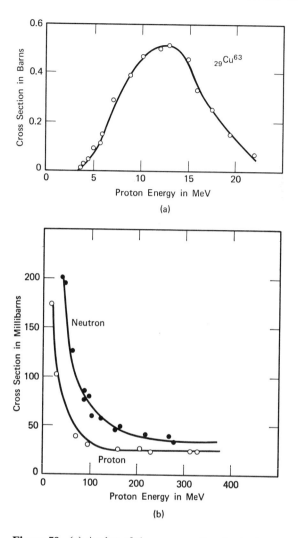

Figure 59. (a) A plot of the cross section for inelastic proton scattering by a copper nucleus against proton energy. [After S. N. Ghoshal, An experimental verification of the theory of the compound nucleus, *Phys. Rev.*, **80**: 939–942 (1950).] (b) The cross section for protons scattered elastically from protons (o) and neutrons (•) as a function of proton energy. [The data are from W. N. Hess, Summary of high-energy nucleon-nucleon cross-section data, *Rev. Mod. Phys.* **30**: 369–401 (1958).]

315

dN_Ω refers to the number of particles scattered per unit time into the unit solid angle element $d\Omega$. The relation between equations (11.1) and (11.3) is not hard to see:

$$\int_\Omega \sigma(\Omega)\, d\Omega = \sigma_t \tag{11.4}$$

11.2. THE COLLISION OF TWO PARTICLES

The form of equation (11.1) suggests that the host of scattering processes which take place when a beam of particles invades a target may be viewed as N two-particle collisions. It follows that the differential scattering cross section must be interpreted, in a mechanical sense, as the area available to an incident particle for striking a target particle and subsequently being scattered into a given element of solid angle. The magnitude of this area is not unique, however, but depends on the way the scattering process is envisioned, which is to say that it depends on the coordinate system chosen for describing the event. Two coordinate systems enjoy wide use; one is more suitable for an experimental description of two-particle scattering, while the other lends itself better to the theoretical analysis. For these reasons we must consider both reference systems as well as the relation between them.

In the *laboratory coordinate system*, a scattering process involves a particle of mass m_1 being fired with speed v_1 at another particle, of mass m_2, which is considered to be at rest. After collision, the first particle moves off at speed v_1' along a trajectory whose orientation in spherical polar space is expressed by the angles $(\vartheta_{L_1}, \varphi_{L_1})$. (See Figure 60.) The target particle recoils from the

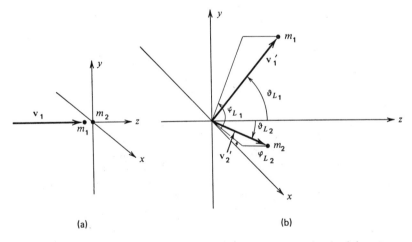

Figure 60. The elastic collision of two particles, as seen in the laboratory frame of reference. (a) Before the collision. (b) After the collision.

collision and so moves off with speed v_2' along a direction specified by the angles $(\vartheta_{L_2}, \varphi_{L_2})$. The differential scattering cross section in this coordinate system may be written, in terms of $(\vartheta_{L_1}, \varphi_{L_1})$,

$$\sigma_{L_1}(\vartheta_{L_1}, \varphi_{L_1}) \sin \vartheta_{L_1} \, d\vartheta_{L_1} \, d\varphi_{L_1}$$

for scattering into the solid angles between Ω_{L_1} and $\Omega_{L_1} + d\Omega_{L_1}$.

In the *center-of-mass coordinate system* a scattering process involves two particles moving toward the center of mass, which is itself considered to be at rest. The velocity of the incident particle is related to those in the laboratory frame by

$$\mathbf{u}_1 = \mathbf{v}_1 - \mathbf{v}_{CM} = \frac{m_2 \mathbf{v}_1}{M}$$

where $\mathbf{v}_1$ is the velocity of the particle in the laboratory reference frame,

$$\mathbf{v}_{CM} = \frac{m_1 \mathbf{v}_1 \oplus m_2 \mathbf{v}_2}{M} = \frac{m_1 \mathbf{v}_1}{M}$$

is the velocity of the center of mass in terms of the velocities of the incident and target particles in the laboratory ($\mathbf{v}_2 = \mathbf{0}$ by hypothesis), and $M = m_1 + m_2$. Similarly, the velocity of the target particle is

$$\mathbf{u}_2 = \mathbf{v}_2 - \mathbf{v}_{CM} = -\frac{m_1 \mathbf{v}_1}{M}.$$

After the collision process has occurred, the speeds of the two particles remain unchanged. (Energy and momentum are conserved!) What has changed is the line of centers of the pair; it has been rotated so that it has an orientation, relative to the original line of centers, expressed by the angles $(\vartheta_{CM}, \varphi_{CM})$. (See Figure 61.) The differential scattering cross section is now

$$\sigma_{CM}(\vartheta_{CM}, \varphi_{CM}) \sin \vartheta_{CM} \, d\vartheta_{CM} \, d\varphi_{CM}.$$

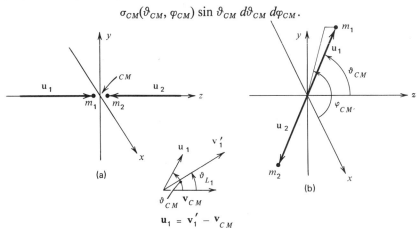

Figure 61. The elastic collision of two particles, as seen in the center-of-mass frame of reference. (a) Before the collision. (b) After the collision.

Scattering Theory

It is evident that the two scattering cross sections must be equivalent. We can find the equivalence relation by writing down the equations relating laboratory coordinates to center-of-mass coordinates. They are:

$$\varphi_{L_1} = \varphi_{CM}$$
$$v_1' \sin \vartheta_{L_1} = u_1 \sin \vartheta_{CM} \qquad (11.5)$$
$$v_1' \cos \vartheta_{L_1} = u_1 \cos \vartheta_{CM} + v_{CM}.$$

The last two equations may be rewritten as

$$\tan \vartheta_{L_1} = \frac{\sin \vartheta_{CM}}{[(v_{CM}/u_1) + \cos \vartheta_{CM}]}. \qquad (11.6)$$

Now, by the first of equations (11.5),

$$d\varphi_{L_1} = d\varphi_{CM}.$$

Also, by equation (11.6) and the relation derived from the last two of equations (11.5),

$$v_1'^2 = u_1^2 + v_{CM}^2 + 2u_1 v_{CM} \cos \vartheta_{CM},$$

we can write

$$\cos \vartheta_{L_1} = \frac{(v_{CM}/u_1) + \cos \vartheta_{CM}}{[1 + (v_{CM}/u_1)^2 + 2(v_{CM}/u_1) \cos \vartheta_{CM}]^{\frac{1}{2}}}. \qquad (11.7)$$

It follows that

$$\sin \vartheta_{L_1} \, d\vartheta_{L_1} = \frac{[1 + (V_{CM}/u_1) \cos \vartheta_{CM}]}{[1 + (v_{CM}/u_1)^2 + 2(v_{CM}/u_1) \cos \vartheta_{CM}]^{\frac{3}{2}}} \sin \vartheta_{CM} \, d\vartheta_{CM}$$

and, therefore, that

$$\sigma_{L_1}(\vartheta_{L_1}, \varphi_{L_1}) = \left\{ \frac{[1 + (v_{CM}/u_1)^2 + 2(v_{CM}/u_1) \cos \vartheta_{CM}]^{\frac{3}{2}}}{|1 + (v_{CM}/u_1) \cos \vartheta_{CM}|} \right\} \sigma_{CM}(\vartheta_{CM}, \varphi_{CM}). \qquad (11.8)$$

The absolute value sign is to preserve the positive value essential to the definition of the differential scattering cross section. When the scattered particle and the target particle are of the same mass, equation (11.8) becomes

$$\sigma_{L_1}(\vartheta_{L_1}, \varphi_{L_1}) = 4 \cos \left(\frac{\vartheta_{CM}}{2} \right) \sigma_{CM}(\vartheta_{CM}, \varphi_{CM}).$$

In this case, as ϑ_{CM} varies from zero to π radians, the laboratory colatitudinal angle may vary from zero to $\pi/2$ radians, as indicated by equation (11.7) when v_{CM} is put equal to u_1.

When (v_{CM}/u_1) is less than unity, equation (11.7) shows that $(\vartheta_{CM}/2) < \vartheta_{L_1} < \vartheta_{CM}$ for all ϑ_{CM} between zero and π radians. When (v_{CM}/u_1) is greater than one equation (11.7) has a maximum at $\vartheta_{CM} = \cos^{-1}(-u_1/v_{CM})$ corresponding to

$$\vartheta_{L_1}(\max) = \sin^{-1}\frac{u_1}{v_{CM}}.$$

Evidently, ϑ_{L_1} increases with ϑ_{CM} to $\vartheta_{L_1}(\max)$ and then falls off to zero once more with further increases in ϑ_{CM}.

11.3. PARTIAL WAVE ANALYSIS: THE PHASE SHIFT

The classical mechanical aspects of the scattering process have provided us with a fundamental quantity—the scattering cross section. Now we must turn to the quantum physical problem of deducing what must be the value of this quantity for a given collision situation. With regard to the relative motion of the incident particle, the quantum problem involves the solution of the eigenvalue equation (neglecting spin forces)

$$-\frac{\hbar^2}{2\mu}\nabla_r^2\psi + V(\mathbf{r})\psi(\mathbf{r}) = E\psi(\mathbf{r}). \tag{11.9}$$

where μ is the reduced mass of the incident particle and $V(\mathbf{r})$ describes the interaction between that particle and the target. We shall see now that the solution of equation (11.9) for very large values of $\mathbf{r}$ is closely related to the differential scattering cross-section. When the incident particle is very far from the target we expect on physical grounds that $\psi(\mathbf{r})$ will be a superposition of two states: the state of the unscattered particle and that of the particle long since scattered. The former state is a solution of equation (11.9) for $V(\mathbf{r})$ equal to zero. Its form, as we know from section 3.5, is

$$\psi(\mathbf{r})_{\,V(\mathbf{r})\to 0} \exp[i(\mathbf{k}\cdot\mathbf{r}),$$

where

$$k^2 = \frac{2\mu E}{\hbar^2}$$

and $\mathbf{k}$ is construed to be the wavenumber vector for the incident particle, the direction of $\mathbf{k}$ being the direction of (de Broglie) wave propagation. After the incident particle has been scattered, the solution of equation (11.9) will be *specified* to be the product of the solution of the radial equation

$$\frac{d^2R}{dr^2} + k^2R(r) = 0$$

and a function of the spherical polar angles ϑ and φ which reflects the

319

direction of propagation of the scattered particle. The form of the radial equation given is meaningful to the scattering problem only if

$$\operatorname*{Lim}_{r\to\infty} rV(\mathbf{r}) = 0,$$

a condition we now impose upon the potential function $V(\mathbf{r})$. The asymptotic solution to equation (11.9) is, therefore,

$$\psi(\mathbf{r}) \underset{r\to\infty}{\sim} \exp\left[i(\mathbf{k}\cdot\mathbf{r})\right] + \frac{\exp(ikr)}{r} f(\vartheta, \varphi) \tag{11.10}$$

where we have put[3]

$$R(r) = \frac{\exp(ikr)}{r}.$$

Now, the number of scattered particles crossing in unit time an element of spherical surface a distance R from the target particle is equal to

(density of *scattered* particles at R) × (speed of scattered particles) × (element of spherical surface)

$$= \left|\frac{\exp(ikR)f(\vartheta, \varphi)}{R}\right|^2 \times \frac{\hbar k}{\mu} \times R^2 \, d\Omega = \frac{\hbar k}{\mu} |f(\vartheta, \varphi)|^2 \, d\Omega,$$

so long as R is very large compared with the range of $V(r)$. Also, by equation (11.3),

$$\sigma(\vartheta, \varphi) \, d\Omega = \frac{\text{number of scattered particles crossing } R^2 \, d\Omega \text{ in unit time}}{\text{number of incident particles crossing unit area in unit time}}$$

$$= \frac{(\hbar k/\mu)|f(\vartheta, \varphi)|^2 \, d\Omega}{(\hbar k/\mu)} = |f(\vartheta, \varphi)|^2 \, d\Omega$$

since the density of incident particles, $|\psi(\mathbf{r})|^2$ for $r \to \infty$, is unity. Therefore

$$\sigma(\vartheta, \varphi) = |f(\vartheta, \varphi)|^2. \tag{11.11}$$

If the complete form of the asymptotic state expressed by equation (11.10) can be found for a given scattering process, the differential scattering cross section can be calculated immediately. The problem of finding that state, of course, depends entirely on the form of the potential function $V(r)$, since equation (11.10) is supposed to represent a solution of equation (11.9). Usually the problem cannot be solved exactly and one of a number of available approximation techniques must be used to ferret out the differential

---

[3] The solution $\exp(ikr)$ has been chosen instead of $\exp(-ikr)$ because the former represents an *outgoing* (spherical) wave as is essential to the physical interpretation of the second term in equation (11.10)—the state of a *scattered* particle. See footnote 12 of Chapter 3 for more remarks.

scattering cross section. These special techniques will not concern us here, as we are after principle, not practice.[4]

Let us return to the "homogeneous" form of equation (11.9), namely,

$$(\nabla_r^2 + k^2)\varphi(\mathbf{r}) = 0. \tag{11.12}$$

In spherical polar space this equation may be separated into the two differential equations

$$\frac{1}{r^2} \frac{d}{dr}\left(r^2 \frac{dR}{dr}\right) + \left[k^2 - \frac{l(l+1)}{r^2}\right]R(r) = 0 \tag{11.13}$$

and

$$\frac{1}{\sin\vartheta} \frac{\partial}{\partial\vartheta}\left(\sin\vartheta \frac{\partial F}{\partial\vartheta}\right) + \frac{1}{\sin^2\vartheta} \frac{\partial^2 F}{\partial\varphi^2} + l(l+1)F(\vartheta, \varphi) = 0. \tag{11.14}$$

The solutions of equation (11.14) are of course the spherical harmonics. If the substitution

$$x = kr$$

is made in equation (11.13), that expression is brought into the form

$$x^2 \frac{d^2 R}{dx^2} + 2x \frac{dR}{dx} + [x^2 - l(l+1)]R(x) = 0. \tag{11.15}$$

The solutions of this eigenvalue problem, which are not infinite at $x = 0$, are the *spherical Bessel functions*

$$j_l(x) = \left(\frac{\pi}{2x}\right)^{\frac{1}{2}} J_{l+\frac{1}{2}}(x), \tag{11.16}$$

where

$$J_{l+\frac{1}{2}}(x) = \sum_{k=0}^{\infty} \frac{(-1)^k}{k!\,(k+l+\frac{1}{2})!} \left(\frac{x}{2}\right)^{2k+l+\frac{1}{2}}$$

is the solution of *Bessel's equation*

$$x^2 \frac{d^2 J_n}{dx^2} + x \frac{dJ_n}{dx} + (x^2 - n^2)J_n(x) = 0,$$

n being any number. The general solution of equation (11.12) is a linear combination of the basis vectors made up of a spherical Bessel function multiplied by a spherical harmonic:

$$\varphi(\mathbf{r}) = \sum_{l=0}^{\infty} \sum_{m=-l}^{l} c_{lm} j_l(kr) P_l^m(\cos\vartheta) \exp(im\varphi). \tag{11.17}$$

[4] See, however, H. S. W. Massey, Theory of atomic collisions, *Handbuch der Physik* **XXXVI**: 232–306 (1956), for a discussion of approximation methods.

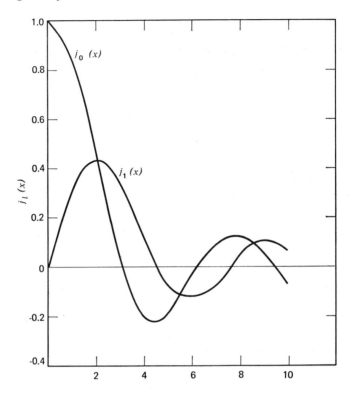

Figure 62. A plot of the first two spherical Bessel functions.

(We have absorbed the normalization constants for the spherical harmonics into the c_{lm}). Equation (11.17) must be compatible with the solution of equation (11.9) for $V(\mathbf{r}) = 0$. Therefore,

$$\exp\,[i(\mathbf{k}\cdot\mathbf{r})] = \sum_{l=0}^{\infty}\sum_{m=-l}^{l} c_{lm}j_l(kr)P_l^{m}(\cos\vartheta)\exp\,(im\varphi). \qquad (11.17')$$

Moreover, we can put the fact that the spherical harmonics form a set of basis vectors to work and write

$$f(\vartheta,\,\varphi) = \sum_{l=0}^{\infty}\sum_{m=-l}^{l} f_{lm}P_l^{m}(\cos\vartheta)\exp\,(im\varphi).$$

By the two foregoing equations, it follows that the asymptotic solution of equation (11.9) may be written

$$\psi(\mathbf{r})\underset{r\to\infty}{\sim}\sum_{l=0}^{\infty}\sum_{m=-l}^{l}\left(c_{lm}j_l(kr) + \frac{\exp\,(ikr)}{r}f_{lm}\right)P_l^{m}(\cos\vartheta)\exp\,(im\varphi). \qquad (11.18)$$

In order to arrive at equation (11.18) we have had to assume that $V(\mathbf{r})$ varies more rapidly than does r^{-1} and that the scattered particle state is uniquely given by the second term on the right-hand side of equation (11.10). We can simplify matters considerably if the potential function is postulated to be independent of ϑ and φ, in which case

$$\psi(\mathbf{r}) \underset{r \to \infty}{\sim} \sum_{l=0}^{\infty} \left(c_l j_l(kr) + \frac{\exp(ikr)}{r} f_l \right) P_l(\cos \vartheta) \tag{11.19}$$

may be written down instead of equation (11.18) by choosing the direction of $\mathbf{k}$ to be that of the z-axis. [This done, $(\mathbf{k} \cdot \mathbf{r})$ becomes $kr \cos \vartheta$, which is independent of φ. We may also choose the scattered particle state to be φ-independent in this case, as is evident from the analysis carried out in the foregoing section.]

Equation (11.19) was achieved by considering equation (11.12). It should also be attainable by looking at the asymptotic form of the general solution of equation (11.9), since it should not matter whether we solve an asymptotic form of equation (11.9) or solve equation (11.9) and then let r get large. Ignoring the azimuthal angle, we may write the incident particle state as

$$\psi(r, \vartheta) = \sum_{l=0}^{\infty} \frac{P_l(r)}{r} P_l(\cos \vartheta), \tag{11.20}$$

where $P_l(r)$ is the solution of

$$\left[\frac{d^2}{dr^2} + (k^2 - U(r)) \right] P_l(r) = 0, \tag{11.21}$$

the effective potential $U(r)$ being

$$U(r) = \frac{2\mu}{\hbar^2} V(r) + \frac{l(l+1)}{r^2}.$$

Now, when r is large enough, equation (11.21) moves toward the form

$$\left(\frac{d^2}{dr^2} + k^2 \right) P_l(r) = 0$$

which has the general solution

$$P_l(r) = A_l' \sin(kr + \delta_l'),$$

where A' and δ_l' are arbitrary constants. Picking up the hint, let us define two functions $A_l(r)$ and $\delta_l(r)$ such that

$$P_l(r) = A_l(r) \sin(kr + \delta_l(r))$$

$$\frac{dP_l}{dr} = A_l(r) k \cos(kr + \delta_l(r)) \tag{11.22}$$

for *all* values of r. Upon putting these expressions into equation (11.21) we find

$$k \cos [kr + \delta_l(r)] \frac{dA_l}{dr} - kA_l(r) \sin [kr + \delta_l(r)] \frac{d\delta_l}{dr}$$

$$- U(r)A_l(r) \sin [kr + \delta_l(r)] = 0.$$

But equations (11.22) imply

$$\sin [kr + \delta_l(r)] \frac{dA_l}{dr} + A_l(r) \cos [kr + \delta_l(r)] \frac{d\delta_l}{dr} = 0,$$

which permits us to reduce equation (11.21) to the two first-order differential equations

$$\frac{dA_l}{dr} = \frac{A_l(r)U(r)}{2k} \sin 2[kr + \delta_l(r)]$$

$$\frac{d\delta_l}{dr} = -\frac{U(r)}{k} \sin^2 [kr + \delta_l(r)]. \tag{11.23}$$

The formal solutions of these two equations are

$$A_l(r) = A_l(0) \exp \left\{ \int_0^r \frac{U(r')}{2k} \sin 2[kr' + \delta_l(r')] \, dr' \right\}$$

$$\delta_l(r) = \delta_l(0) - \int_0^r \frac{U(r')}{k} \sin^2 [kr' + \delta_l(r')] \, dr',$$

respectively. As r approaches infinity, $A_l(r)$ approaches a finite limit because $U(\infty)$ is zero; $\delta_l(r)$ does the same since $U(r)$ goes to zero with increasing r faster than does r^{-1}, by hypothesis. We may conclude, then, that the asymptotic form of $P_l(r)$ is

$$P_l(r) \underset{r \to \infty}{\sim} A_l \sin [kr - \tfrac{1}{2}l\pi + \delta_l] \tag{11.24}$$

where

$$A_l = A_l(\infty)$$
$$\delta_l = \delta_l(\infty) + \tfrac{1}{2}\pi l,$$

the additive term $\tfrac{1}{2}l\pi$ being for the purpose of making δ_l zero when $V(\mathbf{r})$ is zero. [When $V(\mathbf{r})$ is zero and the substitution $R(r) = P_l(r)/r$ has been made, equation (11.21) becomes identical with equation (11.15), whose asymptotic solution is[5]

$$j_l(kr) \underset{kr \to \infty}{\sim} \frac{\sin (kr - \tfrac{1}{2}\pi l)}{kr}. \tag{11.25}$$

[5] See G. Arfken, *Mathematical Methods for Physicists*, Academic Press, New York, 1966, Chapter 11.

Hence the term $\frac{1}{2}\pi l$ in equation (11.24).] Equation (11.20) may now be written

$$\psi(r, \vartheta) \underset{r \to \infty}{\sim} \sum_{l=0}^{\infty} \frac{A_l}{r} \sin (kr - \tfrac{1}{2}l\pi + \delta_l)P_l(\cos \vartheta). \qquad (11.26)$$

This expression and equation (11.19) should say the same thing. However, in the two equations there are four constants to be deduced—something which cannot be done uniquely. To alleviate the problem somewhat, we can calculate the coefficients c_l as follows. If equation (11.17') is differentiated with respect to kr, we find

$$i \cos \vartheta \exp (ikr \cos \vartheta) = \sum_{l=0}^{\infty} c_l \frac{dj_l}{d(kr)} P_l(\cos \vartheta). \qquad (11.27)$$

But, taking note of the recurrence relation

$$(2l + 1) \cos \vartheta P_l(\cos \vartheta) = (l + 1)P_{l+1}(\cos \vartheta) + lP_{l-1}(\cos \vartheta),$$

we can also write

$$i \cos \vartheta \exp (ikr \cos \vartheta) = i \sum_{l=0}^{\infty} c_l j_l(kr) \cos \vartheta P_l(\cos \vartheta)$$

$$= i \sum_{l=0}^{\infty} c_l j_l(kr) \left[\frac{l+1}{(2l+1)} P_{l+1}(\cos \vartheta) \right.$$

$$\left. + \frac{l}{(2l+1)} P_{l-1}(\cos \vartheta) \right]$$

$$= i \sum_{l=0}^{\infty} \left(c_{l+1} j_{l+1}(kr) \frac{l+1}{2l+3} \right) P_l(\cos \vartheta)$$

$$+ i \sum_{l=1}^{\infty} \left(c_{l-1} j_{l-1}(kr) \frac{l}{2l-1} \right) P_l(\cos \vartheta). \qquad (11.28)$$

With the recurrence relations[5]

$$(2l + 1)j_l(kr) = kr[j_{l+1}(kr) + j_{l-1}(kr)]$$

$$j_{l-1}(kr) = \frac{1}{(kr)^{l+1}} \frac{d}{d(kr)} [(kr)^{l+1}j_l(kr)]$$

we can write equation (11.27) as

$$i \cos \vartheta \exp (ikr \cos \vartheta)$$

$$= \sum_{l=0}^{\infty} c_l \left(\frac{l}{2l+1} j_{l-1}(kr) - \frac{l+1}{2l+1} j_{l+1}(kr) \right) P_l(\cos \vartheta). \qquad (11.29)$$

Equations (11.28) and (11.29) must be the same. This means

$$ic_{l+1}j_{l+1}(kr)\frac{l+1}{2l+3} + ic_{l-1}j_{l-1}(kr)\frac{l}{2l-1}$$

$$= c_l\frac{l}{2l+1}j_{l-1}(kr) - c_l\frac{l+1}{2l+1}j_{l+1}(kr)$$

or

$$l\left(\frac{c_l}{2l+1} - \frac{i}{2l-1}c_{l-1}\right)j_{l-1}(kr)$$

$$= (l+1)\left(\frac{c_l}{2l+1} + \frac{i}{2l+3}c_{l+1}\right)j_{l+1}(kr) \qquad (l \geqslant 1).$$

The equation is not generally valid unless the expressions in parentheses are zero identically:

$$c_l = i\frac{(2l+1)}{(2l-1)}c_{l-1} \qquad (l = 1, 2, \ldots)$$

or

$$c_l = (2l+1)i^l c_0 \qquad (l = 1, 2, \ldots).$$

Equation (11.19) is, therefore,

$$\psi(\mathbf{r}) \underset{r\to\infty}{\sim} \sum_{l=0}^{\infty}\left[(2l+1)i^l j_l(kr) + \frac{\exp(ikr)}{r}f_l\right]P_l(\cos\vartheta) \qquad (11.30)$$

if we put c_0 equal to one [as we must to make equation (11.17') true for $r = 0$]. By equation (11.25), moreover,

$$\psi(\mathbf{r}) \underset{r\to\infty}{\sim} \sum_{l=0}^{\infty}\left[\frac{(2l+1)i^l}{kr}\sin(kr - \tfrac{1}{2}l\pi) + \frac{\exp(ikr)}{r}f_l\right]P_l(\cos\vartheta),$$

so that we must have in equation (11.26)

$$\left.\begin{array}{l} A_l = i^l\dfrac{2l+1}{k}\exp(i\delta_l) \\[2ex] f_l = \dfrac{2l+1}{k}\sin\delta_l\exp(i\delta_l) \qquad (l = 0, 1, \ldots) \end{array}\right\} \qquad (11.31)$$

upon writing the sine functions as exponentials and separating incoming and outgoing spherical waves.

Now that we know the coefficients f_l $(l = 0, 1, \ldots)$ we can write down $f(\vartheta)$:

$$f(\vartheta) = \frac{1}{k}\sum_{l=0}^{\infty}(2l+1)\sin\delta_l\exp(i\delta_l)P_l(\cos\vartheta). \qquad (11.32)$$

The sense of this equation hinges upon a physical interpretation for the constants δ_l $(l = 0, 1, \ldots)$. This can be obtained most easily by noting that

$$\frac{P_l(r)}{r} \underset{r \to \infty}{\sim} \frac{(2l + 1)}{2ikr} [(-1)^{l+1} \exp(-ikr) + \exp(ikr) \exp(2i\delta_l)]$$

$$(2l + 1)i^l j_l(kr) \underset{r \to \infty}{\sim} \frac{(2l + 1)}{2ikr} [(-1)^{l+1} \exp(-ikr) + \exp(ikr)].$$

The first expression refers to one of the partial waves making up the asymptotic state of the incident-scattered particle, while the state of a free particle is represented by the second expression. Notice that the incoming wave is the same for both particles. The outgoing waves differ in that the scattered particle has had its phase shifted by $2\delta_l$ relative to that of the free particle. Evidently the presence of the target particle is felt through the appearance of the *phase shift* δ_l in each of the scattered particle state's partial waves.

From equation (11.32) we easily deduce

$$\sigma(\vartheta) = \left| k^{-2} \sum_{l=0}^{\infty} \sum_{l'=0}^{\infty} (2l + 1)(2l' + 1) \right.$$

$$\left. \times \exp[i(\delta_l - \delta_{l'})] \sin \delta_l \sin \delta_{l'} P_l(\cos \vartheta) P_{l'}(\cos \vartheta) \right| \quad (11.33)$$

and

$$\sigma_t = \int_0^{2\pi} \int_0^{\pi} \sigma(\vartheta) \sin \vartheta \, d\vartheta \, d\varphi = \frac{4\pi}{k^2} \sum_{l=0}^{\infty} (2l + 1) \sin^2 \delta_l, \quad (11.34)$$

since

$$\int_{-1}^{1} P_l(z) P_{l'}(z) \, dz = \frac{2}{2l + 1} \delta_{ll'}.$$

The scattering cross sections have now been put into forms dependent on the phase shift. The latter quantity can in principle be calculated once the potential function $V(\mathbf{r})$ is known. One general statement can be made, however. The contribution to the scattering cross section σ_t from the state of angular momentum $[l(l + 1)]^{1/2}\hbar$ is σ_l, given by

$$\sigma_l = \frac{4\pi}{k^2} (2l + 1) \sin^2 \delta_l. \quad (11.35)$$

This quantity is a maximum when the phase shift is a half-integral multiple of π.

11.4. AN EXAMPLE: SCATTERING BY A HARD SPHERE

Suppose that the interaction between the incident particle and the target may be expressed by

$$V(r) = \begin{cases} \infty & r < r_0 \\ 0 & r > r_0 \end{cases} \tag{11.36}$$

where r_0 may be interpreted as the distance of closest approach between the two particles. This kind of potential function represents a "hard sphere" in the sense that the interaction is infinitely repulsive in the interior of the region $0 \leqslant r < r_0$; it is easy to see that the function satisfies our condition

$$\operatorname*{Lim}_{r \to \infty} rV(r) = 0.$$

Equation (11.36) restricts the incident particle from entering the region $r < r_0$. Therefore, we can rewrite the solution of equation (11.15) as

$$j_l(x) + n_l(x),$$

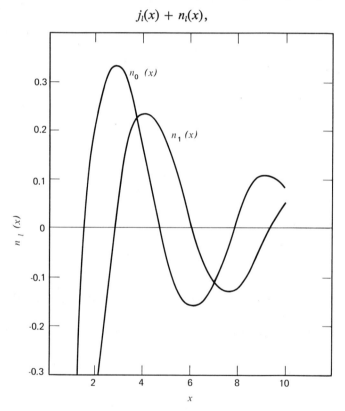

Figure 63. A plot of the first two spherical Neumann functions.

where $n_l(x)$ is the *spherical Neumann function*[5]

$$n_l(x) = (-1)^{l+1}\left(\frac{\pi}{2x}\right)^{\frac{1}{2}} J_{l-\frac{1}{2}}(x),$$

$J_{l-\frac{1}{2}}(x)$ being a solution of Bessel's equation, mentioned previously. Unlike the spherical Bessel function, $n_l(x)$ is singular at the origin:

$$n_l(x) \underset{x \to 0}{\sim} O(x^{-l-1}).$$

This fact prevented us from writing it down before, as we had no guarantee that the form of $V(r)$ would in general insure the finiteness of $\varphi(\mathbf{r})$ as $r \to 0$. With the addition of the spherical Neumann functions, equation (11.24) may be rewritten as

$$P_l(r) \underset{r \to \infty}{\sim} \sin\left(kr - \tfrac{1}{2}l\pi + \delta_l\right)$$

$$= kr[\cos \delta_l j_l(kr) + \sin \delta_l n_l(kr)], \qquad (r > r_0) \quad (11.37)$$

where we have chosen the normalization of $P_l(r)$ to be such that the coefficient A_l drops out and have taken advantage of[5]

$$\sin\left(kr - \tfrac{1}{2}l\pi + \delta_l\right) = \sin\left(kr - \tfrac{1}{2}l\pi\right)\cos\delta_l + \cos\left(kr - \tfrac{1}{2}l\pi\right)\sin\delta_l,$$

$$j_l(x) \underset{x \to \infty}{\sim} \frac{\sin\left(x - \tfrac{1}{2}l\pi\right)}{x}.$$

$$n_l(x) \underset{x \to \infty}{\sim} \frac{\cos\left(x - \tfrac{1}{2}l\pi\right)}{x}.$$

That we can write equation (11.37) as an equality follows from the disappearance of $V(r)$ for $r > r_0$. If equation (11.37) is written in terms of exponentials in the phase shift, we find easily

$$\exp(2i\delta_l) = -\frac{j_l(kr_0) + in_l(kr_0)}{j_l(kr_0) - in_l(kr_0)} \qquad (11.38)$$

since $P_l(r)$ must vanish at r_0. Now,

$$\exp(2i\delta_l) = \cos 2\delta_l + i\sin 2\delta_l = -\frac{j_l(kr_0) + in_l(kr_0)}{j_l(kr_0) - in_l(kr_0)}$$

or

$$\cos 2\delta_l j_l(kr_0) + \sin 2\delta_l n_l(kr_0) = -j_l(kr_0)$$

upon equating real and imaginary parts of equation (11.38). By employing the trigonometric identities for $\cos 2\delta_l$ and $\sin 2\delta_l$ we find with no trouble

$$\sin^2 \delta_l = \frac{j_l^2(kr_0)}{j_l^2(kr_0) + n_l^2(kr_0)}$$

Scattering Theory

and, therefore,

$$\sigma_l = \frac{4\pi}{k^2} (2l + 1) \frac{j_l{}^2(kr_0)}{j_l{}^2(kr_0) + n_l{}^2(kr_0)} \tag{11.39}$$

by equation (11.35). The scattering cross-section is just a sum of terms like that given by equation (11.39). If the primary contribution to σ_t is from the state of lowest angular momentum ("s-wave scattering") we can write

$$\sigma_t = \sigma_0 = 4\pi r_0{}^2 \left[\frac{\sin (kr_0)}{kr_0} \right]^2 \tag{11.40}$$

since[5]

$$j_0(x) = \frac{\sin x}{x}, \qquad n_0(x) = -\frac{\cos x}{x}.$$

A plot of equation (11.40) is given in Figure 64. We notice that the s-wave scattering cross section drops off sharply as the incident particle's (de Broglie) wavelength decreases. It should not be inferred that the scattering cross section *in toto* drops off as the wavelength decreases, because the contributions of higher-order terms in equation (11.34) become important as the energy increases. It follows that equation (11.40) is inaccurate under these circumstances. At quite high energies, in fact, it is found that[6]

$$\sigma_t \underset{k \to \infty}{\sim} 2\pi r_0{}^2 \tag{11.41}$$

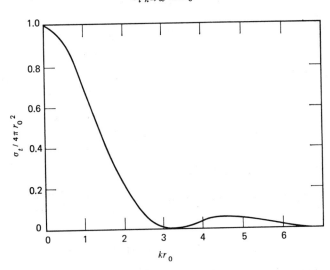

Figure 64. A plot of the cross-section for s-wave scattering by a hard sphere.

[6] A simple derivation of equation (11.41) is given by A. Messiah, *Quantum Mechanics*, John Wiley and Sons, New York, 1961, Vol. I, p. 394*f*.

a result which differs from what might be expected classically (πr_0^2). Thus, even in the limit of every short de Broglie wavelength, the quantum aspects of the scattering process may not be neglected. This is because the potential function (11.36) is never slowly varying in the region of the target and so does not permit a classical-mechanical analysis of the scattering problem. [If $V(r)$ *were* slowly-varying we might neglect it and work only with the kinetic energy portion of the Hamiltonian—which, of course, always admits a "classical" analysis in infinite free space.] Indeed, the sudden change in $V(r)$ at $r = r_0$ in effect "diffracts" the de Broglie wave of the incident particle. It is only to be expected that the quantum aspects of the process will be important at all times.

11.5. THE BORN APPROXIMATION

There is a number of methods available for calculating phase shifts when $V(r)$ is not so simple a function as considered in the previous section. Most of these depend upon the establishment of an integral representation for the phase shift, in the sense that some function of δ_l is made to depend upon an integral over some function of $V(r)$. One of the most useful of these representations is that which demonstrates how the phase shift changes when the interaction between the target and the incident particle changes. We should like to derive it now.

Consider two equations of the form of equation (11.21):

$$\left\{\frac{d^2}{dr^2} + \left[k^2 - U_1(r) - \frac{l(l+1)}{r^2}\right]\right\}P_l^{(1)}(r) = 0 \qquad (11.42)$$

$$\left\{\frac{d^2}{dr^2} + \left[k^2 - U_2(r) - \frac{l(l+1)}{r^2}\right]\right\}P_l^{(2)}(r) = 0 \qquad (11.43)$$

where now

$$U(r) = \frac{2\mu}{\hbar^2} V(r)$$

and it is to be clearly understood that $P_l^{(1)}(r)$ and $P_l^{(2)}(r)$ refer to states of the *same* total energy. We now define the function

$$W(P_l^{(1)}, P_l^{(2)}) \equiv P_l^{(1)} \frac{dP_l^{(2)}}{dr} - P_l^{(2)} \frac{dP_l^{(1)}}{dr}$$

and note that, upon multiplying equation (11.42) by $P_l^{(2)}(r)$, equation (11.43), by $P_l^{(1)}(r)$, and subtracting the latter from the former, we get

$$P_l^{(2)} \frac{d^2 P_l^{(1)}}{dr^2} - P_l^{(1)} \frac{d^2 P_l^{(2)}}{dr^2} + (U_2(r) - U_1(r))P_l^{(1)}P_l^{(2)} = 0$$

or

$$\frac{dW}{dr} - (U_2(r) - U_1(r))P_l^{(1)}P_l^{(2)} = 0.$$

If this equation is integrated between two values of r, say, a and b, we find

$$W(P_l^{(1)}, P_l^{(2)})\big|_a^b = \int_a^b (U_2(r) - U_1(r))P_l^{(1)}P_l^{(2)}\, dr. \tag{11.44}$$

The point of all this manipulation is that, since

$$P_l^{(i)}(r) \underset{r \to \infty}{\sim} \sin(kr - \tfrac{1}{2}l\pi + \delta_l^{(i)}) \qquad (i = 1, 2),$$

we have

$$\begin{aligned} W(P_l^{(1)}, P_l^{(2)}) \underset{r \to \infty}{\sim} {} & k[\sin(kr - \tfrac{1}{2}l\pi + \delta_l^{(1)}) \cos(kr - \tfrac{1}{2}l\pi + \delta_l^{(2)}) \\ & - \cos(kr - \tfrac{1}{2}l\pi + \delta_l^{(1)}) \sin(kr - \tfrac{1}{2}\pi l + \delta_l^{(2)})] \\ = {} & k\sin(\delta_l^{(1)} - \delta_l^{(2)}). \end{aligned}$$

It follows from this result and equation (11.44) that

$$\sin(\delta_l^{(1)} - \delta_l^{(2)}) = \frac{2\mu}{\hbar^2 k}\int_0^\infty P_l^{(2)}(V_2(r) - V_1(r))P_l^{(1)}\, dr \tag{11.45}$$

which is what we were looking for. It must be remembered, of course, that the validity of the derivation requires that the potential functions be integrable, that they vanish more rapidly than r^{-1} for large r, and that they be no more singular than is r^{-2} at the origin. (Quite a lot of restrictions!) In the special case that $V_2(r)$ is equal to zero, equation (11.45) becomes

$$\sin \delta_l = -\frac{2\mu}{\hbar^2}\int_0^\infty j_l(kr)V(r)P_l(r)r\, dr \tag{11.46}$$

since then

$$P_l^{(2)}(r) = krj_l(kr).$$

Returning to equation (11.45), we note that if the difference between potential functions is small we can write

$$\Delta\delta_l = -\frac{2\mu}{\hbar^2 k}\int_0^\infty P_l^2(r)\, \Delta V(r)\, dr \tag{11.47}$$

since $P_l^{(1)}$ and $P_l^{(2)}$ should be virtually the same. This result tells us that the phase shift varies in the direction opposite that of the (monotonic) variation of the potential function. Therefore, if the potential becomes more repulsive, the phases of the scattered partial waves are made more nearly the same as those of a free particle. (Recall the discussion in section 11.3.) In other words, the more a target repels an incident particle, the less will be its effect upon the particle after scattering—a conclusion which seems quite reasonable, speaking physically. On the other hand, the phase shift and, therefore, the influence of the target, becomes more pronounced as the potential function gets more attractive. This, again, seems quite reasonable.

In general, equation (11.46) is not particularly easy to solve because a knowledge of the wavefunction for the incident particle is required. Such knowledge is obtained through the solution of equation (11.21) and so is tantamount to knowing the differential scattering cross section without the benefit of phase shifts. However, we need not solve the equation exactly if the interaction between target and incident particle is weak; for, in that case $P_l(r)$ will differ very little from the free-particle wavefunction and we may write

$$\sin \delta_l \doteq -\frac{2\mu}{\hbar^2} k \int_0^\infty \{j_l(kr)\}^2 V(r) r^2 \, dr. \tag{11.48}$$

Equation (11.48) is known as the *Born approximation*, after Max Born, who devised it in 1926. The expression is valid if the interaction is "weak"; that is, if

$$V(r) \ll E - \frac{l(l+1)}{2\mu r^2} \hbar^2.$$

This condition should obtain at high energies or large values of the total angular momentum [provided $V(r)$ is small enough at infinity].

Some idea of the validity of the Born approximation is given in Table 11.1, where phase shifts for the scattering of electrons by the (Hartree) average central field of the helium atom are listed. The Born approximation is seen to be only fair at low wavenumbers and angular momentum quantum numbers, but is found to be in good agreement with the results of accurate numerical integration of equation (11.9), when energy and angular momentum are large.

Table 11.1. Exact and Born-approximation phase shifts for the scattering of electrons by the spherically-averaged potential field of helium. The data are taken from ref. 4, p. 270.

Wavenumber in in units of a_0^{-1}	δ_0		δ_1		δ_2	
	Exact	Born	Exact	Born	Exact	Born
1	1.40	0.57	0.07	0.04	0.006	0.005
2	1.07	0.74	0.19	0.15	0.041	0.033
3	0.90	0.75	0.27	0.24	0.095	0.077
4	0.78	0.70	0.30	0.27	0.130	0.113
5	0.69	0.64	0.31	0.29	0.152	0.138

11.6. AN EXAMPLE: SCATTERING BY A SCREENED COULOMB POTENTIAL

A proton flying toward an atom is unlikely to be subjected to the full effect of the nuclear charge unless it penetrates deeply and safely into the array of electronic orbitals. Indeed, the interaction should be nil even at fairly close distances because of the strict charge neutrality of the atom; even when the interaction is strong it will probably resemble the coulomb potential only in an asymptotic sense. It follows that the potential function representing the interaction between a proton and an atom should be coulomb-like near the origin and should vanish much more rapidly than does r^{-1} for large values of r. An expression which has these properties is the *screened coulomb potential*

$$V(r) = \frac{Ze^2}{r} \exp(-\alpha r),$$

α being called the screening constant and providing a measure of the effective range of $V(r)$. Let us see what the Born approximation gives in this case.

First, it should be noted that in the Born approximation the phase shift decreases rapidly with increasing angular momentum quantum number. To see this we need only note that[7]

$$j_l(kr) \underset{l \to \infty}{\sim} \tfrac{1}{2}(lkr)^{-\frac{1}{2}} \left(\frac{lkr}{2l+1}\right)^{l+\frac{1}{2}}$$

and, therefore, that

$$\sin \delta_l \underset{l \to \infty}{\sim} -\frac{\mu}{2\hbar^2} \left(\frac{lk}{2l+1}\right)^{2l+1} \int_0^{1/\alpha} \frac{r^{2l+2}}{l} V(r)\, dr.$$

In the present case

$$r^2 |V(r)| < \frac{Ze^2}{\alpha}, \qquad 0 \leqslant r \leqslant \alpha^{-1},$$

and so

$$|\sin \delta_l| < \frac{Ze^2 \mu}{2\hbar^2} \left(\frac{lk}{2l+1}\right)^{2l+1} \frac{(1/\alpha)^{2l+2}}{l(2l+1)}.$$

We see that the phase shift is quite small for large l. (This behavior is evident also from Table 11.1.) Thus only δ_0 will be computed for the screened coulomb potential.

[7] See H. and B. S. Jeffereys, *Methods of Mathematical Physics*, Cambridge University Press, London, 1956, p. 527.

By equation (11.48) we have

$$|\sin \delta_0| = \frac{2Ze^2\mu}{\hbar^2} k \int_0^\infty \{j_0(kr)\}^2 r e^{-\alpha r}\, dr$$

$$= \frac{2Ze^2\mu}{k\hbar^2} \int_0^\infty \frac{e^{-\alpha r}}{r} \sin^2 kr\, dr = \frac{2Ze^2\mu}{k\hbar^2} \int_0^\infty \frac{e^{-\alpha r}}{r} \sin kr \int_0^k \cos k'r\, dk'\, dr$$

$$= \frac{Ze^2\mu}{k\hbar^2} \int_0^k \int_0^\infty e^{-\alpha r}[\sin (k' + k)r - \sin (k' - k)r]\, dr\, dk'$$

$$= \frac{Ze^2\mu}{k\hbar^2} \int_0^k \left[\frac{(k' + k)}{(k' + k)^2 + \alpha^2} - \frac{(k' - k)}{(k' - k)^2 + \alpha^2} \right] dk'$$

$$= \frac{Ze^2\mu}{2\hbar^2 k} \ln \left(1 + \frac{4k^2}{\alpha^2} \right) \doteq \frac{Ze^2\mu}{\hbar^2} \frac{2k}{\alpha^2 + (2k)^2}$$

where, in the last step, it has been assumed that the range of the screened coulomb potential is much smaller than the de Broglie wavelength of the electron: $k \ll \alpha$. Using equation (11.34) and the approximation

$$(\alpha^2 + 2k)^2 \doteq \alpha^4 + 4\alpha^2 k^2,$$

we finally arrive at the scattering cross section

$$\sigma_t = \frac{16\pi\mu^2 Z^2 e^4}{\alpha^2\hbar^4} \frac{1}{(\alpha^2 + 4k^2)}. \tag{11.49}$$

It is not an obvious jump from here to the differential scattering cross section. To calculate that quantity we must take advantage of the smallness of the phase shift and rewrite equation (11.32) as

$$|f(\vartheta)| = \frac{2Ze^2\mu}{\hbar^2} \sum_{l=0}^\infty (2l + 1) \int_0^\infty [j_l(kr)]^2 r e^{-\alpha r}\, dr P_l(\cos \vartheta)$$

$$= \frac{2Ze^2\mu}{\hbar^2} \int_0^\infty \left\{ \sum_{l=0}^\infty (2l + 1)[j_l(kr)]^2 P_l(\cos \vartheta) \right\} r e^{-\alpha r}\, dr$$

$$= \frac{2Ze^2\mu}{\hbar^2} \int_0^\infty \frac{\sin ar}{ar} r e^{-\alpha r}\, dr = \frac{2Ze^2\mu}{\hbar^2} \frac{1}{\alpha^2 + a^2},$$

where

$$a = 2k \sin \frac{\vartheta}{2}$$

and we have used the identity[8]

$$\frac{\sin ar}{ar} = \sum_{l=0}^\infty (2l + 1)[j_l(kr)]^2 P_l(\cos \vartheta).$$

[8] See G. N. Watson, *Theory of Bessel Functions*, Cambridge University Press, London, 1944, p. 363.

It follows that

$$\sigma(\vartheta) = \left(\frac{2Ze^2\mu}{\hbar^2}\right)^2 \left[\alpha^2 + 4k^2 \sin^2 \frac{\vartheta}{2}\right]^{-2}. \tag{11.50}$$

The accuracy of equation (11.50) may be assessed readily by looking at Table 11.2, which lists $\sigma(\vartheta)$ for different energies and scattering angles corresponding to the potential function

$$V(r) = \frac{11.63 \times 10^{-5}}{r} \exp(-0.7407r)$$

where $V(r)$ is in ergs and r is in femtometers ($1f = 10^{-13}$ cm). In the first column of cross sections the results of the Born approximation are shown; the second column lists the results of an accurate numerical integration of equation (11.9) using $V(r)$ as given above. We see that in this instance the Born approximation reproduces only the trends in the values of $\sigma(\vartheta)$ and not the values themselves. Evidently this result may be attributed to our

Table 11.2. Comparison between the Born approximation and the exact method for calculating differential cross sections appropriate to the screened coulomb potential.[9]

Energy MeV	Wavenumber (femtometers)$^{-1}$	ϑ (degrees)	$\sigma(\vartheta)$ (Born)	$\sigma(\vartheta)$ (Exact)
20	0.4911	0	5.60	3.09
		90	1.58	2.28
		180	0.74	2.27
50	0.7763	0	5.60	4.58
		90	0.546	0.752
		180	0.192	0.531
90	1.041	0	5.60	5.07
		90	0.227	0.309
		180	0.071	0.151
120	1.203	0	5.60	5.25
		90	0.142	0.190
		180	0.028	0.048
150	1.345	0	5.60	5.31
		90	0.097	0.127
		180	0.028	0.048

[9] The data are taken from E. Gerjuoy and D. S. Saxon, Application of variational methods to intermediate and high-energy scattering, *Phys. Rev.* **94**: 478–491(1954).

assuming the phase shift is small and that the energy of the incident particle is large. Indeed, as the energy increases, the table indicates that the Born approximation improves considerably in its accuracy.

11.7. RUTHERFORD SCATTERING

In the year 1909 Hans Geiger, working under the direction of Ernest Rutherford at the University of Manchester, made a remarkable discovery as regards the scattering of helium ions by thin metal foils. Geiger found that although most of these particles were little deflected in traversing the foil, occasionally one of them would be scattered through a very large angle, approaching 90°, and that a few of them could be scattered *backwards*. This result was quite unexpected because of the high speed and great bulk of the helium ions, which together should have given them the relative inertia of a freight train. Indeed, at the time it was believed, largely through the work of J. J. Thomson on electron scattering, that particles entering matter were deflected many times through small angles, the observed angle of scattering being the resultant of these. However, small-angle scattering, no matter how many times compounded, could not explain the appearance of scattered particles at angles of deflection greater than 90°. It is no wonder, then, that Rutherford was astounded when Geiger told him of the large-angle deflections.

Admittedly, the number of helium ions scattered from the incident beam at larger than right angles was small. For example, only about 1 particle in 8,000 was so deflected when the beam struck platinum foil. But, on the other hand, the fraction of backwards-scattered particles should have been far smaller than 1/8,000 were the theory of multiple small-angle scattering adequate to describe the penetration of helium ions into matter. Rutherford puzzled over this dilemma for two years, deciding in the meanwhile that the large-angle deflections *had* to be explained and that the explanation could *not* be in terms of multiple scattering. Early in 1911 the answer seemed to have come. It would have to be that the helium particles encountered but *one* atom in the metal foil and after this encounter appeared at a point far removed from the line of its original path. This being the case, it follows that an atom must be the seat of a very strong electric field—one strong enough to stop a heavy particle moving one-tenth the speed of light. If the atom were construed as a structure containing a large, positive, central charge surrounded by a "sphere of electrification" (to use Rutherford's words) of negative charge, we could expect the positive nucleus to create a strong field. A fast-moving particle could easily penetrate the outer sphere, but might be stopped entirely by the central core.

Of course, we have been assuming all along that the atom has a kind of planetary structure, and Rutherford's model is nothing new. On the other hand, our enlisting of this structure was based on notions of *simplicity* rather than on experimental data. Obviously, the former are without much significance in physics without the latter. It was Rutherford's work, epitomized in a paper published in 1911,[10] that provided the first concrete evidence for a nuclear atom. The theoretical studies of Bohr, and later Schrödinger and Heisenberg, properly may be said to have been the culmination of Rutherford's experimental enquiries.

Rutherford's model of the atom leads directly to an estimate of the differential scattering cross section. We can see this by returning to the formalism of section 11.3. In the present case, equation (11.9) becomes

$$\left[-\frac{\hbar^2}{2\mu} \nabla_\mathbf{r}^2 + \frac{ZZ'e^2}{r} \right] \psi(\mathbf{r}) = E\psi(\mathbf{r}), \tag{11.51}$$

where Z is the atomic number of the target and Z' is that of the incident particle. Upon substituting equation (11.20) into equation (11.51) and putting

$$k^2 = \frac{2\mu}{\hbar^2} E, \qquad \alpha = \frac{ZZ'e^2\mu}{\hbar^2 k}, \qquad \rho = 2kr, \qquad S_l(\rho) = \frac{P_l(\rho)}{\rho},$$

we get the set of differential equations

$$\frac{1}{\rho^2} \frac{d}{d\rho} \left(\rho^2 \frac{dS_l}{d\rho} \right) + \left\{ \frac{1}{4} - \frac{\alpha}{\rho} - \frac{l(l+1)}{\rho^2} \right\} S_l(\rho) = 0 \quad (l = 0,1,\ldots) \tag{11.52}$$

each of which is similar to equation (4.33), save for the signs of the first two terms inside the brackets. Following the procedure of Chapter 4, we substitute

$$S_l(\rho) = \rho^l \exp(i\rho/2) L_l(\rho)$$

into equation (11.52) to find

$$\rho \frac{d^2 L_l}{d\rho^2} + [2(l+1) + i\rho] \frac{dL_l}{d\rho} + (i(l+1) - \alpha) L_l(\rho) = 0$$

or, upon putting

$$\rho = iZ,$$

$$Z \frac{d^2 L_l}{dZ^2} + [2(l+1) - Z] \frac{dL_l}{dZ} - (i\alpha + l + 1) L_l(Z) = 0. \tag{11.53}$$

This expression looks just like equation (4.35), except for the term in α. Accordingly, its solution must be the associated Laguerre polynomial

$$L_l(Z) \equiv L_{i\alpha+l}^{2l+1}(Z) = \frac{d^{2l+1}}{dZ^{2l+1}} \sum_{k=0}^{\infty} \frac{(-1)^k \Gamma(i\alpha + l + 1)}{(k!)^2 \Gamma(i\alpha + l - k)} Z^k,$$

[10] E. Rutherford, The scattering of α and β particles by matter and the structure of the atom, *Phil. Mag.* **21**: 669–688 (1911).

where[11]

$$\Gamma(x) \equiv \int_0^\infty e^{-t} t^{x-1} dt \qquad (\text{Re } (x) > 0)$$

is the gamma function, a generalization of the factorial for non-integer, complex x.[12] What is necessary to us is the behavior of $L_l(\rho)$ for large values of its argument. The asymptotic form of $L_l(\rho)$ is[13]

$$L_l(\rho) \underset{\rho \to \infty}{\sim} \frac{2}{\rho} \sin \left(\frac{\rho}{2} - \tfrac{1}{2} l\pi + \delta_l - \alpha \operatorname{lu} \rho \right), \qquad (11.54)$$

where[11]

$$\delta_l = \text{Im } [\Gamma(i\alpha + l + 1)]$$

and the Laguerre polynomials have been normalized such that

$$\psi(r, \vartheta) \underset{r \to \infty}{\sim} \sum_{l=0}^\infty \left[\frac{(2l + 1)}{kr} i^l \right.$$

$$\times \exp (i\delta_l - i\alpha \ln 2kr) \sin (kr - \tfrac{1}{2} l\pi + \delta_l - \alpha \ln 2kr) \Big] P_l(\cos \vartheta). \quad (11.55)$$

This expression is the analog of equation (11.26) (which is valid only for potentials satisfying

$$\lim_{r \to \infty} rV(r) = 0).$$

We see that the only difference between the two is that the coulomb potential introduces a further distortion of the phase of the scattered particle, represented by the term $\alpha \ln 2kr$. Evidently a necessary criterion for the phase shift method developed in section 11.3 is that α be small; that is,

$$\frac{ZZ'e^2\mu}{\hbar^2 k} = \frac{ZZ'e^2}{\hbar v} \ll 1,$$

where v is the speed of the particle and de Broglie's relation has been used.

Equation (11.55) does not lend itself readily to a calculation of the differential scattering cross section. A simpler method is to set up equation (11.51) in parabolic coordinates and find the asymptotic solution. Doing so, we discover[14]

$$f(\vartheta) = \frac{\alpha}{2k} \csc^2 \frac{\vartheta}{2} \exp \left[-i\alpha \ln (1 - \cos \vartheta) + i\pi + 2i\delta_0 \right] \qquad (11.56)$$

[11] Re (x) and Im (x) mean "real part of x" and "imaginary part of x," respectively.
[12] See, for example, Chapter 10 of reference 5 for details.
[13] See p. 272f. of reference 4 for details.
[14] Reference 4, pp. 271ff.

and

$$\sigma_R(\vartheta) = \left(\frac{ZZ'e^2\mu}{2\hbar^2 k^2}\right)^2 \csc^4 \frac{\vartheta}{2}, \tag{11.57}$$

an expression derived by Rutherford from purely classical mechanical considerations. We notice that the cross section does not permit the determination of the sign of the nuclear charge; Rutherford was aware of this as well and so did not insist at the beginning that the cores of atoms were positively charged. However, the remainder of his model for the atom was well verified experimentally by Geiger, who found that equation (11.57) was followed by helium ions deflected from gold foil at scattering angles between 30° and 150°.

PROBLEMS

1. When the nucleus Al^{27} is bombarded by a proton, the result may be a change in the former to Si^{27} with the concomitant release of a neutron. If K_i is the total kinetic energy of $(Al^{27} + \text{proton})$ and K_f is that of $(Si^{27} + \text{neutron})$, show that the emission angle of the neutron in the laboratory reference frame is related to that in the center-of-mass reference frame by

$$\cos \vartheta_{L_n} = \frac{[(v_{CM}/u_n) + \cos \vartheta_{CM}]}{[1 + (v_{CM}/u_n)^2 + 2(v_{CM}/u_n) \cos \vartheta_{CM}]^{1/2}}$$

where u_n is the speed of the neutron in the center-of-mass frame. Show also that

$$\frac{v_{CM}}{u_n} = \left(\frac{m_p m_n K_i}{m_{Al} m_{Si} K_f}\right)^{1/2}$$

where the m's are masses.

2. The general nuclear reaction at low energy may be expressed

$$a + A \rightarrow b + B + Q$$

where a and b are incident and product particles, A and B are target and product nuclei, and Q is the energy transferred through the reaction. In symbols, the reaction is noted by $A(a, b)B$. Now, if Q is negative, it follows that $(a + A)$ must have certain minimum kinetic energy in order that the reaction occur:

$$K_i \geqslant |Q|,$$

where K_i is the total kinetic energy of $(a + A)$. In the center-of-mass reference frame we have

$$\tfrac{1}{2}m_a u_a^2 + \tfrac{1}{2}m_A u_A^2 = |Q|$$

as the least condition for $A(a, b)B$ to take place. Show that, in the laboratory reference frame,

$$\tfrac{1}{2}\mu v_a{}^2 = |Q|$$

is the same condition, where μ is the reduced mass of $(a + A)$.

3. Carry out the analysis of equation (11.7) which will yield the behavior of ϑ_{L_1} as a function of ϑ_{CM} for (v_{CM}/u_1) less than, equal to, and greater than one, respectively.

4. Show that equation (11.11) implies the *optical theorem*

$$\sigma_t = \frac{4\pi}{k} \operatorname{Im}(f(0)).$$

(*Hint*: Use equations (11.32) and (11.34) and the fact that $P_l(1) = 1$.)

5. Work out the Born approximation for the potential function expressed by

$$V(r) = \begin{cases} -V_0 & r < r_0 \\ 0 & r > r_0 \end{cases}$$

in order to calculate $\sigma(\vartheta)$. [The best method is to use equation (11.32), noting that $\exp(i\delta_l) \doteq 1$ in this case, along with the identity

$$\frac{\sin ar}{ar} = \sum_{l=0}^{\infty} (2l + 1)[j_l(kr)]^2 P_l(\cos \vartheta)$$

where

$$a = 2k \sin \frac{\vartheta}{2}\Big]$$

6. Calculate the value of σ_t for the screened coulomb potential discussed in section 11.6, using the Born approximation, at the wavenumber-values listed in Table 11.2. Compare your results with the following table, which was obtained by numerical integration (from Gerjuoy and Saxon, reference 9). Take $\mu = 0.83 \times 10^{-24}$ gm; this is the reduced mass of a proton-neutron system. (A potential of the screened-coulomb form is appropriate here and is called in this case a *Yukawa potential*.)

Wavenumber (femtometers)$^{-1}$	$\dfrac{\sigma_t}{(\pi/\alpha^2)}$ femtometers^2
0.4911	10.1
0.7763	4.53
1.041	2.64
1.203	2.01
1.345	1.63

Scattering Theory

7. Show by direct integration of equation (11.50) that the assumption $k \ll \alpha$ is not *necessary* to the derivation of equation (11.49). (Note that this assumption is *not* made in getting equation (11.50).)

8. Use the Born approximation to get equation (11.57) by considering a screened coulomb potential in the limit of vanishing α. Compare the Born expression for $f(\vartheta)$ in this case with the exact expression, given by equation (11.56). What must be the condition on α so that the Born scattering amplitude is nearly exact?

9. Compute σ_t for Rutherford scattering and explain its value in physical terms.

10. Equation (11.32) can be derived from very general considerations of the conservation of matter applied to the beam of incident particles in a scattering process. Let us see how this is done.

(a) Write down the Schrödinger equation and its complex conjugate. Multiply each equation by the solution complex conjugate to its own and show that

$$\frac{\partial}{\partial t} |\psi|^2 = -\frac{\hbar}{2mi} (\bar{\psi}\nabla^2\psi - \psi\nabla^2\bar{\psi}).$$

(Remember that $|\psi|^2 = \bar{\psi}\psi$.)

(b) Define a *current density vector*

$$\mathbf{j} \equiv \frac{\hbar}{2mi} (\bar{\psi}\nabla\psi - \psi\nabla\bar{\psi})$$

and show that

$$\frac{\partial\rho}{\partial t} + \nabla\cdot\mathbf{j} = 0$$

where

$$\rho \equiv |\psi|^2$$

is the *particle density function*. This is an expression of the conservation of matter; it has the general form of a "continuity equation."

(c) As we know already, the incident and (elastically) scattered particle wavefunction is at large distances from the target

$$\psi_l \underset{r \to \infty}{\sim} \frac{(2l+1)i^l}{kr} \sin\left(kr - \tfrac{1}{2}l\pi\right) + \frac{\exp(ikr)}{r} f_l$$

in terms of its lth component $\psi_l(r)$. Show that for this wavefunction

$$j_l = \frac{\hbar}{mr^2} [\tfrac{1}{2}i(2l+1)(\bar{f}_l - f_l) - k|f_l|^2].$$

342

(d) Now we invoke the conservation principle. If the only process occurring is scattering, j_l—which represents the net flux of particles into the vicinity of the target—must be zero. (Remember that j_l represents both incoming and outgoing waves.) Therefore,

$$|f_l|^2 = -(\bar{f}_l - f_l)\frac{(2l + 1)}{2ik}.$$

Now define

$$f_l \equiv (c_l - 1)\frac{2l + 1}{2ik}$$

where c_l is some complex number whose absolute value must be one:

$$|c_l - 1|^2 \equiv 2 - \bar{c}_l - c_l.$$

If we put

$$c_l \equiv \exp(2i\delta_l),$$

show that

$$f_l = \frac{2l + 1}{2ik}[\exp(2i\delta_l) - 1]$$

$$= \frac{2l + 1}{k}\sin\delta_l \exp(i\delta_l),$$

as given in the second of equations (11.31).

CHAPTER 12

The Relativistic Electron

I ain't got no body.

12.1. COMPTON SCATTERING AND LORENTZ INVARIANCE

One very important kind of collision problem which has not yet been considered involves the scattering of radiation by free electrons. This process, as is well known, owes its theoretical explanation to the photon concept. Indeed, the work of A. H. Compton and a number of others on the scattering of X-rays by electrons is regarded as some of the best empirical evidence for the existence of light quanta. We certainly do not require such evidence now, but it is still quite worthwhile to look at the problem of Compton scattering. The reason for this will become apparent a little later.

Suppose a photon of frequency ν_0 strikes a resting free electron and imparts to it a momentum $\mathbf{p}$. For kinetic energy and momentum to be conserved in this process, we must have

$$h\nu_0 = \frac{p^2}{2m_e} + h\nu \tag{12.1}$$

$$\mathbf{p} = \hbar\mathbf{k}_0 - \hbar\mathbf{k} \tag{12.2}$$

where ν is the frequency of the scattered photon and $\mathbf{k}$ is its wavenumber vector. If we call θ the angle between the directions of $\mathbf{k}_0$ and $\mathbf{k}$, the insertion of equation (12.2) into equation (12.1) yields

$$h\nu_0 = \frac{\hbar^2}{2m_e c^2}(\nu_0{}^2 - 2\nu\nu_0 \cos\theta + \nu^2) + h\nu$$

or

$$\lambda - \lambda_0 = \frac{\lambda\lambda_0}{\lambda_c} - \left[\left(\frac{\lambda\lambda_0}{\lambda_c}\right)^2 - 2\lambda\lambda_0(1 - \cos\theta)\right]^{\frac{1}{2}} \tag{12.3}$$

344

in terms of the wavelength shift upon scattering, where

$$\lambda_C \equiv \frac{h}{m_e c} = 2.42621 \pm 0.00006 \times 10^{-10}\, \text{cm}$$

is the *Compton wavelength*. (The corresponding frequency is called the Compton frequency, $\nu_C = 1.23564 \times 10^{20}\, \text{sec}^{-1}$.) Equation (12.3) predicts that the wavelength shift depends on both the incident and scattered wavelengths, and that the shift increases as the incident wavelength increases. These predictions are in qualitative *disagreement* with experiment, which shows the wavelength shift to be independent of either λ or λ_0, and equation (12.3) itself is in quantitative disagreement, unless the ratio (λ_C/λ_0) is small. We see immediately that our theory is absolutely *incorrect*.

What has gone wrong? The problem is *not* with the quantum physical aspect of the analysis; the photon concept has been correctly applied to the radiation-electron interaction. The difficulty lies with equation (12.1), which does not properly account for the energy transformation at very high photon frequencies. In other words, we have not taken the special theory of relativity into account during our discussion! To do so means rewriting equation (12.1) as[1]

$$h\nu_0 = mc^2 - m_e c^2 + h\nu \tag{12.1'}$$

where

$$m = m_e \left(1 - \left(\frac{u}{c}\right)^2\right)^{-\frac{1}{2}},$$

u being the speed of the electron. Upon noting that

$$p^2 = m^2 c^2 - m_e{}^2 c^2,$$

equations (12.1') and (12.2) yield

$$\lambda - \lambda_0 = \lambda_C(1 - \cos\theta). \tag{12.4}$$

We see immediately that equation (12.4) is in good qualitative agreement with experiment. In the limit of low incident photon energies, equation (12.3) takes on the form

$$\lambda - \lambda_0 = \frac{\lambda \lambda_0}{\lambda_C} \left\{ 1 - \left[1 - \frac{2\lambda_C{}^2}{\lambda \lambda_0}(1 - \cos\theta) \right]^{\frac{1}{2}} \right\}_{\lambda_C/\lambda_0 \to 0} \lambda_C(1 - \cos\theta), \tag{12.5}$$

showing that the relativistic and non-relativistic theories agree in this case. Otherwise, equation (12.4) is quantitatively the only correct expression.

[1] Those who need to refresh their memories on elementary relativity might consult C. Kittel, W. D. Knight, and M. Ruderman, *Mechanics*, McGraw-Hill Book Co., New York, 1965; or R. Resnick, *Introduction to Special Relativity*, John Wiley and Sons, New York, 1968.

The Relativistic Electron

The point of these remarks is not really to show that radiation may be thought of as aggregates of photons—we already know that—but rather to demonstrate that a correct quantum physical analysis is not necessarily a correct *kinematical* analysis. In the present case, the application of Einstein's hypothesis about radiation is not sufficient to describe Compton scattering even though it *is* sufficient to describe the collision phenomenon [in the sense of equation (12.2)]. The reason for this, of course, is that the structure of Newtonian mechanics does not admit the Lorentz transformations as a symmetry group. It follows that equation (12.1), which is a result of Newtonian mechanics, will not correctly describe the behavior of the electron at all possible values of its kinetic energy, but only for those which are small enough to make the Lorentz and Galilean transformations indistinguishable. (The latter form the symmetry group in Newtonian mechanics.) As a simple example, suppose we calculate the force on an electron moving in one dimension. According to Newtonian mechanics, the force is

$$\frac{d}{dt}(m_e u_x),$$

where u_x is the speed of the electron. The appropriate Lorentz transforms are

$$du_x = \left(1 - \left(\frac{v}{c}\right)^2\right) du_{x'}$$

$$dt = \left(1 - \left(\frac{v}{c}\right)^2\right)^{-\frac{1}{2}} dt' \tag{12.6}$$

where v is the speed of the frame of reference. It follows from equations (12.6) that

$$\frac{d}{dt}(m_e u_x) = m_e\left(1 - \left(\frac{v}{c}\right)^2\right)^{\frac{3}{2}} \frac{du_{x'}}{dt'} x'. \tag{12.7}$$

We see that it is only when $(v/c) \ll 1$ that Newton's force expression is invariant under a Lorentz transform. On the other hand, the Galilean transforms

$$du_x = du_{x'}$$

$$dt = dt' \tag{12.8}$$

lead immediately to an expression of the same form in both frames Quite clearly, then, Newtonian physics does not mix well with special relativity. But perhaps this does not matter. It is not the Newtonian equation of motion we are concerned with, anyway, but, in fact, the Schrödinger equation of motion. Now we come to the rub: *in writing down the Schrödinger*

346

equation we used the form of the Hamiltonian valid in non-relativistic mechanics. Indeed, we merely applied the observable postulate to

$$\frac{p^2}{2m_e} + V(\mathbf{r}) = H,$$

which obviously violates the principle of special relativity, except at low values of $\mathbf{p}$. It is beginning to look as if the equation

$$\mathcal{H}\psi = i\hbar \frac{\partial \psi}{\partial t} \tag{12.9}$$

will not be invariant under Lorentz transformations. Let us see if this is so by checking equation (12.9) for the free electron in one dimension. The appropriate transformations here are:

$$\frac{\partial}{\partial t} = \left(1 - \left(\frac{v}{c}\right)^2\right)^{-\frac{1}{2}} \left(\frac{\partial}{\partial t'} - v\frac{\partial}{\partial x'}\right)$$

$$\frac{\partial^2}{\partial x^2} = \left(1 - \left(\frac{v}{c}\right)^2\right)^{-1} \left[\frac{\partial^2}{\partial x'^2} - \frac{2(v/c)}{c}\frac{\partial^2}{\partial x'\,\partial t'} + \frac{(v/c)^2}{c^2}\frac{\partial^2}{\partial t'^2}\right] \tag{12.10}$$

$$\psi(x, t) = S^{-1}\psi'(x', t')$$

where S is a unitary operator which transforms the state vector in consonance with the coordinate and time transformations. When equations (12.10) are applied to equation (12.9) for the free electron, we get

$$-\frac{\hbar^2}{2m_e}\left(1 - \left(\frac{v}{c}\right)^2\right)^{-1}\frac{\partial^2\psi'}{\partial x'^2} + i\hbar v\left(1 - \left(\frac{v}{c}\right)^2\right)^{-\frac{1}{2}}\frac{\partial\psi'}{\partial x'}$$

$$-\frac{\hbar^2 v}{2m_e c^2}\left(1 - \left(\frac{v}{c}\right)^2\right)^{-1}\left[\left(\frac{v}{c^2}\right)\frac{\partial^2\psi'}{\partial t'^2} - \frac{2\partial^2\psi'}{\partial x'\,\partial t'}\right] = \left(1 - \left(\frac{v}{c}\right)^2\right)^{-\frac{1}{2}}i\hbar\frac{\partial\psi'}{\partial t'}$$

or

$$\frac{1}{2m_e}\left[-i\hbar\left(1 - \left(\frac{v}{c}\right)^2\right)^{-\frac{1}{2}}\frac{\partial}{\partial x'} - m_e v\right]^2 \psi'(x', t')$$

$$-\frac{\hbar v}{2m_e c^2}\left(1 - \left(\frac{v}{c}\right)^2\right)^{-\frac{1}{2}}\left[\left(\frac{v}{c^2}\right)\frac{\partial^2\psi'}{\partial t'^2} - 2\frac{\partial^2\psi'}{\partial x'\,\partial t'}\right]$$

$$= \left(1 - \left(\frac{v}{c}\right)^2\right)^{-\frac{1}{2}}i\hbar\frac{\partial\psi'}{\partial t'} + \tfrac{1}{2}m_e v^2\psi'(x', t') \tag{12.11}$$

where equation (12.11) has been operated on by S. When $(v/c) \ll 1$, equation (12.11) becomes

$$\frac{1}{2m_e}\left(-i\hbar\frac{\partial}{\partial x'} - m_e v\right)^2\psi'(x', t') = \left(i\hbar\frac{\partial}{\partial t'} + \tfrac{1}{2}m_e v^2\right)\psi'(x', t').$$

This expression represents the equation of motion for an electron moving with speed $-v$ relative to its speed as measured in the original frame of reference, provided that Galilean transforms are taken as the appropriate symmetry group. Thus, equation (12.11) does not have the expected form of a free-particle Schrödinger equation unless $(v/c) \ll 1$. As we had feared, the Schrödinger picture is not relativistically invariant. Evidently, we must extend it in some way if we wish to have a theory that is dynamically sound in every physical situation.

12.2. THE DIRAC EQUATION OF MOTION

As with the Schrödinger postulate, we cannot expect to determine the correct form of a relativistic equation of motion just by a thorough sifting of available experimental data. The best we can do is to list certain conditions which experience would seem to impose and leave the rest to inspiration. The requirements which a relativistic equation of motion must fulfill are necessarily more numerous than those for a non-relativistic equation of motion, since the latter must be a special case of the former. To be precise, we expect the following conditions to be satisfied:

(a) *The equation of motion must be intrinsically non-deterministic, applicable to all pure states, and compatible with the probability postulate.*

(b) *The equation of motion must be a differential equation of the first order with respect to the time.* This requirement is most important to the uniqueness of the wavefunction and to the last part of the foregoing condition.

(c) *The equation of motion must be relativistically invariant, and must lead to results, in the limit of low energies, that agree with those obtained through the Schrödinger postulate. In the limit of short de Broglie wavelengths it must be compatible with classical relativistic mechanics.* This condition, of course, is essential to the consistency of the quantum theory.

An equation of motion for coordinate representatives of state vectors, valid for all energies, was first written down by Paul A. M. Dirac in 1928. It may be given as:

The Dirac Postulate. The equation of motion for the wavefunction representing a single particle is

$$i\hbar \frac{\partial \psi}{\partial t} = -i\hbar c \left(\alpha_1 \frac{\partial \psi}{\partial x} + \alpha_2 \frac{\partial \psi}{\partial y} + \alpha_3 \frac{\partial \psi}{\partial z} \right) + \beta m_0 c^2 \psi + V(r)\psi \quad (12.12)$$

where $\psi(x, y, z, t)$ is an N-rowed column matrix and α_1, α_2, α_3, and β are N by N Hermitian matrices satisfying

$$\alpha_i\alpha_j + \alpha_j\alpha_i = 2\delta_{ij}I \quad (i,j = 1, 2, 3) \quad (12.13)$$

$$\alpha_i\beta + \beta\alpha_i = 0 \quad (i = 1, 2, 3) \quad (12.14)$$

$$\alpha_i^2 = \beta^2 = I \quad (i = 1, 2, 3) \quad (12.15)$$

This is quite an unusual postulate! On the face of it, there is little resemblance between equation (12.12) and the Schrödinger equation of motion; and the requirement that the wavefunction be the column matrix

$$\psi(\mathbf{r}, t) = \begin{pmatrix} \psi_1(\mathbf{r}, t) \\ \vdots \\ \psi_N(\mathbf{r}, t) \end{pmatrix}$$

seems very unusual. We shall have much to do, it seems, in giving a physical interpretation to the Dirac equation of motion.

Let us begin by checking equation (12.12) against the conditions (a) through (c) mentioned earlier. We see immediately that the first two are fulfilled, since the Dirac equation prescribes the spacial and temporal dependence of a wavefunction and is a linear, homogeneous differential equation of the first order with respect to the time. The third condition, however, is a complicated one and requires careful attention. First we note that equation (12.12) is consistent with the observable postulate and the Schrödinger postulate in the sense that (in the absence of potential energy) it is the quantum physical analog of

$$E^2 = p^2c^2 + m_0^2c^4. \quad (12.16)$$

To see this, we must apply equation (12.12) twice:

$$-\hbar^2 \frac{\partial^2\psi}{\partial t^2} = -\hbar^2c^2 \sum_{k=1}^{3}\sum_{j=1}^{3} \frac{\alpha_k\alpha_j + \alpha_j\alpha_k}{2} \frac{\partial^2\psi}{\partial x_k\,\partial x_j}$$

$$- i\hbar^2 m_0 c^3 \sum_{k=1}^{3} (\alpha_k\beta + \beta\alpha_k) \frac{\partial\psi}{\partial x_k} + \beta^2 m_0^2 c^4\psi$$

$$= -\hbar^2c^2 \sum_{k,j} \delta_{kj} \frac{\partial^2\psi}{\partial x_k\,\partial x_j} + m_0^2c^4\psi$$

$$= -\hbar^2c^2\nabla^2\psi + m_0^2c^4\psi, \quad (12.17)$$

the second and third steps resulting from the application of equations (12.13), (12.14), and (12.15). (We have written $x = x_1$, etc., to simplify the notation.) Equation (12.17) follows from equation (12.16) directly if we employ the Schrödinger prescription for the total energy and momentum operators. Thus we find equation (12.12) to be consistent with our general notions about the relation between operators and dynamical quantities.

Continuing in this vein, we should like to see if equation (12.12) reduces to the Schrödinger equation in the low-energy limit. In order to do this, we must represent the matrices α_i ($i = 1, 2, 3$) and β in a specific basis and make N a definite integer. Now, equation (12.15) indicates that the eigenvalues of α_i ($i = 1, 2, 3$) and β are ± 1. Moreover, according to equation (12.14),

$$\alpha_i \beta^2 = \alpha_i = -\beta \alpha_i \beta \qquad (i = 1, 2, 3)$$

and, by equation (12.15),

$$\text{Tr}\,(\alpha_i) = \text{Tr}\,(\beta^2 \alpha_i) = \text{Tr}\,(\beta \alpha_i \beta) = -\text{Tr}\,(\alpha_i) \qquad (i = 1, 2, 3)$$

because the trace of a matrix is invariant under a transposition. (See problem 2 in Chapter 6.) It follows that

$$\text{Tr}\,(\alpha_i) = 0 \qquad (i = 1, 2, 3)$$

and that N must be an *even* number. The smallest value of N, then, is two. However, if the α_i and β are two by two matrices, they must be equivalent to the Pauli matrices and the unit matrix. (See problem 14 in Chapter 6.) But this prescription violates equation (12.14); so, two by two matrices are ruled out. The next possibility is $N = 4$. The matrices α_i and β can then be represented as

$$\alpha_i = \begin{pmatrix} 0 & \sigma_i \\ \sigma_i & 0 \end{pmatrix} \qquad \beta = \begin{pmatrix} I' & 0 \\ 0 & -I' \end{pmatrix} \qquad (i = 1, 2, 3) \qquad (12.18)$$

where $\sigma_1 = \sigma_x$, etc., are the Pauli matrices and I' refers to a two-by-two unit matrix.

Now let us consider equation (12.12) as applied to an electron in a constant magnetic field and go to the limit of low kinetic energies. (The reason for selecting this particular situation will become apparent quite soon.) The Schrödinger equation in this instance has the same form as it took in section 5.5,

$$-\frac{\hbar^2}{2m_e} \nabla^2 \psi + \frac{i\hbar e}{m_e c} \mathbf{A} \cdot \nabla \psi = i\hbar \frac{\partial \psi}{\partial t}, \qquad (12.19)$$

except that the vector potential $\mathbf{A}(\mathbf{r})$ is no longer a function of the time. Evidently, equation (12.12) should have the form of equation (12.19) in the low-energy limit. To see if this is so, we write down the appropriate form of equation (12.12),

$$i\hbar \frac{\partial \psi}{\partial t} = c \sum_{k=1}^{3} \alpha_k \left(-i\hbar \frac{\partial}{\partial x_k} - \frac{e}{c} A_k \right) \psi + \beta m_e c^2 \psi,$$

and operate on both sides with $\partial/\partial t$ to get

$$-\hbar^2 \frac{\partial^2 \psi}{\partial t^2} = c^2 \sum_{j,k} \alpha_j \alpha_k \left(i\hbar \frac{\partial}{\partial x_j} + \frac{e}{c} A_j \right) \left(i\hbar \frac{\partial}{\partial x_k} + \frac{e}{c} A_k \right) \psi + m_e^2 c^4 \psi \qquad (12.20)$$

upon using equation (12.14). The summation in this expression is reduced as follows. We note that

$$\sum_{j,k} \alpha_j \alpha_k \left(i\hbar \frac{\partial}{\partial x_j} + \frac{e}{c} A_j \right) \left(i\hbar \frac{\partial}{\partial x_k} + \frac{e}{c} A_k \right)$$

$$= \sum_{j=1}^{3} \left(i\hbar \frac{\partial}{\partial x_j} + \frac{e}{c} A_j \right)^2 + \sum_{j \neq k} \alpha_j \alpha_k \left(i\hbar \frac{\partial}{\partial x_j} + \frac{e}{c} A_j \right) \left(i\hbar \frac{\partial}{\partial x_k} + \frac{e}{c} A_k \right)$$

and that the second sum reduces to

$$\frac{i\hbar e}{c} \sum_{j \neq k} \alpha_j \alpha_k \left(A_j \frac{\partial}{\partial x_k} + \frac{\partial}{\partial x_j} A_k \right) = \frac{i\hbar e}{c} \sum_{j \neq k} \alpha_j \alpha_k \frac{\partial A_k}{\partial x_j}$$

$$= \frac{i\hbar e}{2c} \sum_{j \neq k} \alpha_j \alpha_k \left(\frac{\partial A_k}{\partial x_j} - \frac{\partial A_j}{\partial x_k} \right)$$

by virtue of equation (12.13), the identity

$$\frac{\partial}{\partial x_j} (A_k \varphi) = \frac{\partial A_k}{\partial x_j} \varphi + A_k \frac{\partial \varphi}{\partial x_j}$$

(φ being any scalar), and the fact that

$$\mathbf{\nabla} \cdot \mathbf{A} = \sum_{j=1}^{3} \frac{\partial A_j}{\partial x_j} = 0.$$

If we drop the term in A^2 and note that

$$\left(\frac{\partial A_k}{\partial x_j} - \frac{\partial A_j}{\partial x_k} \right) = (\mathbf{\nabla} \times \mathbf{A})_l = (\mathbf{H}_e)_l \qquad (j, k, l \text{ in cyclic order})$$

$$\alpha_j \alpha_k = \sigma_j \sigma_k I = i\sigma_l I$$

[the second relation following easily from the first of equations (12.18)], we can write equation (12.20) as

$$-\hbar^2 \frac{\partial^2 \psi}{\partial t^2} = \left\{ c^2 \left[-\hbar^2 \nabla^2 + \frac{2i\hbar e}{c} (\mathbf{A} \cdot \mathbf{\nabla}) - \frac{e\hbar}{c} \mathbf{\sigma} \cdot \mathbf{H}_e \right] + m_e^2 c^4 \right\} \psi \quad (12.21)$$

where $\mathbf{\sigma}$ refers to the "vector" with components σ_1, σ_2, σ_3. Finally, we rearrange equation (12.21) to have the form

$$\left(i\hbar \frac{\partial}{\partial t} + m_e c^2 \right) \left(i\hbar \frac{\partial}{\partial t} - m_e c^2 \right) \psi = c^2 \left(-\hbar^2 \nabla^2 + \frac{2ie\hbar}{c} (\mathbf{A} \cdot \mathbf{\nabla}) - \frac{e\hbar}{c} \mathbf{\sigma} \cdot \mathbf{H}_e \right) \psi$$

and notice that, for small kinetic energies, equation (12.20) reduces to

$$-\hbar^2 \frac{\partial^2 \psi}{\partial t^2} \doteq m_e^2 c^4 \psi,$$

so that

$$\left(i\hbar \frac{\partial}{\partial t} + m_e c^2 \right) \doteq 2m_e c^2.$$

The Relativistic Electron

Thus equation (12.21) may be expressed, relative to the rest energy, as

$$i\hbar \frac{\partial \psi}{\partial t} = -\frac{\hbar^2}{2m_e} \nabla^2 \psi + \frac{ie\hbar}{m_e c} (\mathbf{A} \cdot \nabla)\psi - \frac{e\hbar}{2m_e c} \boldsymbol{\sigma} \cdot \mathbf{H}_e \psi. \qquad (12.22)$$

Equation (12.22) differs from equation (12.19) only in the term

$$-\frac{e\hbar}{2m_e c} \boldsymbol{\sigma} \cdot \mathbf{H}_e = \frac{2\mu_B \mathbf{S} \cdot \mathbf{H}_e}{\hbar}$$

where μ_B is the Bohr magneton and $\mathbf{S} = \{S_1, S_2, S_3\}$ is the total spin angular momentum operator. (See problem 7 in Chapter 7.) The Dirac equation, in the low-energy limit, does not yield the Schrödinger equation after all, it seems, but instead gives the Schrödinger equation *as modified by the spin hypothesis*. Indeed, the extra term in equation (12.22) is *exactly* that representing the interaction between a spinning electron and a magnetic field, with the spin magnetic moment correctly given by twice the Bohr magneton.[2] This is a most unexpected and spectacular result! We see that the Dirac postulate does not replace the Schrödinger postulate, but, in fact, replaces the latter *and* the spin hypothesis: the dependence of the wavefunction upon the spin variable no longer appears as a semiempirical part of the quantum theory, but *arises naturally from a relativistically sound equation of motion*.

One last thing we must do is establish the invariance of equation (12.12) under Lorentz transformations. To simplify matters we shall consider once again the free particle in one spacial dimension.[3] Equation (12.12) is then

$$i\hbar \frac{\partial \psi}{\partial t} = -i\hbar c \alpha_1 \frac{\partial \psi}{\partial x_1} + \beta m_e c^2 \psi(x_1, t). \qquad (12.23)$$

If we put

$$\frac{\partial}{\partial x_1} = \left(1 - \left(\frac{v}{c}\right)^2\right)^{-\frac{1}{2}} \frac{\partial}{\partial x_1'} - \left(1 - \left(\frac{v}{c}\right)^2\right)^{-\frac{1}{2}} \frac{v}{c^2} \frac{\partial}{\partial t'}$$

$$\equiv \cosh\omega \frac{\partial}{\partial x_1'} - \frac{\sinh\omega}{c} \frac{\partial}{\partial t'}$$

$$\frac{\partial}{\partial t} = \cosh\omega \frac{\partial}{\partial t'} - c \sinh\omega \frac{\partial}{\partial x_1'}$$

$$S^{-1}\psi'(x_1', t') \equiv \beta \exp\left(-\omega\alpha_1/2\right)\beta\psi'(x_1', t') = \psi(x_1, t)$$

[2] Very careful measurements have shown this statement to be slightly in error. The presently accepted value is not 2.00, but

$$2.0031923 \pm 0.0000002.$$

The difference finds its explanation in quantum electrodynamics. For an interesting discussion see P. Kusch, The electron dipole moment, *Physics Today* **19**: 23–35 (1966).

[3] The problem is considered in detail in J. D. Bjorken and S. D. Drell, *Relativistic Quantum Mechanics*, McGraw-Hill Book Co., New York, 1964, Chapter 2.

where ω is defined by

$$\tanh\omega = \frac{v}{c}$$

and $\mathcal{S}$ is the (matrix) operator defined in the foregoing section, equation (12.23) becomes [4]

$$i\hbar(\cosh\omega - \alpha_1 \sinh\omega)\beta \exp\left(-\omega\alpha_1/2\right)\beta \frac{\partial\psi'}{\partial t'}$$

$$= -i\hbar c(\alpha_1 \cosh\omega - \sinh\omega)\beta \exp\left(-\omega\alpha_1/2\right)\beta \frac{\partial\psi'}{\partial x_1'}$$

$$+ \beta m_e c^2 \beta \exp\left(-\omega\alpha_1/2\right)\beta\psi'(x_1', t')$$

or upon operating with $\exp\left(\omega\alpha_1/2\right)$,

$$i\hbar \exp\left(\omega\alpha_1/2\right)(\cosh\omega - \alpha_1 \sinh\omega)\beta \exp\left(-\omega\alpha_1/2\right)\beta \frac{\partial\psi'}{\partial t'}$$

$$= -i\hbar c \exp\left(\omega\alpha_1/2\right)(\alpha_1 \cosh\omega - \sinh\omega)\beta \exp\left(-\omega\alpha_1/2\right)\beta \frac{\partial\psi'}{\partial x_1'}$$

$$+ \beta m_e c^2 \psi'(x_1', t'). \tag{12.24}$$

Now, by equations (12.14) and (12.15),

$$\beta \exp\left(-\omega\alpha_1/2\right) = \beta \sum_{k=0}^{\infty} \frac{(-\omega\alpha_1/2)^k}{k!} = \beta \sum_{m=0}^{\infty} \frac{(\omega/2)^{2m}}{(2m)!} - \beta\alpha_1 \sum_{n=0}^{\infty} \frac{(\omega/2)^{2n+1}}{(2n+1)!}$$

$$= \beta \cosh\frac{\omega}{2} + \alpha_1\beta \sinh\frac{\omega}{2} = \exp\left(\omega\alpha_1/2\right)\beta,$$

$$\exp\left(\omega\alpha_1\right) = (\cosh\omega + \alpha_1 \sinh\omega),$$

so that equation (12.24) becomes

$$i\hbar(\cosh\omega - \alpha_1 \sinh\omega)(\cosh\omega + \alpha_1 \sinh\omega)\beta^2 \frac{\partial\psi'}{\partial t'}$$

$$= -i\hbar c(\alpha_1 \cosh\omega - \sinh\omega)(\cosh\omega + \alpha_1 \sinh\omega)\beta^2 \frac{\partial\psi'}{\partial x_1'} + \beta m_0 c^2 \psi'(x_1', t')$$

or

$$i\hbar \frac{\partial\psi'}{\partial t'} = -i\hbar c \frac{\partial\psi'}{\partial x_1'} + \beta m_e c^2 \psi'(x_1', t') \tag{12.25}$$

which establishes invariance.

[4] α_1' and β' can be shown to be equivalent to α_1 and β. Thus we lose nothing by dropping the primes in equation (12.24). See Chapter 2 of the reference cited in note 3 for more details.

12.3. SOLUTIONS OF THE DIRAC EQUATION FOR A FREE ELECTRON

In the case of a free electron, equation (12.12) takes on the form

$$i\hbar \frac{\partial \psi}{\partial t} = -i\hbar c \sum_{k=1}^{3} \alpha_k \frac{\partial \psi}{\partial x_k} + \beta m_e c^2 \psi(\mathbf{r}, t). \tag{12.26}$$

The solutions of this equation are, if we use the representation of the α_i ($i = 1, 2, 3$) and β given by equation (12.18),

$$\psi_1(\mathbf{r}, t) = N \exp\left[(i/\hbar)(\mathbf{p}\cdot\mathbf{r}) - (i/\hbar)|E_p|t\right] \begin{pmatrix} 1 \\ 0 \\ cp_3/\pi_p \\ cp_+/\pi_p \end{pmatrix}$$

$$\psi_2(\mathbf{r}, t) = N \exp\left[(i/\hbar)(\mathbf{p}\cdot\mathbf{r}) - (i/\hbar)|E_p|t\right] \begin{pmatrix} 0 \\ 1 \\ cp_-/\pi_p \\ -cp_3/\pi_p \end{pmatrix}$$

$$\tag{12.27}$$

$$\psi_3(\mathbf{r}, t) = N \exp\left[-(i/\hbar)(\mathbf{p}\cdot\mathbf{r}) + (i/\hbar)|E_p|t\right] \begin{pmatrix} cp_3/\pi_p \\ cp_+/\pi_p \\ 1 \\ 0 \end{pmatrix}$$

$$\psi_4(\mathbf{r}, t) = N \exp\left[-(i/\hbar)(\mathbf{p}\cdot\mathbf{r}) + (i/\hbar)|E_p|t\right] \begin{pmatrix} cp_-/\pi_p \\ -cp_3/\pi_p \\ 0 \\ 1 \end{pmatrix}$$

where E_p is the energy eigenvalue,

$$p_\pm = p_1 \pm ip_2,$$
$$\pi_p = (|E_p| + m_e c^2),$$

$p_1 = p_x$, etc., and N is a normalization constant. The solutions are obtained in the most straightforward way by solving equation (12.26) for the case $|\mathbf{p}| = 0$ and operating upon the solution with S, whose columns, in fact, are just the matrix portions of the $\psi_i(\mathbf{r}, t)$ ($i = 1, 2, 3, 4$).[5]

[5] See, for example, Chapter 3 of the reference in note 3.

The question of interest, of course, is the physical interpretation of equations (12.27). The most obvious peculiarity is that there are four solutions rather than two, as we obtained in Chapter 3 for the Schrödinger equation. This is the direct result of having chosen the α_i and β to be four by four matrices, so that equation (12.26) becomes four differential equations instead of one. Evidently, any state of the electron will be, in general, a linear combination of these four state vectors. Another thing which seems odd is that ψ_1 and ψ_2 appear to correspond to momentum $-\mathbf{p}$ and energy $-E_\mathbf{p}$. Since $E_\mathbf{p}$ is entirely kinetic energy here, the latter pair of solutions are highly unphysical, implying a particle of negative rest mass! The reason that such solutions arise has to do with equation (12.16). In solving the expression for $E_\mathbf{p}$ we have to write

$$E_\mathbf{p} = \pm(p^2c^2 + m_e^2c^4)^{1/2}.$$

Ordinarily, we would throw away the negative square root; but we cannot do that here and still maintain our notion that the superposition of solutions of equation (12.26) gives the most general free-electron state.

If we have to keep the whole of equations (12.27), let us see what kind of electron the negative-energy solutions describe. When $|\mathbf{p}| = 0$, the equations become

$$\psi_1(t) = N \exp\left[-(i/\hbar)m_ec^2t\right]\begin{pmatrix}1\\0\\0\\0\end{pmatrix},$$

$$\psi_2(t) = N \exp\left[-(i/\hbar)m_ec^2t\right]\begin{pmatrix}0\\1\\0\\0\end{pmatrix},$$

(12.28)

$$\psi_3(t) = N \exp\left[(i/\hbar)m_ec^2t\right]\begin{pmatrix}0\\0\\1\\0\end{pmatrix},$$

$$\psi_4(t) = N \exp\left[(i/\hbar)m_ec^2t\right]\begin{pmatrix}0\\0\\0\\1\end{pmatrix}.$$

Let us operate upon these with the four by four matrix (generalized Pauli matrix)

$$\hat{\sigma}_3 = \sigma_3 I = \begin{pmatrix} \sigma_3 & 0 \\ 0 & \sigma_3 \end{pmatrix}$$

where $\sigma_3 \equiv \sigma_z$. We find

$$\hat{\sigma}_3 \psi_1 = \psi_1 \qquad \hat{\sigma}_3 \psi_3 = \psi_3$$
$$\hat{\sigma}_3 \psi_2 = -\psi_2 \qquad \hat{\sigma}_3 \psi_4 = -\psi_4$$

which suggests that ψ_1 and ψ_3 describe electrons with spin along the positive vertical axis and that ψ_2 and ψ_4 describe electrons with the opposite spin. Thus, the negative-energy solutions refer to a certain kind of electron, with both of the usual spin possibilities, and the positive-energy solutions to another kind of electron.

What are we to make of this? No one has ever seen a particle with negative rest mass, yet the Dirac equation, which is a correct generalization of the Schrödinger equation, requires such particles to exist. If we admit that they do exist in some non-detectable way, the principle of superposition is maintained; but then we must deal in a believable way with possibility of an electronic transition from the positive groundstate ($E_0 = m_e c^2$) to the highest negative energy level, $-m_e c^2$. This amounts to prescribing the conditions for the destruction (and creation, for the reverse transition) of matter—a rather unusual process at best! In effect we have come to the extreme consequence of de Broglie's hypothesis: the relativistic quantum theory of the electron demands that this particle be created and absorbed just as are photons.

12.4. HOLE THEORY: THE POSITRON

The physical interpretation of negative energy electrons was given by Dirac himself two years after he postulated his equation of motion. The idea is as follows. First we assert that negative-energy electrons cannot be detected directly, in agreement with experience, and, therefore, that the totality of such particles forms the *vacuum state*. What we usually call a perfect vacuum is now construed as the set of all negative energy levels. Moreover, it is postulated that these levels, in the absence of perturbations, are completely filled. It follows from the exclusion postulate that a positive-energy electron cannot disappear into the vacuum state under ordinary circumstances. This explains the observed stability of matter relative to transitions into the negative-energy state.

Now, the negative-energy electron can be detected indirectly if it absorbs radiation of sufficient energy and enters a positive energy state. When this happens, an ordinary electron appears out of the vacuum, leaving a *hole* in the set of negative-energy states. The hole represents the *absence* of a particle of charge $-|e|$ and negative energy, and so is itself a particle of charge $-e$ and positive energy! Therefore, if a photon whose frequency satisfies the inequality

$$\nu \geqslant 2\nu_C = 2m_e c^2$$

is absorbed by a negative-energy electron, *two* particles will appear out of the vacuum state in its place. (See Figure 65.) We see that the relativistic wavefunction, unlike its low-energy counterpart, describes a many-particle state automatically. It makes no sense to speak of a one-particle solution of the Dirac equation, as the theory is not complete unless both kinds of electron are accounted for simultaneously. It is in this sense that relativistic quantum mechanics is a *field theory*: material particles are created (and destroyed) just as are photons in the electromagnetic field.

As is well known, Dirac's interpretation of the vacuum state was amply verified by experiment. In 1932, C. D. Anderson reported that, during his

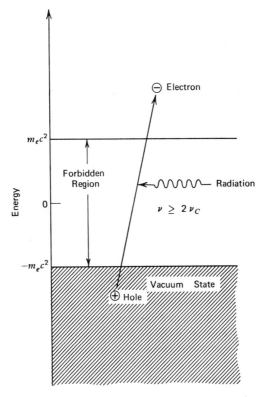

Figure 65. A schematic view of electron-positron creation.

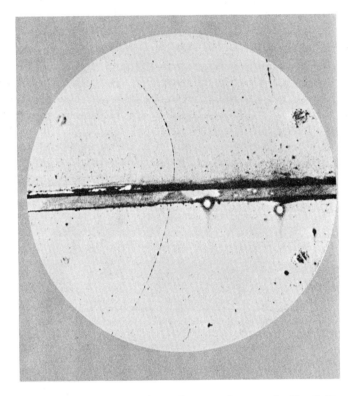

Figure 66. The track of the positron as photographed by C. D. Anderson.

study of the energy spectrum of secondary electrons produced in the atmosphere by cosmic radiation, he photographed the track of a very light, positively charged particle moving through a magnetic field applied to the interior of his cloud chamber. (See Figure 66. The magnetic field is *into* the plane of the figure.) The track was more sharply curved above the lead barrier inserted into the chamber, indicating that the particle was moving upward.[6] The direction of travel, then, was just opposite that appropriate to an electron in

[6] The curvature of the path of a charge particle moving in a magnetic field is inversely proportional to its speed and thus to its kinetic energy. To be moving downward, the particle would have had to be accelerated while in the lead plate.

the applied field and could only represent the motion of a positively charged particle. Measurements of the radius of curvature of the path indicated the new particle had a mass essentially the same as that of the electron. This was Dirac's "hole"! Today we know it as the *positron*.

Hole theory makes it quite clear that a certain kind of symmetry exists between electrons and positrons, in that to each positive-energy electron there corresponds a positive-energy "hole" electron differing from its counterpart only by the sign of its charge. Indeed, equation (12.12) stands as it is if interpreted as the equation of motion for a positron, rather than an electron. When a magnetic field is present, we need only require that the components of the vector potential change sign as the charge does to apply the equation of motion

$$ i\hbar \frac{\partial \psi}{\partial t} = c \sum_{k=1}^{3} \alpha_k \left(i\hbar \frac{\partial}{\partial x_k} - \frac{e}{c} A_k \right) \psi + \beta m_e c^2 \psi $$

unaltered to positrons. We see, then, that *the transformation*

(a) *change the sign of the charge*

(b) *change the sign of the vector potential*

is a realization of a member of the group of the Dirac equation. The operator corresponding to this transformation is called the *charge conjugation operator.* Its effect on the wavefunction is defined by

$$ \psi_c = C\beta\bar{\psi} \equiv -i\alpha_2\beta\bar{\psi} \tag{12.29} $$

where α_2 and β are given by equations (12.18) and ψ_c is the charge-conjugated ψ. For example, if C operates on a resting, spin-down electron of negative energy we find, according to equations (12.18), (12.28), and (12.29),

$$ C\beta\bar{\psi}_4 = -i\alpha_2\beta\bar{\psi}_4 = -i \begin{pmatrix} 0 & \sigma_2 \\ \sigma_2 & 0 \end{pmatrix} \begin{pmatrix} I' & 0 \\ 0 & -I' \end{pmatrix} N \exp\left[-(i/\hbar)m_e c^2 t \right] \begin{pmatrix} 0 \\ 0 \\ 0 \\ 1 \end{pmatrix} $$

$$ = i \begin{pmatrix} 0 & 0 & 0 & -i \\ 0 & 0 & i & 0 \\ 0 & i & 0 & 0 \\ -i & 0 & 0 & 0 \end{pmatrix} \begin{pmatrix} 0 \\ 0 \\ 0 \\ 1 \end{pmatrix} N \exp\left[-(i/\hbar)m_e c^2 t \right] $$

$$ = \begin{pmatrix} 1 \\ 0 \\ 0 \\ 0 \end{pmatrix} N \exp\left[-(i/\hbar)m_e c^2 t \right] = \psi_1 $$

The Relativistic Electron

A charge-conjugated, spin-down, negative-energy electron at rest is a resting, spin-up, positive-energy electron. Both of these particles arise from the solutions of the same equation of motion when no force fields are present, in agreement with our earlier remark concerning the charge-independence of the Dirac equation for free electrons. However, if a magnetic field is present, the conjugated wavefunction is a solution of the equation of motion for *positrons*. (Remember that the conjugated wavefunction satisfies the conjugated equation of motion.) It follows that a positive-energy, spin-up positron corresponds to a conjugated, spin-down, negative-energy electron, in exact agreement with the ideas of hole theory.

PROBLEMS

1. We may reasonably wonder why it is not appropriate to look at the relativistic Hamiltonian function

$$H^2 = p^2 c^2 + m_0^2 c^4$$

for a free particle and apply the Schrödinger postulate *directly* to obtain the equation of motion

$$-\hbar^2 \frac{\partial^2 \psi}{\partial t^2} = -\hbar^2 c^2 \nabla_r^2 \psi + m_0^2 c^4 \psi(\mathbf{r}, t).$$

Let us see why we cannot easily do this.

(a) Multiply $\bar{\psi}$ times the equation of motion and ψ times the complex conjugate equation of motion, then subtract the latter from the former to obtain the *equation of continuity*

$$\frac{\partial}{\partial t} \left[\frac{i\hbar}{2m_0 c^2} \left(\bar{\psi} \frac{\partial \psi}{\partial t} - \psi \frac{\partial \bar{\psi}}{\partial t} \right) \right] + \nabla \cdot \frac{\hbar}{2im_0} (\bar{\psi}\nabla\psi - \psi\nabla\bar{\psi}) = 0.$$

(b) Ordinarily, this equation is written in the form

$$\frac{\partial \rho}{\partial t} + \nabla \cdot \mathbf{j} = 0$$

where ρ is the *probability density* and $\mathbf{j}$ is the *probability current density*. The equation expresses the "conservation of probability" within a small volume. Adopting this point of view, show why it is difficult to suppose that ρ in the present case is a probability distribution function.

(*Hint*: It may be of help to work the problem through for the non-relativistic Schrödinger equation first. For that equation ρ is an acceptable probability function.)

2. Carry out the same steps as in the foregoing problem for the Dirac equation. Show that ρ in this case is acceptable.

3. Quite often in the literature of relativistic quantum mechanics the so-called γ-matrices are encountered. These are defined by

$$\gamma_0 = \beta \qquad \gamma_i = \beta\alpha_i \qquad (i = 1, 2, 3)$$

where β and α are the Dirac matrices.

(a) Show that γ_0 is an Hermitian matrix, but that the γ_i are anti-Hermitian:

$$\gamma_i^\dagger = -\gamma_i.$$

(b) Show that

$$\gamma_i^2 = -1 \qquad (i = 1, 2, 3).$$

4. Show that, by using the definition of the γ-matrices and writing $x_0 = ct$, the Dirac equation may be expressed as

$$i\hbar \sum_{k=0}^{3} \gamma_k \frac{\partial \psi}{\partial x_k} - m_0 c\psi(\mathbf{x}) = 0$$

for a free particle, where $\mathbf{x} = \{x, y, z, ct\}$.

5. Show that, in the representation of equation (12.18),

$$\gamma_i = \begin{pmatrix} 0 & \sigma_i \\ -\sigma_i & 0 \end{pmatrix} \qquad \gamma_0 = \begin{pmatrix} I' & 0 \\ 0 & -I' \end{pmatrix}. \qquad (i = 1, 2, 3).$$

6. Show that the generalized Pauli matrices $\hat{\sigma}_j$ can be written

$$\hat{\sigma}_j = i\gamma_k\gamma_l \qquad (j, k, l = 1, 2, 3)$$

where the j, k, l are in cyclic order.

7. As we know from Chapter 5, the interaction between a free electron and a magnetic field is of the form

$$H' = -\frac{e}{c}\mathbf{v}\cdot\mathbf{A}$$

where $\mathbf{v}$ is the velocity of the electron. By comparison with the comparable Dirac Hamiltonian operator,

$$\mathcal{H}' = -e\boldsymbol{\alpha}\cdot\mathbf{A},$$

this suggests that

$$c\boldsymbol{\alpha} = \mathbf{v}_{op}.$$

The Relativistic Electron

Show that such a correspondence also follows from the Ehrenfest relation

$$\frac{d\mathbf{r}_{\text{op}}}{dt} = -\frac{i}{\hbar}[\mathbf{r}_{\text{op}}, \mathcal{H}] = \mathbf{v}_{\text{op}}$$

where $\mathcal{H}$ is the Dirac Hamiltonian operator

$$\mathcal{H} = -i\hbar c\boldsymbol{\alpha}\cdot\boldsymbol{\nabla} + \beta m_0 c^2 + V(\mathbf{r}).$$

8. The solutions of the Dirac equation for a free particle at rest are

$$\psi_j = x_j(0)\exp\left(-i\epsilon_j m_0 c^2 t/\hbar\right)$$

where

$$x_1(0) = \begin{pmatrix} 1 \\ 0 \\ 0 \\ 0 \end{pmatrix} \quad x_2(0) = \begin{pmatrix} 0 \\ 1 \\ 0 \\ 0 \end{pmatrix} \quad x_3(0) = \begin{pmatrix} 0 \\ 0 \\ 1 \\ 0 \end{pmatrix} \quad x_4(0) = \begin{pmatrix} 0 \\ 0 \\ 0 \\ 1 \end{pmatrix}$$

and

$$\epsilon_j = \begin{cases} 1 & j = 1, 2 \\ -1 & j = 3, 4 \end{cases}$$

We can find ψ_j for the particle moving with speed v in the x-direction by transforming to a reference frame moving with speed v in the $-x$-direction. This is accomplished by writing $(px - |E_p|t)$ for $m_0 c^2 t$ and putting

$$x_j(p) = \exp\left(-\omega\alpha_1/2\right)x_j(0)$$

where $\tanh\omega = -v/c$. Show that, in this case,

$$x_j(p) = \cosh\left(\frac{\omega}{2}\right) \begin{vmatrix} 1 & 0 & 0 & -\tanh\left(\frac{\omega}{2}\right) \\ 0 & 1 & -\tanh\left(\frac{\omega}{2}\right) & 0 \\ 0 & -\tanh\left(\frac{\omega}{2}\right) & 1 & 0 \\ -\tanh\left(\frac{\omega}{2}\right) & 0 & 0 & 1 \end{vmatrix} x_j(0),$$

where

$$\cosh\left(\frac{\omega}{2}\right) = \left(\frac{|E_p| + m_0 c^2}{2m_0 c^2}\right)^{1/2}$$

$$\tanh\left(\frac{\omega}{2}\right) = \frac{-pc}{|E_p| + m_0 c^2}.$$

362

9. Using the commutation relation between $\mathscr{L}_3$ and the components of the momentum operator, show that

$$[\mathscr{L}_3, \mathscr{H}] = ich(\alpha_1 p_2 - \alpha_2 p_1).$$

Also, show that

$$[\hat{\sigma}_3, \mathscr{H}] = -2ic(\alpha_1 p_2 - \alpha_2 p_1)$$

and, therefore that

$$[j_3, \mathscr{H}] = 0$$

where the p_i ($i = 1, 2$) are components of the momentum operator,

$$j_3 = \mathscr{L}_3 + \tfrac{1}{2}\hbar\hat{\sigma}_3,$$

and $\mathscr{H}$ is the Dirac Hamiltonian for a free particle. What do you immediately conclude from this analysis as regards the nature of a Dirac electron?

10. Show that the Dirac equation

$$i\hbar \frac{\partial \psi}{\partial t} = -i\hbar c \sum_{k=1}^{3} \alpha_k \left(\frac{\partial}{\partial x_k} - \frac{ie}{\hbar c^2} A_k \right)\psi + \beta m_e c^2 \psi$$

is invariant under space inversion $\mathscr{P}$, provided that

$$P\psi(\mathbf{r}, t) \equiv e^{i\varphi}\beta\psi(\mathbf{r}, t) = \psi'(-\mathbf{r}, t),$$

and

$$\mathbf{A}'(-\mathbf{r}, t) = -\mathbf{A}(\mathbf{r}, t),$$

where φ is an arbitrary, real constant.

11. Show that the Dirac equation as given just above is also invariant under time reversal $\mathscr{T}$, provided that

$$T\psi(\mathbf{r}, t) \equiv -i\alpha_1\alpha_3\bar{\psi}(\mathbf{r}, t) = \psi'(\mathbf{r}, -t),$$

and

$$A'(\mathbf{r}, -t) = -A(\mathbf{r}, t).$$

12. Show that the Dirac equation is invariant under $\mathscr{P}\mathscr{C}\mathscr{T}$, where $\mathscr{C}$ represents charge conjugation, and that

$$\psi'(-\mathbf{r}, -t) = PCT\psi(\mathbf{r}, t) = e^{i\varphi}\alpha_1\alpha_2\alpha_3\psi(\mathbf{r}, t).$$

It follows from this that we may interpret a positron corresponding to positive energy as an electron corresponding to negative energy and *running backward in space and time*. This interpretation is the basis for R. P. Feynman's approach to relativistic quantum theory.

CHAPTER 13

Nuclear Phenomena

The heart of the matter.

13.1. NUCLEAR STRUCTURE: THE DEUTERON

The experiments which Ernest Rutherford carried out on the scattering of helium nuclei led us to a simple but very useful model of the atom. Indeed, once we have insisted that the atom is composed of electrons whirling about a central nucleus, we can fully understand the microstructure of matter simply by making inquiries as to the precise meanings of the terms "whirling" and "nucleus." Until now we have been attempting to define "whirling" and have come to a rather complete understanding of the quantum physical meaning of this word! We should be able to complete our work by saying "point charge" for "nucleus" if it were not for one remarkable fact. The problem is that atoms can emit matter and radiation without prompting from a concomitant change in their electronic configurations. This phenomenon—called *radioactivity* by Marie Curie—allows us no other course than to accuse the nucleus of its cause and thus to impart to the nucleus some kind of structure. And so our investigation goes on.

Let us begin by recalling some of the empirical information about nuclei which can be gleaned from a casual knowledge of nuclear physics.

(a) Nuclei are generally characterized by the number of protonic charges on them (*atomic number*) and by their masses expressed in units of the protonic mass (*mass number*). A nucleus of an atom of chemical element X is thus denoted by

$$_ZX^A$$

where Z is the atomic number and A is the mass number. The atomic number is unique and is an integer because atoms contain whole numbers of electrons and possess no electric charge, as has been demonstrated by careful experiments with atomic beams put through electric fields. The mass number is not so unique as the atomic number since nuclear masses are not really proportional to the atomic number. However, the mass number is sufficient to aid in distinguishing one chemical element from another. A given nucleus is usually called a *nuclide*. Nuclides having common values of Z are called *isotopes*; those with common values of A are called *isobars*. Thus, the mass number determines an isobaric group while the atomic number determines an isotopic group.

(b) On the grossest scale, nuclei are thought to be composed of protons and neutrons, as suggested by Werner Heisenberg in 1932. The basic properties of these two particles are listed in Table 13.1. We note that the neutron is electrically neutral, and has the same spin but greater mass than the proton. The mass difference is 1.3888×10^{-3} amu, where

$$1 \text{ amu} = 1.66042 \times 10^{-24} \text{ gm}.$$

The spin magnetic moments of the two particles have been expressed in units of the *nuclear magneton*, defined by

$$\mu_N \equiv \frac{|e|\hbar}{2m_p c} = 5.0505 \pm 0.0004 \times 10^{-24} \text{ erg/gauss}.$$

where m_p is the mass of the proton. This definition is quite analogous to that for the Bohr magneton, given in Chapter 4.

Table 13.1. Some properties of the proton and the neutron.

Particle	Mass (amu)	Charge (10^{-10} esu)	Spin (units of $\hbar$)	Magnetic Moment (units of μ_N)	Mean Life
proton	1.0072766	+4.80298	$\pm\frac{1}{2}$	+2.79282	stable
neutron	1.0086654	0	$\pm\frac{1}{2}$	-1.91315	1.01×10^3 sec $(_0n^1 \rightarrow {}_1H^1$ $+ {}_{-1}e^0 + \bar{\nu})$

It is clear that the neutron-proton structure for the nucleus is necessary because protons alone cannot be the sole nuclear constituents and because electrons have the wrong charge and are too large to fit inside the nucleus in any stable way. To see this, we note that, for nearly every nuclide, A is just about equal to 2Z—which implies that Z protonic masses reside in the nucleus along with matter of equal mass. Electrons are ruled out as members of the nucleus because their radii are larger than the radii of many nuclei and

365

because their presence is inconsistent with a large amount of data on nuclear angular momenta and nuclear spectra. The former are fairly compelling in this regard. The basic facts are:[1]

(i) all stable nuclides with even A and even Z have $I = 0$, where I is the total *nuclear* angular momentum quantum number,

(ii) all stable nuclides with even A and odd Z have $1 \leqslant I \leqslant 7$, and

(iii) all stable nuclides with odd A have $\frac{1}{2} \leqslant I \leqslant \frac{9}{2}$.

If electrons were in the nucleus, an even A-odd Z nuclide, for example, would comprise A protons and $(A - Z)$ electrons. But this means that there are $(2A - Z)$ particles—an odd number—and, therefore, a half-integral total angular momentum. Since the even A-odd Z nuclides have integral total angular momenta, we conclude that electrons are not in the nucleus.

(c) A systematic examination of the radioactive behavior of nuclei shows that

(i) isotopes corresponding to odd values of Z are very seldom non-radioactive,

(ii) nuclides with odd Z- and even A-values are radioactive, with the exception of $_1H^2$, $_3Li^6$, $_5B^{10}$, $_7N^{14}$, and $_{73}Ta^{180}$,

(iii) nuclides with both odd Z- and A-values are very seldom stable against electron emission (*beta decay*), and

(iv) nuclides with both even Z- and A-values are often non-radioactive.

(d) An assembly of nucleons (protons and neutrons) always suffers a decrease in total energy as it forms a nucleus. This change in energy is called the *binding energy* of the nucleus. The *binding energy per nucleon* is the binding energy divided by the number of neutrons and protons in the nuclide. The binding energy per nucleon is observed experimentally to increase erratically with mass number, then, for $A > 11$, to remain within 10 per cent of 8.1 MeV. (See Figure 67.) Nuclides with both Z and A equal to even integers always have greater binding energies per nucleon than those with both Z and A equal to odd integers. When either Z or $(A - Z)$ are near in value to any of the numbers in the set $\{2, 8, 20, 28, 50, 82, 126\}$, the corresponding nuclides are found to be unusually stable and quite abundant, relative to their fellows. For this reason, these seven numbers have come to be called "magic numbers." As examples of the magic numbers at work, we may cite the abnormally high binding energies per nucleon of the doubly magic $_2He^4$ and $_8O^{16}$ and the great abundance of nuclides possessing either 50 protons or 50 neutrons.

[1] For a complete introductory discussion of the method for getting the data, see E. Segrè. *Nuclei and Particles*, W. A. Benjamin, New York, 1964, pp. 217–235.

Nuclide	Binding Energy Per Nucleon	Nuclide	Binding Energy Per Nucleon
	(MeV)		(MeV)
$_1H^3$	2.8283	$_8O^{15}$	7.4676
$_2He^3$	2.5733	$_7N^{16}$	7.3748
$_2He^4$	7.0743	$_8O^{16}$	7.9759
$_2He^5$	5.4680	$_7N^{17}$	7.2790
$_3Li^5$	5.3000	$_8O^{17}$	7.7506

These are the basic notions. With them in mind, we should like to investigate the structure of the nucleus. It would seem that the first question to ask is: "How large is the edifice with which we are dealing?" Rutherford attempted to give a solution to this problem by estimating the nuclear radius from the results of his scattering experiments. He suggested that the "distance of closest approach" corresponding to the scattering angle at which equation (11.57) fails be called the nuclear radius. While this is certainly not a quantum physical definition, it does have an intuitive physical appeal and should give us something to work with. In doing a series of experiments, Rutherford and James Chadwick, discoverer of the neutron, found that the nuclear radius was directly proportional to the cube root of the mass number:

$$R = r_0 A^{1/3} \tag{13.1}$$

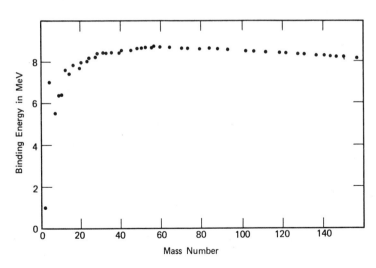

Figure 67. The binding energy per nucleon as a function of mass number.

where r_0 is a constant whose value in *fermis* (femtometers) may be taken to be

$$r_0 = 1.20 \pm 0.01 \text{ f}.$$

It is encouraging that many experiments done subsequent to Rutherford's work have supported the validity of equation (13.1). Evidence from high-energy electron scattering, in fact, has provided the value of r_0 just given. When low-energy electrons are scattered, we find that r_0 is still of the order of 1 fermi.

Nucleus	Electron Energy (MeV)	r_0 (fermi)
Silver	15.7	1.13 ± 0.03
Gold	15.7	1.18 ± 0.03
Tungsten	30–60	1.1 ± 0.1

Another source for the estimate of r_0 comes from the semiempirical expression for the nuclide mass[2] as a function of A and Z. There is a term in this equation proportional to $A^{-\frac{1}{3}}$ which can be used to compute r_0. The most recent statistical fitting of the equation to nuclide mass data indicates that

$$r_0 = 1.22 \pm 0.01$$

which is in fine agreement with the estimates from scattering data. Other methods of getting at the nuclear radius, such as the interpretation of radioactive decay and nuclear reaction experiments, provide values of r_0 within 0.3 fermi (but systematically high) of those suggested above. We certainly can conclude, then, that the atomic nucleus is of the order of a few fermis in diameter.

The fact that Rutherford's expression for the differential scattering cross section fails at a certain point permits us to define the nuclear radius and thus to infer that some structure is present, even in the absence of evidence from radioactive decay experiments. Since we know that the nucleus is charged, but yet contains uncharged matter, it would be worthwhile to know what is the distribution of nuclear charge and, therefore, "where the protons are." The simplest way to picture the nuclear charge distribution is to write down the differential scattering cross section in terms of a variable charge density and then calculate this function by fitting the cross-section expression to the observed values of $\sigma(\vartheta)$ for a chosen nucleus. Let us see how to do this.

[2] See, for example, E. Fermi, *Nuclear Physics*, University of Chicago Press, Chicago, 1950, p. 6f.

One easy way is to use Born's approximation, which, as suggested in problem 8 of Chapter 11, is fortuitously exact for coulomb interactions. We can modify equation (11.50) of that chapter to read

$$\sigma(\vartheta) = \left(\frac{2Z'e\mu}{\hbar^2 q^2}\right)^2 \left| \int_0^\infty 4\pi\rho(r) \frac{\sin qr}{qr} r^2 \, dr \right|^2$$

$$= \sigma_R(\vartheta) \left| \frac{1}{Ze} \int_0^\infty 4\pi\rho(r) \frac{\sin qr}{qr} r^2 \, dr \right|^2$$

$$\equiv \sigma_R(\vartheta) |F(q)|^2 \tag{13.2}$$

where

$$q = 2k \sin (\tfrac{1}{2}\vartheta),$$

in the previous notation, $\rho(r)$ is the charge distribution function, and $F(q)$ is known as the *form factor*. Equation (13.2) represents the effect of replacing $Ze\, \delta(r)/4\pi r^2$ by an arbitrary distribution of charge. Whenever $\rho(r)$ represents a charge concentrated at the origin, equation (13.2) then reproduces equation (11.57), as it should. In all realistic cases, however, $\rho(r)$ will have a definite functional form dictated by the empirical values of $\sigma(\vartheta)$. Some results of this kind of calculation are shown in Figure 68. There we see

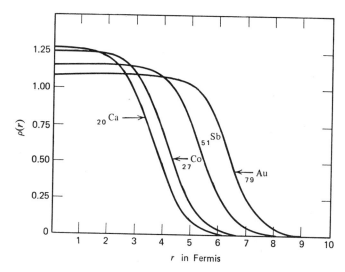

Figure 68. Charge distributions for several nuclei as deduced from the results of electron scattering experiments. [The data are from B. Hahn, D. G. Ravenhall, and R. Hofstadter, High-energy electron scattering and the charge distributions of selected nuclei, *Phys. Rev.* **101**: 1131–1142 (1956).]

that the charge density in the nucleus does not fall off rapidly in value, but persists at least for a distance R as given by equation (13.1). It would appear from this behavior that the distribution of nuclear charge is rather uniform, that protons and neutrons are "well mixed" in the nucleus. (Actually, we might have expected such a result from an elementary consideration of the behavior of a swarm of positive charges confined to a finite region of space. The neutrons would naturally be expected to reduce the repulsive coulomb interactions by interspersing themselves between protons.) We should note in passing that the diffuse "tail" of the charge distribution gives the lie to any attempt to define an absolute nuclear radius. It is clear now that this quantity will vary somewhat, depending on just how the nuclear charge distribution is disturbed when its extent is ascertained.

Thus far our picture of the nucleus includes the notion of a fairly uniform distribution of neutrons and protons over a very small volume. The next step in our investigation would seem to be an inquiry into the geometric arrangement of nuclear matter. There are two ways of answering this question. In the first, we note that

$$\frac{\text{mass density of nucleus}}{\text{mass density of a solid}} \doteq \frac{\text{volume of an atom}}{\text{volume of nucleus}} \doteq \frac{(10^{-8})^3}{(10^{-12})^3} = 10^{12}$$

which shows that the mass density of nuclear matter is enormous. But this result does not really have a bearing upon the question of geometric arrangement—it only informs us that nucleons are relatively heavy and quite small. What we must know is the *volume occupied per nucleon* in the nucleus. We can estimate that quantity by writing[3]

$$\frac{(\text{collision radius of nucleon})^3}{(\text{radius of nucleus})^3} \doteq \frac{(0.4)^3}{(1.2)^3}\left(\frac{1}{A}\right) = \frac{0.037}{A}$$

which is quite small. It follows that each nucleon occupies very little of the nuclear volume in most cases and, therefore, that we might suppose the nucleons to be *independently moving particles*, in the zeroth approximation.

Our model of the nucleus has now developed to the point where some of the mathematical formalism of quantum physics can be applied to it. If we assume that a nucleon moves independently of its fellows, yet is confined to the nuclear volume, we have nothing more than the problem of a free particle in a finite region of space. This problem was discussed in section 3.5. There we discovered that a particle subject to the potential

$$V(q) = \begin{cases} \infty & q < 0 \\ 0 & 0 \leqslant q \leqslant L \qquad (q = x, y, z) \\ \infty & q > L \end{cases}$$

[3] The estimate of the collision radius (or hard core radius) is from L. Hulthén and M. Sugawara, The two-nucleon problem, *Handbuch der Physik* **XXXIX**: 27–30 (1957).

could be represented by sinusoidal wavefunctions and possessed the total energy

$$E_N = \left(\frac{2\pi\hbar}{L}\right)^2 \frac{N^2}{8m} \qquad (N = 3, 6, \ldots) \qquad (13.3)$$

where L is the length of the (cubical) region in which the particle moves, m is the particle mass, and

$$N^2 = N_x{}^2 + N_y{}^2 + N_z{}^2$$

in the notation of Chapter 3. Equation (13.3) evidently is valid for either protons or neutrons in the present approximation. However, because of the exclusion postulate, no more than one proton and one neutron can be characterized by a given $\{N_x, N_y, N_z, m_s\}$. Otherwise we should have, say, two protons with identical sets of quantum numbers—a clear violation of the simple form of the exclusion postulate. It follows that, in the groundstate of the nucleus, the nucleons will arrange themselves two to an energy state (characterized by $\{N_x, N_y, N_z\}$) until all the needed states are occupied. The total energy of the groundstate nucleus, then, is much greater than would be expected in the absence of the exclusion postulate!

The uppermost energy level filled when a nucleus is in its groundstate is known as the *Fermi energy*. It may be calculated as follows. Let us imagine that there are a great many nucleons, so that we may express the number of quantum numbers between N and $N + dN$ as

$$\rho(N) \, dN = \tfrac{1}{2}\pi N^2 \, dN \qquad (13.4)$$

the right-hand side of equation (13.4) being just the volume of an octant of a shell between spheres of radii N and $N + dN$. Now, by equation (13.3),

$$dE_N = \left(\frac{2\pi\hbar}{L}\right)^2 \frac{N}{4m} \, dN = \frac{\pi\hbar}{L} \left(\frac{2E_N}{m}\right)^{\!\!1/2} dN$$

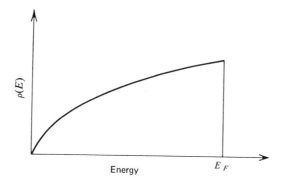

Figure 69. The density of states as a function of energy.

371

and

$$\rho(N)\, dN \equiv \rho(E_N)\, dE_N = \left(\frac{m^3}{2}\right)^{\!1/2} \frac{V}{\pi^2 \hbar^3} E_N^{1/2}\, dE_N \tag{13.5}$$

where we have noted the obvious equivalence of the density of states $\rho(E)$ to $\rho(N)$ and have put L^3 equal to the nuclear volume. Equation (13.5) shows that the density of states increases as $E_N^{1/2}$ between zero and the Fermi energy, where it instantly must fall to zero. (See Figure 69.) The total number of particles in the nucleus must then be

$$\mathcal{N} = 2\int_0^{E_F} \rho(E_N)\, dE_N = \frac{(2m^3)^{1/2}V}{\pi^2 \hbar^3} \int_0^{E_F} E_N^{1/2}\, dE_N$$

$$= \frac{(2m)^{3/2}V}{3\pi^2\hbar^3} E_F^{3/2}$$

or

$$E_F = \frac{(3\pi^2)^{2/3}\hbar^2}{2m}\, \rho^{2/3} \tag{13.6}$$

where $\rho = \mathcal{N}/V$ is the (uniform) nucleon number density. We notice that the Fermi energy determines the average nucleon energy; for,

$$\langle E \rangle = \frac{\int_0^{E_F} E_N \rho(E_N)\, dE_N}{\int_0^{E_F} \rho(E_N)\, dE_N} = \frac{\int_0^{E_F} E_N^{3/2}\, dE_N}{\int_0^{E_F} E_N^{1/2}\, dE_N}$$

$$= \frac{\frac{2}{5}E_F^{5/2}}{\frac{2}{3}E_F^{3/2}} = \frac{3}{5}\, E_F. \tag{13.7}$$

The value of the Fermi energy for neutrons may be calculated, once the nucleon number density has been estimated. This we shall do by writing

$$\rho = \frac{3(A - Z)}{4\pi r_0^3 A} \doteq \frac{3}{8\pi r_0^3}$$

.or $A \doteq 2Z$. Thus

$$E_F \doteq \left(\frac{9\pi}{8}\right)^{2/3} \frac{\hbar^2}{2mr_0^2} = \frac{48.1}{r_0^2}\ \text{MeV} - \text{f}^2 \tag{13.8}$$

If we choose r_0 to be in the range 1.2–1.5 f, we get a Fermi energy in the range 21–33 MeV. To this we must add the average binding energy per nucleon, 8 MeV. The final picture, then, is one wherein the neutron is confined to a potential well of a few fermis in diameter and about 35 MeV in depth. (See Figure 70.) The proton is in very much the same situation, except that its potential well is shallower than that for the neutron and is surmounted by a coulomb potential barrier. The total energy for both nucleons must be the

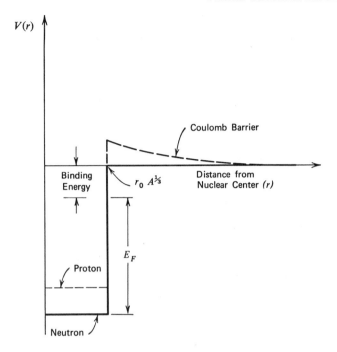

Figure 70. The potential wells for neutrons and protons in a heavy nucleus. The proton well is shown in dashed lines.

same; if it were not, the neutron would be unstable, relative to the proton, in *every* nucleus.

The potential well idea can be extended convincingly to a description of the simplest compound nucleus, that of the hydrogen isotope deuterium. The deuteron, as this nuclide is called, is composed of a proton and a neutron; the binding energy is 2.22452 ± 0.00020 MeV. The nuclide has a unit total angular momentum quantum number, a spin magnetic moment equal to 0.85741 nuclear magnetons, and an electric quadrupole moment equal to 2.74×10^{-27} cm^2. It is not observed to possess any bound excited states. In order to describe the deuteron theoretically, we shall make the following two assumptions:

(a) The relative potential governing the motions of the neutron and the proton is central, short-range, and possesses a minimum.

(b) The relative Hamiltonian operator describing the nucleons is spin-independent.

Both of these assumptions, strictly speaking, are incorrect. Because the deuteron possesses a small electric quadrupole moment, the relative potential cannot be central. Moreover, the non-central character of the potential turns out to arise from spin-spin coupling. But these modifications are not critical when it comes to supplying a quantum mechanical rationale for the deuteron's properties, as given above. Therefore, we shall stay with the assumptions (a) and (b).

In particular, we shall assume that $V(r)$ is a spherical potential well:

$$V(r) = \begin{cases} -V_0 & r < a \\ 0 & r > a \end{cases}$$

where V_0 and r_0 are constants. The appropriate eigenvalue problem is, then, for the groundstate,

$$\frac{d^2 p}{dr^2} + \alpha^2 p(r) = 0 \qquad (r < a),$$

$$\frac{d^2 p}{dr^2} - \beta^2 p(r) = 0 \qquad (r > a),$$

(13.9)

where

$$p(r) = r\psi(r),$$

$$\alpha^2 = \left(\frac{2\mu}{\hbar^2}\right)(V_0 - E_B),$$

$$\beta^2 = \left(\frac{2\mu}{\hbar^2}\right)E_B,$$

$\psi(r)$ being the relative groundstate wavefunction, and E_B is the binding energy, the negative of the groundstate energy. The solution of the eigenvalue problem must vanish at the center of the well and at infinite separation of the nucleons. It follows that

$$p(r) = \begin{cases} A' \sin \alpha r & (r < a) \\ B' \exp(-\beta r) & (r > a) \end{cases}$$

is a solution, subject to the condition that it be continuous and twice-differentiable at a. This condition, in turn, implies

$$A' \sin \alpha a = B' \exp(-\beta a)$$
$$A'\alpha \cos \alpha a = -\beta B' \exp(-\beta a)$$

or

$$(\alpha a) \cot (\alpha a) = -(\alpha \beta), \qquad (13.10)$$

upon combining the two equations. (It is understood, of course, that A' and B' are normalization constants.) We shall choose a to be the nuclear radius, as estimated from equation (13.1), for concreteness. Thus, with $A = 2$, we have $a = 1.5\,\text{f}$ and, using the experimental value of E_B with $\mu = 0.5 m_p$,

$(a\beta) = 0.348$. Now, if $(a\beta)$ were precisely zero, then (αa) would be $\pi/2$ and the bottom of the potential well would be

$$V_0 = E_B + \left(\frac{\hbar^2}{2\mu}\right)\left(\frac{\pi}{2a}\right)^2.$$

Let us suppose that $(\alpha a) = \pi/2 + \epsilon$, where ϵ is a small number. Then we have

$$\left(\frac{\pi}{2} + \epsilon\right) \cot\left(\frac{\pi}{2} + \epsilon\right) = -0.348,$$

or

$$\left(\frac{\pi}{2} + \epsilon\right)\epsilon = 0.348$$

or

$$\epsilon \doteq \frac{0.695}{\pi},$$

upon expanding the cotangent to first order in ϵ. It follows that

$$(\alpha a) \doteq \frac{\pi}{2}\left(1 + \left(\frac{2}{\pi}\right)^2 0.348\right) = 1.14\left(\frac{\pi}{2}\right) = 1.79$$

and that

$$V_0 \doteq E_B + \frac{\hbar^2}{2\mu}\left(\frac{1.79}{a}\right)^2. \tag{13.11}$$

If the empirical values of E_B, μ, $\hbar$, and a are put into equation (13.11), we find

$$V_0 \doteq 27.2 E_B.$$

We see that the groundstate energy level is quite near the top of the potential well, indicating that the nucleons are rather loosely bound. Moreover, we see that $\alpha^2 \doteq 2\mu V_0/\hbar^2$ here, so that

$$\frac{\pi}{2} < ak_0 < \pi \tag{13.12}$$

is implied by our results, where

$$k_0{}^2 = \frac{2\mu V_0}{\hbar^2}.$$

On the other hand, the zeroth approximation to the first excited energy level, $-E'$, is

$$a\alpha = \frac{3\pi}{2}$$

which suggests

$$a\left[\left(\frac{2\mu}{\hbar^2}\right)(V_0 - E')\right]^{1/2} \geq \frac{3\pi}{2}$$

in direct contradiction with equation (13.12). We conclude that no excited bound states corresponding to zero orbital angular momentum are possible, in agreement with experiment. A very similar argument[4] shows that no bound state is possible for orbital angular momenta greater than zero, either.

Because the groundstate of the deuteron corresponds to zero orbital angular momentum, the total angular momentum of the nucleus must arise solely from the spins of the two nucleons. This conclusion is in agreement with the facts that $I = 1$ for the deuteron and that the spin magnetic moment is 0.85741 nuclear magneton, while the sum of nucleon spin moments is $(2.79282 - 1.91315) = 0.87967$ nuclear magneton. Evidently, then, the groundstate of the deuteron is a triplet state.

In Figure 71 the groundstate wavefunction is plotted against the inter-nucleon separation. One striking aspect of the plot is that the wavefunction has an extremely long "tail" outside the potential well. In fact, it turns out

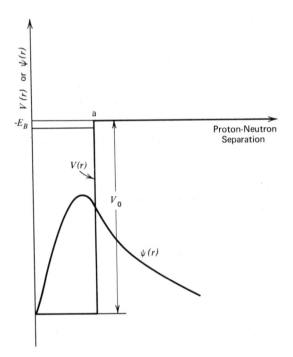

Figure 71. A sketch of the groundstate wavefunction for the deuteron, in the potential-well approximation.

---

[4] See, for example, L. R. B. Elton, *Introductory Nuclear Theory*, W. B. Saunders, Philadelphia, 1966, pp. 64–66.

that there is only about one chance in three that the nucleons will be closer together than the distance a. This is just one more result of the very weak binding of the deuteron, as we have found in our analysis.

13.2. ALPHA DECAY

The "alpha particle" is the name Ernest Rutherford gave to the nucleus of helium of mass number four. Its very great importance to our present investigation, aside from the fact that it may be used as a scattering probe, lies with the observation that some 170 nuclides produce this particle in the decomposition

$$_ZX^A \rightarrow {}_{Z-2}Y^{A-4} + {}_2He^4 \tag{13.13}$$

where X is termed the "parent" and Y, the "daughter." Equation (13.13) may be called simply *alpha decay*. It is characterized by two parameters, the decay constant, λ, and the decay energy, Q. The first of these appears in the well-known expression for the rate of radioactive decay

$$\frac{dN}{dt} = -\lambda N(t) \tag{13.14}$$

where $N(t)$ is the number of X-nuclides at time t. The physical significance of λ is made transparent by noting that the time, $t_{1/2}$, for $N(t)$ to become equal to half its value at time zero is a solution of the integrated form of equation (13.14)

$$N(t_{1/2}) = \tfrac{1}{2}N(0) = N(0)\exp{(-\lambda t_{1/2})}$$

such that

$$t_{1/2} = \frac{\ln 2}{\lambda}. \tag{13.15}$$

The quantity $t_{1/2}$ is known as the *half-life*. For the nuclides engaging in alpha decay, it takes on values between 10^{-21} second (${}_2He^5$) and 10^{12} years (${}_{78}Pt^{190}$) —a fantastic range of variation! Thus, the half-life, while characteristic of a given alpha decay process, in no way represents the general aspects of helium nucleus emission.

The decay energy, Q, fares somewhat better in this regard. Possibly because it is only the heavy nuclei which emit alpha particles to any extent (Figure 72), the primary variation of the alpha decay energy is from 4 to 8 MeV. (See Figure 73.) Now, Q is the sum of the kinetic energies of the daughter nuclide and the alpha particle in the laboratory frame of reference, by definition. It is not hard to see from this (or, indeed, to show rigorously by using energy and momentum conservation principles) that most of Q is made up from the kinetic energy of the alpha particle when the daughter is a heavy nuclide. Therefore, a measurement of the speed of the alpha particle as it

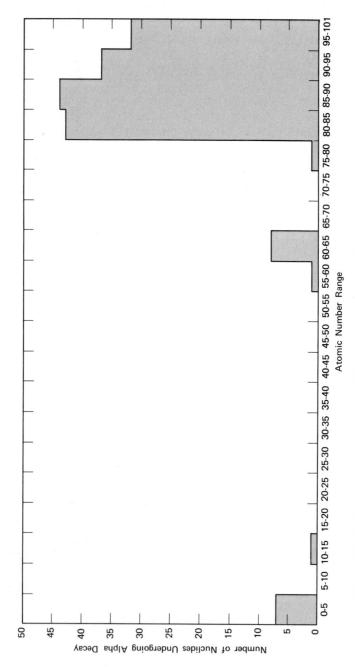

Figure 72. A plot of the number of nuclides undergoing alpha decay against the atomic number.

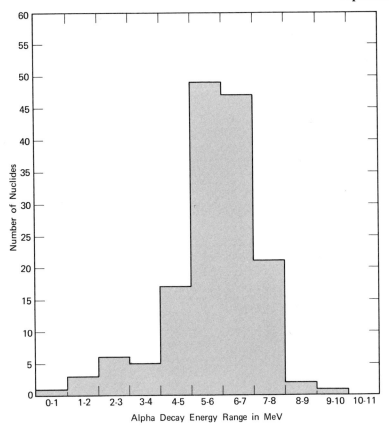

Figure 73. A plot of the number of nuclides with alpha decay energy in a given range against alpha decay energy.

escapes the nucleus should yield the value of Q reliably. This may be done by deflecting the alpha particle in a magnetic field and using the ideas developed in section 2.1 in order to relate the speed of the particle to its angle of deflection and the strength of the magnetic field. In this way we find speeds of about 2×10^9 cm/sec, as indicated in this table:

Nuclide	Alpha Particle Speed (cm/sec)	Alpha Decay Energy (MeV)
$_{86}Em^{222}$	1.62×10^9	5.49
$_{84}Po^{216}$	1.80×10^9	6.77
$_{84}Po^{212}$	2.05×10^9	8.78

Another, but less exact, way to discover the energy of the alpha particle is to measure its *range* in air. The range, by definition, is the distance from the source of alpha particles to a point where they cannot be detected in significant numbers.[5] This quantity is quite easy to measure, much more so than is the alpha particle speed. Therefore, if the ranges of the alpha particles whose speeds are known are tabulated, a correlation graph can be drawn which allows the computation of the energy of a particle of known range but unknown speed. Better still, an empirical formula might be derived from the correlation graph so that energies can be calculated directly. As an example of this, for ranges in air at 15°C and atmospheric pressure, the range-energy relation

$$\text{mean range} = \text{constant} \times (\text{energy})^{3/2} \qquad (13.16)$$

seems to be valid.

As odd as it may seem, the widely varying decay constant and the relatively stationary Q-value are not independent quantities. In 1911, Geiger and Nuttall published a correlation between the decay constant and the range in the form

$$\text{decay constant} = \text{constant} \times (\text{range})^{\text{constant}},$$

which, in view of equation (13.16), may be written

$$\text{decay constant} = \text{constant} \times (\text{energy})^{\text{constant}}. \qquad (13.17)$$

Equation (13.17) appears to conform to experiment rather well. (See Figure 74.) In general, we may conclude from the Geiger-Nuttall relation that relatively stable nuclides emit the slowest-moving alpha particles, and vice versa.

Before attempting to account for alpha decay in terms of our picture of the atomic nucleus, we must pause to consider one more empirical observation. Quite some time ago, it was discovered that the energy of certain alpha particles, such as those emitted by $_{83}\text{Bi}^{212}$, was not of a certain value, but, in fact, was a *multi-valued* quantity. The range of values, to be sure, was in general quite small—so small, in fact, as to be masked by the insensitivity of the range-energy method of getting at Q-values. But the magnetic deflection method showed clearly that the energy spectrum for alpha emission possessed

---

[5] This definition is rather sensitive to the experimental situation for measuring ranges. A quantity called the *mean range*, the distance where the number of alpha particles falls off most rapidly, is less sensitive and so is more often used in determining range-energy relationships.

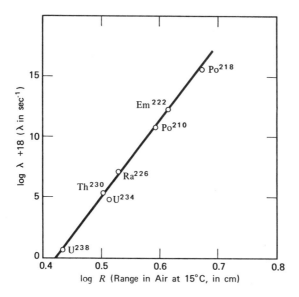

Figure 74. An empirical verification of the Geiger-Nuttall relation.

fine structure. The situation for $_{94}Pu^{238}$ is indicated in Figure 75. Looking at the whole picture, we find that alpha particle spectra can be arranged in three groups:

(a) those which are but a single line (examples are $_{86}Em^{222}$, $_{84}Po^{210}$, $_{84}Po^{218}$),

(b) those which are a set of closely spaced lines (examples are $_{83}Bi^{212}$, $_{86}Em^{219}$), and

(c) those which are widely spaced sets of closely-spaced lines ($_{84}Po^{214}$, $_{84}Po^{212}$).

These observations we must fit into whatever theoretical scheme we devise for explaining alpha decay.

Our concept of the atomic nucleus is one which has it that the nucleons move about rather independently of one another. It is obvious that this picture is too simple to explain alpha decay. Alpha particles, after all, are groups of nucleons in very close association! On the other hand, if we embellish our view only to the extent of acknowledging the existence of magic numbers, we can postulate that, at least for the cases where alpha decay can take place, the substructure of two neutrons with two protons will be an especially stable one in the nucleus. Thus we see the alpha-disintegrating nuclide as an assembly of independently moving *alpha particles* rather than

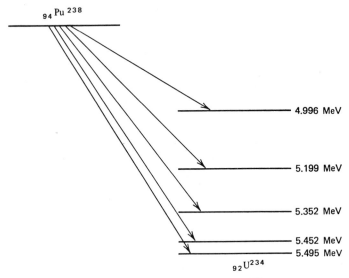

Figure 75. The alpha decay scheme for $_{94}Pu^{238}$, indicating the existence of a gamma radiation spectrum.

nucleons. Moreover, we may visualize the nucleus as a potential well similar in form to that for protons. (See Figure 76.) Inside the nucleus, the alpha particle resides in a well of appropriate depth; to escape the well, it must surmount a barrier of height V_B, where

$$V_B \doteq \frac{2Ze^2}{R + R_{He}}$$

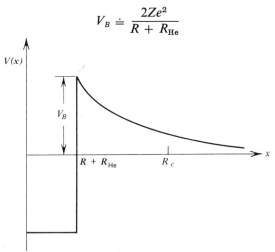

Figure 76. The nucleus as a potential well in alpha decay.

R is the radius of the daughter nuclide and R_{He} is the radius of the alpha particle. If we use equation (13.1) and the "typical" values 90 and 225 for Z and A, respectively, we discover

$$V_B \doteq 28 \text{ MeV},$$

which is much higher than the alpha decay energies indicated in Figure 73. Evidently the alpha particle is able to surmount the coulomb barrier to its escape with an energy of one-seventh the barrier height. Speaking in terms of classical mechanics, this is impossible. However, we are not restricted to classical mechanics! Recall that in section 3.5 we found that a particle with a total energy too small to get over a rectangular potential barrier could "tunnel" *through* the barrier: the wavefunction representing the particle was not zero inside the barrier region. In fact, so long as the product of

$$k_1 = \frac{[2m(V_B - E)]^{\frac{1}{2}}}{\hbar}$$

with the barrier thickness is quite large,

$$T = \frac{|\psi(\text{outside barrier})|^2}{|\psi(\text{incident on barrier})|^2} \underset{k_1 d \to \infty}{\sim} \exp(-2k_1 d) \qquad (13.18)$$

where T is the transmitted fraction of the alpha particle wave and d is the barrier thickness. We see that there is always a small but non-zero chance of getting the alpha particle outside potential barriers of finite height and width. This seems to be the explanation of alpha decay: the alpha particle tunnels out of the nucleus as often as equation (13.18) will allow.

Of course, equation (13.18) is applicable only to the case of a rectangular potential barrier, and not to a coulomb barrier. But this is no serious problem. We can write equation (13.18) in the modified form[6]

$$T \sim \exp\left[\frac{-2(2m)^{\frac{1}{2}}}{\hbar} \int_{R+R_{He}}^{R_c} (V(x) - E)^{\frac{1}{2}} \, dx\right] \equiv \exp(-2G) \quad (13.19)$$

where R_c is the "classically allowed" alpha particle-nucleus separation, the solution of

$$Q - \frac{2Ze^2}{R_c} = 0$$

Q being the alpha decay energy. Now, if we adopt the simplest approach and write

$$V(x) = \frac{2Ze^2}{x} \qquad (x > R + R_{He})$$

$$E = Q = \frac{2Ze^2}{R_c}$$

[6] This action is justifiable. See, for details, A. Messiah, *Quantum Mechanics*, John Wiley and Sons, New York, 1961, Vol. I, p. 237*ff*.

then equation (13.19) may be rewritten as

$$G = \frac{(2m)^{\frac{1}{2}}}{\hbar} \int_{R+R_{He}}^{R_c} \left(\frac{2Ze^2}{x} - \frac{2Ze^2}{R_c} \right)^{\frac{1}{2}} dx$$

$$= \frac{[4mZe^2R_c]^{\frac{1}{2}}}{\hbar} \{ \cos^{-1}(y^{\frac{1}{2}}) - (y - y^2)^{\frac{1}{2}} \}$$

$$\equiv \frac{4Ze^2}{\hbar v} \gamma(y) \tag{13.20}$$

where

$$v = \left(\frac{2E}{m} \right)^{\frac{1}{2}} = \left(\frac{4Ze^2}{mR_c} \right)^{\frac{1}{2}}$$

$$y = \frac{R + R_{He}}{R_c} = \frac{E}{V_B}$$

and

$$\gamma(y) \equiv \cos^{-1}(y^{\frac{1}{2}}) - y^{\frac{1}{2}}(1 - y)^{\frac{1}{2}}.$$

In the present case we are assuming the (E/V_B) is small, so that

$$\gamma(y) = \frac{\pi}{2} - 2y^{\frac{1}{2}} + \cdots$$

and equation (13.20) becomes, to first order,

$$G \underset{y \to 0}{\sim} \frac{2\pi Ze^2}{\hbar v} - \frac{4|e|}{\hbar} [Zm(R + R_{He})]^{\frac{1}{2}}. \tag{13.21}$$

Equation (13.21) indicates that the barrier transparency increases with an increase in the speed of the alpha particle or in the size of the daughter nuclide. The first of these trends we already noted in our discussion of the Geiger-Nuttall relation.

We can pose a reasonable estimate of the decay constant by writing

$$\lambda = \frac{v_0}{R + R_{He}} \exp(-2G), \tag{13.22}$$

which is to say that λ is the product of the frequency of escape attempts and the barrier transparency. According to our previous information, v_0, the speed of the escaped alpha particle, is about 10^9 cm per second and $(R + R_{He})$ is about 9 fermis; therefore, the barrier transparency must range from about 10^0 to 10^{-40} in order to account for the observed variation in the decay constant!

If equation (13.21) is put into equation (13.22), we get

$$\ln \lambda = \ln \frac{v_0}{(R + R_{\text{He}})} - \frac{(2mV_B)^{1/2}}{\hbar} (R + R_{\text{He}}) \left[\pi \left(\frac{V_B}{E} \right)^{1/2} - 4 \right] \quad (13.23)$$

or, in parameter form,

$$\log \lambda = A - BE^{-1/2}. \quad (13.24)$$

Equation (13.24) may be thought of as a "theoretical form" of the Geiger-Nuttall relation in that it, too, expresses a relationship between the decay constant and the alpha decay energy. As a matter of fact, equation (13.24), with empirically deduced values of the parameters A and B, has been found to fit the results of experiment fairly well.[7]

So we can account at least qualitatively for the properties of alpha radioactivity. We have seen that alpha decay comes about as a distinctly quantum physical phenomenon—the "leaking" of the alpha particle's de Broglie wave through the nuclear coulomb barrier—and that the experimentally observed decay constants can be understood in terms of the barrier properties. All that remains is to examine the Q-values. Experiment has it that this quantity may be many-valued, such that a given nuclide may emit alpha particles with a discrete set of kinetic energies. How do we explain this? It is clear that our ideal gas model of the nucleus offers no direct aid since the alpha particle is not a fermion and so is not subject to Pauli selection as regards its occupation of energy levels. (The helium nucleus evidently has spin quantum number zero, corresponding to pairs of neutrons and protons.) It follows that there is no reason to suppose the nuclear alpha particles occupy different energy levels, let alone a given, discrete set of them. The clue to the solution of the problem lies with the word *discrete*. Just as the atom can possess only a discrete set of energy levels, so perhaps is the nucleus similarly restricted. If this were true, the result of alpha decay need not always be a groundstate daughter nuclide and an emitted alpha particle, but instead possibly an *excited* daughter nuclide and an emitted alpha particle. In order to conserve energy, the latter particle would have to be less energetic than it would be if it accompanied a groundstate nuclide, and so would appear with a discrete set of kinetic energies. Evidently the excited daughter nuclide would decay to its groundstate sometime later, with the emission of radiation. A check on our hypothesis, then, would be the observation of radiation from the daughter nuclide—radiation whose energy was just that needed to satisfy the conservation condition

radiated energy $= Q -$ alpha energy $-$ daughter energy

[7] See, for example, I. Kaplan, On the systematics of even-even alpha emitters, *Phys. Rev.* **81**: 962–968 (1951).

Table 13.2. The alpha decay of $_{90}Th^{228}$. Notice the close correspondence between the figures in the fourth and fifth columns.

Q MeV	K_α MeV	K_d MeV	$(Q - K_\alpha - K_d)$ MeV	Observed Radiation Energy MeV
5.516	5.421	0.095	0.000	0.000
5.516	5.338	0.095	0.083	0.084
5.516	5.208	0.095	0.213	0.217
5.516	5.173	0.095	0.248	0.253

In Table 13.2 are shown some data on the alpha decay of $_{90}Th^{228}$:

$$_{90}Th^{228} \rightarrow {}_{88}Ra^{224} + {}_2He^4 + 5.516 \text{ MeV}.$$

We see that, indeed, the emission of alpha particles *is* accompanied by radiation and that the radiation energy makes up the difference between the Q-value and the alpha particle energy, making due allowance for the kinetic energy of the daughter nuclide. It would appear from this result that our hypothesis on the role of the nucleus in alpha decay is basically correct; but our ideal gas model of nuclear structure would need some improving to account quantitatively for nuclear spectra.

13.3. BETA DECAY

The "beta particle" is an electron and is one of the results of the isobaric nuclear decomposition

$$_ZX^A \rightarrow {}_{Z+1}Y^A + {}_{-1}e^0. \tag{13.25}$$

The reaction (13.25) is called *beta decay*. Beta decay fundamentally involves the transformation of a neutron into a proton and an electron. When it occurs outside the nucleus—that is, when a free neutron decays—the process is characterized by a half-life of 11.7 ± 0.3 minutes. Otherwise, the half-life may vary from a fraction of a second to 10^{15} years, as indicated in Table 13.3.

The decomposition (13.25) represents but one of three possibilities for beta decay. The other two are "positron emission,"

$$_ZX^A \rightarrow {}_{Z-1}Y^A + {}_{-1}\bar{e}^0,$$

and "K-capture,"

$$_ZX^A + {}_{-1}e^0 \rightarrow {}_{Z-1}Y^A + \gamma,$$

where the bar denotes the antiparticle and γ refers to gamma radiation. The last-named kind of beta decay involves the absorption, by the nucleus, of an

Table 13.3. Some typical beta decay reactions.

Parent	Daughter	$t_{1/2}$	$Q_{max}(MeV)$
$_0n^1$	$_1H^1$	11.7 ± 0.3 min	0.782
$_1H^3$	$_2He^3$	12.4 years	0.0194
$_2He^6$	$_3Li^6$	0.813 sec	3.50
$_6C^{14}$	$_7N^{14}$	5,568 years	0.155
$_{18}A^{39}$	$_{19}K^{39}$	260 years	0.565
$_{19}K^{40}$	$_{20}Ca^{40}$	1.3×10^9 years	0.63
$_{27}Co^{60}$	$_{28}Ni^{60}$	5.2 years	0.314
$_{49}In^{115}$	$_{50}Sn^{115}$	6×10^{14} years	0.6

electron in the "K-shell" of an atom with the subsequent emission of gamma radiation because of electronic transitions. K-capture and positron emission are equivalent processes, insofar as the state of the daughter nuclide is concerned.

Not long after the turn of the century, James Chadwick provided conclusive evidence that something very strange was going on when beta decay occurred. The electrons emitted from decomposing nuclear matter, by contrast with the situation for alpha decay, carried off energies varying *continuously* from zero to a characteristic maximum. Moreover, *no* concomitant gamma radiation generally was found for the electrons emitted with energies less than the maximum. This was quite surprising! Where did the extra energy go in these cases? It did not take long to find out that the daughter nuclide had nothing to do with the matter. The only other possibility seemed to be that, after the electron was ejected from the nucleus, some kind of interaction occurred which reduced its energy. In other words, the hypothesis was to suppose every expelled electron possessed the maximum energy, but that some of them lost energy by one means or another. An experimental test of this idea could be made by trapping and measuring in a calorimeter the total energy of decomposition per electron of some beta emitter and comparing the result with the average energy, deduced from the continuous beta-particle spectrum, and with Q_{max}. The nuclide $_{83}Bi^{210}$, for example, when undergoing the decay

$$_{83}Bi^{210} \rightarrow {}_{84}Po^{210} + {}_{-1}e^0$$

emits electrons with a maximum energy of 1.05 MeV; from its beta spectrum, the average energy is 0.39 ± 0.06 MeV. A direct calorimetric determination of the total energy per electron yielded 0.35 ± 0.04 MeV, which means that all of the ejected electrons do *not* leave the nucleus with the energy Q_{max}, but, in fact, with all energies from zero to Q_{max}.

To compound the problem, Wolfgang Pauli pointed out in 1930 that beta decay did not appear to conserve angular momentum in addition to not conserving energy. Once again, taking the case of $_{83}Bi^{210}$, we have to reconcile the facts that the nuclear angular momentum quantum number for the parent nuclide is one and for the daughter is zero. It follows that the angular momentum quantum number changes from one to one-half (that for the electron) in this process. Clearly, the situation was a drastic one for theoretical physics. Pauli took up the challenge, however, and postulated that, in fact, energy and angular momentum *were* conserved in beta decay. The reason that they did not seem to be constants was that a third, heretofore undetected particle was emitted when the decay process took place. This particle had gone unnoticed because it was virtually non-interacting: it possessed no charge and no rest mass. Also, its spin quantum number was one-half in order that angular momentum be conserved. The energy not carried off by the electron in beta decay was carried away by this "little neutral one" or *neutrino*, as Enrico Fermi called it.

Quite a postulate! In order to save energy and angular momentum conservation, we have to invent fantastic particles! Nevertheless, we shall adopt the neutrino hypothesis and see if it leads us to a workable theory of beta decay. To begin with, we note that our hypothesis requires us to rewrite the process (13.25) as

$$_zX^A \rightarrow \, _{z+1}Y^A + \, _{-1}e^0 + \bar{\nu} \qquad (13.25')$$

where ν refers to the neutrino. [We have put an antineutrino in expression (13.25') in order to pay heed to another conservation principle: the conservation of leptons.[8] The difference between a neutrino and its antiparticle will be taken up later.] The process (13.25') may be thought of as a transition wherein two particles are emitted whose energies must add up to Q_{max}, but whose momenta are subject to no restriction. The justification for this lies with the very heavy mass of the daughter nucleus, relatively speaking. The daughter is too heavy to carry away any but a neglible fraction of the total disintegration energy; on the other hand, its great mass permits it to carry off an arbitrary momentum. If we consider only the energetics of beta decay, we can ignore the daughter and impose no momentum requirement on the electron and neutrino. Therefore, in this approximation,

$$Q_{max} = E_e + E_{\bar{\nu}} \qquad (13.26)$$

in an obvious notation. Now, the transition rate for emission into a set of

---

[8] See, for example, E. Segré, *Nuclei and Particles*, W. A. Benjamin, New York, 1965, pp. 370–373, for a discussion of this principle.

energy states between E_f and $E_f + dE_f$ is, according to Fermi's "golden rule of time-dependent perturbation theory" (see problem 7 in Chapter 5),

$$W = 4|H'_{fi}|^2 \rho(E_f) \frac{d}{dt} \int_{-\infty}^{\infty} \frac{\sin^2 (\omega_{fi} t/2)}{(E_f - E_i)^2} \, dE_f$$

$$= 2 \frac{|H'_{fi}|^2}{\hbar} \rho(E_f) \int_{-\infty}^{\infty} \frac{\sin^2 x}{x^2} \, dx = \frac{2\pi}{\hbar} |H'_{fi}|^2 \rho(E_f), \quad (13.27)$$

where H'_{fi} is a matrix element of the (unknown) beta-decay perturbation operator and $\rho(E_f)$ is the density of final states of the system, which we can take as independent of the energy in the final state, E_f. The density of states in the present case is equal to the product of the density of states for the electron and the neutrino:

$$\rho(E_f) \, dE_f = \rho(E_e) \, dE_e \rho(E_{\bar{\nu}}) \, dE_{\bar{\nu}}$$

$$= \frac{p_e E_e (Q_{\max} - E_e)^2}{4\pi^4 \hbar^6 c^5} \, dE_{\bar{\nu}} \, dE_e, \quad (13.28)$$

where we have noted that

$$\rho(E_e) \, dE_e = \frac{4\pi p_e^2 \, dp_e}{(2\pi\hbar)^3},$$

$$\rho(E_{\bar{\nu}}) \, dE_{\bar{\nu}} = \frac{4\pi p_{\bar{\nu}}^2 \, dp_{\bar{\nu}}}{(2\pi\hbar)^3},$$

$$c^2 p \, dp = E \, dE, \quad (13.29)$$

and

$$p_{\bar{\nu}} = \frac{E_{\bar{\nu}}}{c}$$

The first two of equations (13.29) are just the result of expressing the density of states in momentum space and, for example, follow easily from equation (13.4) as applied to an entire momentum sphere with unit value for the volume of the system. The third of equations (13.29) is a differential form of equation (12.16), while the last equation is written under the assumption that the neutrino possesses no rest mass. When equations (13.28) and (13.27) are combined, we get

$$W(E_e) \, dE_e = \frac{|H'_{fi}|^2}{2\pi^3 \hbar^7 c^5} p_e E_e (Q_{\max} - E_e)^2 \, dE_e \quad (13.30)$$

as the transition probability per unit time for electron emission, that is, the rate of electron emission in the range of energies between E_e and $E_e + dE_e$. Equation (13.30) represents the energy spectrum for beta decay. As such, it is not quite accurate, since we have neglected entirely the coulomb field of the nucleus and its effect upon the emitted electron. In practice, it is found that this aspect of the problem is adequately represented by a correction factor which depends upon the atomic number of the daughter and upon the electron energy. The correction factor, $F(Z, E_e)$, multiplies the expression we have developed for $W(E_e)$, so that equation (13.30) becomes[9]

$$W(E_e)\, dE_e = \frac{|H'_{fi}|^2 F(Z, E_e)}{2\pi^3 \hbar^7 c^5}\, p_e E_e (Q_{max} - E_e)^2\, dE_e. \qquad (13.30')$$

In order to compare our result with experiment, we define

$$K(E_e) \equiv \left(\frac{W(E_e)}{F(Z, E_e) p_e E_e}\right)^{\!\frac{1}{2}}$$

and notice that, by equation (13.30′),

$$K(E_e) = A(Q_{max} - E_e), \qquad (13.31)$$

where

$$A = \frac{|H'_{fi}|}{2\pi^3 \hbar^7 c^5}.$$

Thus, a plot of the *Kurie function*, $K(E_e)$, against E_e should be a straight line, if our theory is sound. In Figure 77 we have a Kurie plot of the energy

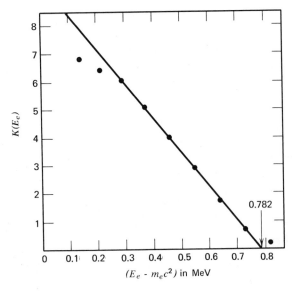

Figure 77. A Kurie plot of the beta-decay spectrum for the free neutron. [From J. M. Robson, The radioactive decay of the neutron, *Phys. Rev.* **83**: 349–358 (1951).]

[9] *Ibid.*, pp. 348–353.

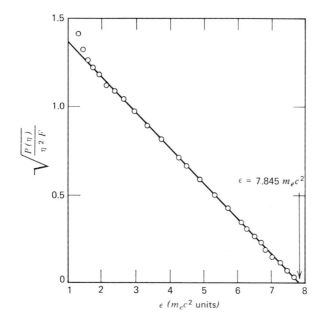

Figure 78. A Kurie plot of the beta-decay spectrum for $_2He^6$. (Reprinted with permission from C. S. Wu and S. A. Moszkowski, *Beta Decay*, Interscience Publishers, New York, 1966.)

spectrum for beta decay by a free neutron. Although there is a fair amount of experimental error associated with the data, it is evident that equation (13.31) is adhered to over the range of energies shown. The x-intercept of the plot occurs at 0.782 ± 0.013 MeV; this *is* the value of Q_{max} for neutron beta decay. In Figure 78 a more accurate Kurie plot is shown for the beta spectrum of He.[6] It is clear that in this instance the theory is rather good. In other cases (for example, $_{17}Cl^{36}$) the Kurie plot is decidedly non-linear. This may be attributed to a number of things, the most prominent among them being a dependence of the perturbation operator's matrix element upon the energy of the final state.

If one measures the momentum of the daughter and the electron, a more sensitive test of the neutrino hypothesis can be made, since one can then compute the neutrino momentum by difference from zero (the parent is at rest) and compare this with the neutrino energy divided by the speed of light. The results of this kind of experiment indicate an upper limit on the neutrino rest mass equal to 0.006 MeV$/c^2$, which is just 1.2 per cent of the electron's rest mass.[10]

All of the evidence we have discussed so far indicates indirectly that the neutrino hypothesis is correct. In 1953 Cowan, Reines, and coworkers at Los Alamos provided direct evidence for the existence of the neutrino by inducing the reaction

$$_1H^1 + \bar{\nu} \to {}_0n^1 + {}_{-1}\bar{e}^0. \tag{13.32}$$

The experimental arrangement involved an intense beam of antineutrinos impinging upon a cadmium chloride solution. The positron produced in the reaction (13.32) annihilates with an electron about 0.001 μsec after its creation, producing a pair of gamma rays. Some 10 μsec after its creation, the neutron in (13.32) is captured by a Cd nucleus; this event also results in gamma radiation. (See Figure 79.) The sequence of radiation described here is quite

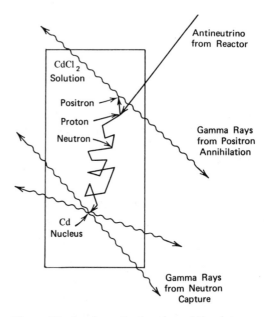

Figure 79. A schematic drawing of the decay

$$_1H^1 + \bar{\nu} \to {}_0n^1 + {}_{-1}\bar{e}^0$$

as it occurred in Cowan and Reines' experiment.

[10] Another possibility is to rewrite equation (13.30) without the assumption of zero rest mass for the neutrino. In that case we have W proportional to

$$[(Q_{max} - E_e) + m_\nu c^2]\{[(Q_{max} - E_e) + m_\nu c^2]^2 - m_\nu^2 c^2\}^{1/2}$$

instead of $(Q_{max} - E_e)^2$. The mass of the neutrino is then inferred by making a Kurie plot fitted to the appropriate value of m_ν. A careful analysis of the beta spectrum for $_1H^3$ leads to $m_\nu < 0.00025$ MeV$/c^2$—less than 0.05 per cent of the electron's rest mass.

characteristic of the process (13.32). Cowan and Reines observed only a few of these sequences per hour, despite the intensity of the incident beam. Nevertheless, their work provides independent verification that Pauli's neutrino really exists.

Having assured ourselves of the neutrino hypothesis, we should like to see the way in which this little particle fits into the scheme of quantum theory. We may begin by writing down the Dirac equation for the free neutrino, which is, evidently,

$$i\hbar \frac{\partial \psi}{\partial t} = -i\hbar c(\boldsymbol{\alpha} \cdot \boldsymbol{\nabla})\psi, \tag{13.33}$$

where $\boldsymbol{\alpha} = \{\alpha_1, \alpha_2, \alpha_3\}$. Equation (13.33) differs from equation (12.12) by the term in β, which vanishes when the rest mass is put equal to zero. Equations (12.13) through (12.15), because of the omission of β, now become

$$\alpha_i\alpha_j + \alpha_j\alpha_i = 2\delta_{ij}, \tag{12.13'}$$

$$\alpha_i^2 = I. \tag{12.15'}$$

Since equation (12.14) *is no longer a condition on equation* (13.33), *we can represent the* α_i $(i = 1, 2, 3)$ *by two by two matrices.* Therefore, we can put, for example,

$$\alpha_i = \sigma_i \qquad (i = 1, 2, 3), \tag{13.34}$$

where σ_i is one of the Pauli matrices. The combination of equations (13.33) and (13.34) yields

$$i\hbar \frac{\partial \psi}{\partial t} = -i\hbar c(\boldsymbol{\sigma} \cdot \boldsymbol{\nabla})\psi \tag{13.35}$$

as the Dirac equation for a free neutrino. This equation is separable, its unnormalized formal solution being

$$\psi(\mathbf{r}, t) = Nu(p, s) \exp\left[-(i/\hbar)(Et - \mathbf{p} \cdot \mathbf{r})\right],$$

where N is a normalization constant and $u(p, s)$ is the solution of the eigenvalue problem

$$c(\boldsymbol{\sigma} \cdot \mathbf{p})u(p, s) = Eu(p, s). \tag{13.36}$$

If $\mathbf{p}$ is chosen to point along the z-axis, then (for $E > 0$)

$$|+\rangle \Rightarrow u(p, s) = \begin{pmatrix} 1 \\ 0 \end{pmatrix},$$

since, in that case, $\boldsymbol{\sigma} \cdot \mathbf{p} = \sigma_3 p$. We see that equation (13.36) describes a neutrino whose spin angular momentum vector points along the direction of motion. This kind of neutrino is called *right-handed*, because its spin is in the sense of a right-handed screw. To create *left-handed* neutrinos, we must write equation (13.34) as

$$\alpha_i = -\sigma_i \quad (i = 1, 2, 3), \tag{13.34'}$$

so that (13.36) becomes

$$-c(\boldsymbol{\sigma}\cdot\mathbf{p})u(p, s) = Eu(p, s) \tag{13.36'}$$

with the solution

$$|-\rangle \Rightarrow u(p, s) = \begin{pmatrix} 0 \\ 1 \end{pmatrix}.$$

We can characterize this "handedness" property of the neutrino by the *helicity operator*,

$$\hbar \equiv \frac{(\boldsymbol{\sigma}\cdot\mathbf{p})}{p}$$

and note that

$$\langle +|\hbar|+\rangle = +1 \quad \text{(right-handed)}$$
$$\langle -|\hbar|-\rangle = -1 \quad \text{(left-handed)}.$$

Experiments on beta decay show that the neutrinos emitted in beta decay are right-handed. Therefore, *we identify the antineutrino with the right-handed neutrino and the neutrino with the left-handed neutrino.*

It is useful to see what is the relation between the two-component column matrices which are the solutions of equations (13.36) and (13.36') and the four-component column matrices we encountered in Chapter 12. To do so, we express equations (12.18) by the equivalent

$$\alpha_i = \begin{pmatrix} \sigma_i & 0 \\ 0 & -\sigma_i \end{pmatrix}, \quad \beta = \begin{pmatrix} 0 & -I' \\ -I' & 0 \end{pmatrix} \quad (i = 1, 2, 3), \tag{13.37}$$

which differ from equations (12.18) by the unitary transformation

$$U = \frac{1}{\sqrt{2}}(I + i\alpha_1\alpha_2\alpha_3\beta).$$

If we write

$$\psi(\mathbf{r}, t) = \begin{pmatrix} \psi^+(\mathbf{r}, t) \\ \psi^-(\mathbf{r}, t) \end{pmatrix}$$

into equation (12.12), then we get the two coupled equations

$$i\hbar \frac{\partial \psi^+}{\partial t} = -i\hbar c\boldsymbol{\sigma}\cdot\nabla\psi^+ - m_0c^2\psi^-,$$

$$i\hbar \frac{\partial \psi^-}{\partial t} = i\hbar c\boldsymbol{\sigma}\cdot\nabla\psi^- + m_0c^2\psi^+.$$

These equations are linked only in the rest-mass terms. If we put m_0 equal to zero, the uncoupled equations are just the Dirac equations for right- and left-handed neutrinos, respectively.

It appears that the theory of the neutrino follows in a very natural way from putting m_0 equal to zero in the Dirac equation. However, we should note that something rather drastic has occurred as well; for, a vanishing rest mass implies a non-existent β-matrix. But, as shown in problem 10 of Chapter 12, the space inversion operation is *defined* in terms of the β-matrix:

$$P\psi(\mathbf{r}, t) \equiv e^{i\varphi}\beta\psi(\mathbf{r}, t),$$

where φ is an arbitrary real number. It follows that *the space-reversal symmetry of the Dirac equation is lost for neutrinos.* The parity operation is no longer a constant of the motion! Indeed, we can see this directly, since the Hamiltonian operator

$$-i\hbar c(\boldsymbol{\sigma}\cdot\boldsymbol{\nabla})$$

in equation (13.35) does not commute with the parity operator P:

$$(\boldsymbol{\sigma}\cdot\boldsymbol{\nabla}) \xrightarrow[\mathbf{r}\to-\mathbf{r}]{} -(\boldsymbol{\sigma}\cdot\boldsymbol{\nabla}).$$

Moreover, the same thing can be said about the charge conjugation operation, since

$$C\psi \equiv i\alpha_2\beta\bar{\psi}$$

according to equation (12.29). It seems quite clear that $\mathscr{P}$ and $\mathscr{C}$ are no longer realizations of the group of the Dirac equation when neutrinos are considered.

Of course, the lack of parity conservation has been demonstrated so far only on paper. It may well be that the Dirac equation (13.35) is simply inaccurate; only an experiment can let us know for sure. In 1957, C. S. Wu and her coworkers reported such an experiment,[11] done on account of the speculations of T. D. Lee and C. N. Yang, who in 1956 had made arguments comparable to those we have just gone through. Wu's investigation involved the beta decay

$$_{27}\mathrm{Co}^{60} \to {}_{28}\mathrm{Ni}^{60} + {}_{-1}e^0 + \bar{\nu}$$

which is diagrammed in Figure 80. The spin magnetic moments of the $_{27}\mathrm{Co}^{60}$ nuclei were oriented at low temperature (about $0.01°\mathrm{K}$) by the Rose-Gorter method, which amounts to adding the beta emitter to cerium-magnesium nitrate, lining up the paramagnetic moments of the ions in that salt with a strong applied magnetic field, and using the oriented ions' field to orient the nuclear moments. The experiment then consists in a determination of whether

[11] C. S. Wu, *et al.*, Experimental test of parity conservation in beta decay, *Phys. Rev.* **105**: 1413–1415 (1957).

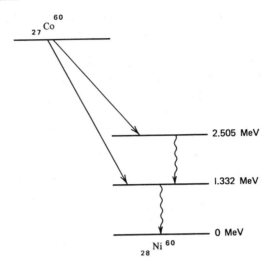

Figure 80. The decay scheme for $_{27}Co^{60}$.

the beta particles are emitted preferentially in one direction or another along the orientation axis. If a preferential direction of emission were discovered, then, obviously, the process would not be invariant under space reversal. A diagram of Wu's experimental apparatus is shown in Figure 81. The anthracene crystal served as the electron detector. The NaI crystals were used to detect gamma rays emitted by the $_{28}Ni^{60}$ nucleus in order to ascertain the degree of orientation of the cobalt nuclei: experiments had shown that oriented nuclei resulted in gamma emission preferentially along an equatorial axis through the sample. In Figure 82 Wu's data are plotted. The upper graph shows the preferential emission of gamma rays as a function of time. After about 6 minutes, we see, thermal agitation has caused the nuclei to become randomly oriented. The lower graph shows the very decided preferential emission of the beta particles, a preferential emission which is along a direction opposite to that in which the nuclear moments point and which disappears when the nuclear moments become disoriented. (Note that the disappearance of gamma anisotropy and beta asymmetry are closely correlated.) We must conclude from this that *beta decay is not invariant under space reversal.* Our theoretical speculation is entirely confirmed by experiment.

Returning to the Dirac equation, it should be pointed out that, although $\mathscr{C}$ and $\mathscr{P}$ are not realizations of the symmetry group, $\mathscr{C}\mathscr{P}$ remains so. Fundamentally the reason for this is that

$$\begin{aligned}
\psi_{CP}(\mathbf{r}, t) &= C(\overline{P\psi}(\mathbf{r}, t)) = C(\exp(-i\varphi)\beta\bar{\psi}(\mathbf{r}, t)) \\
&= i\exp(-i\varphi)\alpha_2\beta^2\bar{\psi}(\mathbf{r}, t) \\
&= i\exp(-i\varphi)\alpha_2\bar{\psi}(\mathbf{r}, t),
\end{aligned}$$

396

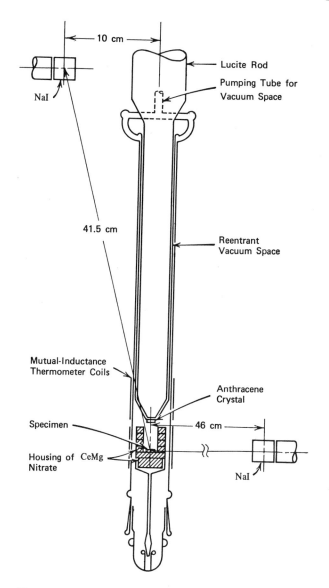

10 cm

Lucite Rod

Pumping Tube for
Vacuum Space

NaI

41.5 cm

Reentrant
Vacuum Space

Mutual-Inductance
Thermometer Coils

Anthracene
Crystal

Specimen

46 cm

Housing of CeMg
Nitrate

NaI

Figure 81. A schematic diagram of the apparatus used
to first detect the violation of the law of conservation of
parity. [From C. S. Wu *et al.*, Experimental test of
parity conservation in beta decay, *Phys. Rev.* **105**: 1413–
1415 (1957).]

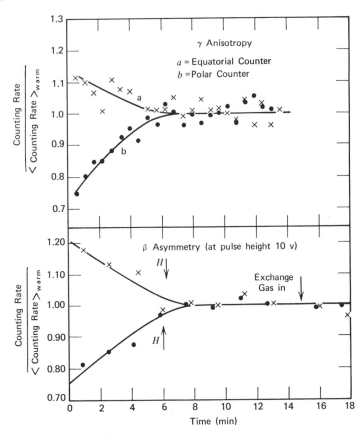

Figure 82. Plots of the gamma anisotropy and the beta asymmetry found in Wu's experiment with $_{27}Co^{60}$. H refers to the applied magnetic field. (After C. S. Wu *et al.*, Experimental test of parity conservation in beta decay, *Phys. Rev.* **105**: 1413–1415 (1957).)

which demonstrates that the operation $\mathscr{CP}$ does not involve the β-matrix. Therefore, $\mathscr{CP}$ is still a good symmetry operation for beta decay. It also follows that beta decay is invariant under the complete set $\mathscr{CPT}$—a consequence most important to the internal consistency of relativistic quantum theory.

PROBLEMS

1. Nuclides corresponding to large values of the atomic number generally possess more neutrons than protons. Show that this is a direct result of the independent-particle model of the nucleus. (You will find Figure 70 to be helpful in your argument.)

2. If electrons existed in the nucleus, they would be confined to within a distance of the order of 10 f.

(a) Calculate the approximate momentum deviation corresponding to this deviation in position.

(b) Calculate the approximate spread in energy. Is this deviation compatible with the values of Q_{max} obtained from studies of beta decay?

3. Derive equation (13.4).

4. In arguing for the neutron-proton model of the nucleus, it was stated that nuclei show a tendency toward $A = 2Z$. Explain why this ought to be in terms of the independent-particle model, presuming that the nucleons arrange themselves so as to minimize the groundstate energy of the nucleus.

5. Explain why the facts that the spin magnetic moment of the deuteron is 0.85741 nuclear magneton and its electric quadrupole moment is not zero force us to conclude that the groundstate of this nuclide cannot be purely a spherically symmetric triplet state.

6. Because $I = 1$ for the deuteron and because of what was stated in problem 5, it must be that the groundstate has some D-state character. (The mixed-in state must have the same parity as the spherically symmetric S-state, but must not correspond to an L-value too large to yield the correct magnitude for I.) Let us see if we can estimate how much D-state is present. The spin magnetic moment of the deuteron can be shown to be representable as

$$\mu_D = (\mu_n + \mu_p)\mathbf{S} + \tfrac{1}{2}\mathbf{L}$$

where $\mathbf{L}$ is the angular momentum of the proton relative to the center of mass of the neutron-proton system and μ_D, μ_n, and μ_p are spin moments for the deuteron, neutron, and proton, respectively. The measured value of μ_D is the projection of μ_D onto $\mathbf{I}$, that is

$$\mu_D = \frac{\mu_D \cdot \mathbf{I}}{|\mathbf{I}|^2}\mathbf{I} \equiv \mu_D\mathbf{I},$$

where $\mathbf{I} = \mathbf{L} + \mathbf{S}$ is construed to be a vector of magnitude $\sqrt{I(I+1)}\,\hbar$.

(a) Show that $\mu_D = 0.87967$ nuclear magneton if there is no D-state character.

(b) Show that

$$\mu_D = (\mu_n + \mu_p)\mathbf{I} - (\mu_n + \mu_p - \tfrac{1}{2})\mathbf{L}.$$

(c) If we write

$$\mathbf{L} = \alpha \mathbf{L}_S + \beta \mathbf{L}_D,$$

399

where α and β represent the fractions of S- and D-state character, respectively, and $\mathbf{L}_S$ is the value of $\mathbf{L}$ in an S-state, $\mathbf{L}_D$, the value of $\mathbf{L}$ in a D-state, show that we may conclude

$$\frac{\mathbf{L} \cdot \mathbf{I}}{|\mathbf{I}|^2} = \tfrac{3}{2}\beta$$

for the groundstate of the deuteron. (Remember the vector model!)

(d) Finally, show that the value of β leading to agreement between the calculated and empirical values of μ_D is about 0.04. Evidently, the groundstate of the deuteron has about 4 per cent D-state character.

7. Show that Q for alpha decay is equal to the kinetic energy of the alpha particle to $O(m_\alpha/m_d)$, where m_α is the mass of the alpha particle and m_d is the mass of the daughter nucleus.

8. The nuclide $_{84}Po^{214}$ undergoes alpha decay with a half-life of 160 μsec and a Q-value of 7.68 MeV.

(a) Calculate the value of $(R + R_{He})$ for this nucleus using the theory of alpha decay in the zeroth approximation: take $\gamma(y) = \pi/2$ and $v_0 = 2 \times 10^9$ cm/sec.

(b) Calculate $(R + R_{He})$ in the first approximation, using $\gamma(y) = (\pi/2) - 2y^{1/2}$. Compare this result to what you got in (a).

(c) Calculate r_0, the constant in equation (13.1). How do your values compare with those given in section 13.1?

9. Why did Pauli have to maintain that the neutrino was virtually non-interacting? Formulate your argument in light of the experiments discussed concerning the beta decay of $_{83}Bi^{210}$.

10. Show that the neutrino must always move with the speed of light. Does this fact violate any of the precepts of relativity?

11. Particles whose spin quantum numbers are half an odd integer obey the exclusion postulate and are called *fermions*. Show that, without the neutrino hypothesis, beta decay does not conserve fermions. (Note that two fermions do not make a fermion, just as two odd numbers do not make an odd number.)

12. Write down the modified expressions for positron emission and K-capture when the neutrino hypothesis is invoked.

13. Consider the K-capture reaction which transmutes $_{18}A^{37}$ into $_{17}Cl^{37}$. The reaction may be supposed to involve a resting parent and a moving daughter whose momentum is 0.806 ± 0.008 MeV/c. Given that the Q-value for the reaction is 0.816 ± 0.004 MeV, estimate the rest mass of the neutrino. Comment on the accuracy of your estimate.

14. A *lepton* is any fermion of smaller mass than a nucleon. The *lepton number*, l, is assigned as follows: $l = +1$ for electrons, muons, and neutrinos;

$l = -1$ for antileptons. Show that the following processes conserve the lepton number. (The muon has the same charge as the electron. The lepton number is zero for non-leptons.)

(a) $_0n^1 \rightarrow {}_1H^1 + e^0 + \bar{\nu}$

(b) $_1H^1 + \bar{\nu} \rightarrow {}_0n^1 + {}_{-1}\bar{e}^0$

(c) $_{-1}\mu^0 + {}_1H^1 \rightarrow {}_0n^1 + \nu$

(d) $_{-1}\pi^0 \rightarrow {}_{-1}\mu^0 + \bar{\nu}$

(e) $_{-1}\mu^0 \rightarrow {}_{-1}e^0 + \nu + \bar{\nu}$

(f) $_{-1}e^0 + {}_{-1}\bar{e}^0 \rightarrow 2\gamma$

At present the basis in symmetry for the law of conservation for leptons is not known.

15. Show that the reaction

$$_{-1}\mu^0 \rightarrow {}_{-1}e^0 + \gamma$$

should be a possible alternative to reaction (e) in problem 14. In reality, this reaction almost *never* occurs. One way out of the dilemma is to write (e) as

$$_{-1}\mu^0 \rightarrow {}_{-1}e^0 + \bar{\nu}_e + \nu_\mu,$$

where ν_e is electron neutrino and ν_μ is the muon neutrino, and to postulate that ν_e and ν_μ are *different* species. Experiment has indicated that this is indeed the case, because the neutrinos arising from the decay of the pion,

$$_{-1}\pi^0 \rightarrow {}_{-1}\mu^0 + \bar{\nu}_\mu$$

will *not* foment the reaction

$$\bar{\nu}_e + {}_1H^1 \rightarrow {}_0n^1 + {}_{-1}\bar{e}^0.$$

Evidently we must introduce a *muon number, m*, which is $+1$ for muons and muon neutrinos, -1 for the antiparticles, and zero otherwise, and postulate that m and l are conserved separately. Using this notion, show that the reaction

$$_{-1}\mu^0 \rightarrow {}_{-1}e^0 + \gamma$$

violates the conservation laws.

16. Consider the function $W(E_e)$, given in equation (13.20) for a massless neutrino and discussed in footnote 10 for a neutrino with rest mass. Show that, as $E_e \rightarrow Q_{max}$, $W(E_e)$ has zero slope in the first case and infinite slope in the second case, assuming that

$$(Q_{max} - E_e) \ll m_\nu c^2.$$

Explain how these facts could be used to estimate m_ν.

17. Prove that equation (13.34) is consistent with equations (12.13′) and (12.15′).

18. Derive equations (13.37) from equation (12.18) and the expression for U.

19. Show that

$$[\mathcal{H}, P] \neq 0,$$

where

$$\mathcal{H} = -i\hbar c(\boldsymbol{\sigma} \cdot \boldsymbol{\nabla})$$

and P is the space reversal operator.

20. Show that the Hamiltonian

$$\mathcal{H} = -i\hbar c(\boldsymbol{\sigma} \cdot \mathbf{p})$$

is invariant under space reversal and charge conjugation taken together.

Mathematical Appendix

A.1. FOURIER TRANSFORMS

The Fourier transform is a linear operator which maps functions $\psi(x)$ in one Hilbert space into functions $\varphi(p)$ in another. The transform is formally defined by

$$\varphi(p) \equiv \mathscr{F}\psi(x) = \left(\frac{a}{2\pi}\right)^{\frac{1}{2}} \int_{-\infty}^{\infty} \psi(x) \exp\left(-iapx\right) dx \qquad \text{(A.1)}$$

where a is any positive real number. To see that $\mathscr{F}$ is a linear operator, we need only note that integration is a linear operator as defined in section 3.2. Equation (A.1) is mathematically rigorous as it stands, since $\varphi(p)$ is simply defined by that equation whenever the integral exists. However, we shall want $\mathscr{F}$ to possess an inverse $\mathscr{F}^{-1}$ so that $\varphi(p)$ and $\psi(x)$ may be considered equivalent vectors and this poses a problem of some complexity. The solution of the problem comes in the form of the

Fourier Inversion Theorem. Let $\psi(x)$ be a complex-valued function on the real line which satisfies the following conditions:
(a) $|\psi(x)|$ has only a finite number of extrema in $(-\infty, \infty)$,
(b) $\psi(x)$ has only a finite number of finite discontinuities in $(-\infty, \infty)$, and
(c) the integral

$$\int_{-\infty}^{\infty} |\psi(x)| \, dx$$

converges absolutely. Then $\psi(x)$ may be written in the form of a Fourier integral,

$$\psi(x) \equiv \mathscr{F}^{-1}\varphi(p) = \left(\frac{a}{2\pi}\right)^{\frac{1}{2}} \int_{-\infty}^{\infty} \varphi(p) \exp\left(iapx\right) dp, \qquad \text{(A.2)}$$

except possibly on a set of isolated points, with $\varphi(p)$ given by equation (A.1).

The Fourier inversion theorem is proved by a straightforward but tedious analysis we shall not reproduce here.[1] Rather, we shall show through examples that equation (A.2) is valid for the appropriate $\psi(x)$.

[1] See I. N. Sneddon, *Fourier Transforms*, McGraw-Hill Book Co., New York, 1951, for a proof.

Mathematical Appendix

Consider, first,

$$\psi(x) = \exp(-\alpha x^2) \qquad (-\infty < x < \infty) \tag{A.3}$$

where α is a positive real number. The Fourier transform of this gaussian function is[2]

$$\varphi(p) = \left(\frac{a}{2\pi}\right)^{\frac{1}{2}} \int_{-\infty}^{\infty} \exp(-\alpha x^2) \exp(-iapx) \, dx$$

$$= \left(\frac{a}{2\alpha}\right)^{\frac{1}{2}} \exp(-a^2 p^2 / 4\alpha) \qquad (-\infty < p < \infty) \tag{A.4}$$

which tells us that the Fourier transform of a gaussian function is still a gaussian function. If we put equation (A.4) into equation (A.2) we get

$$\left(\frac{a}{2}\right)(\alpha\pi)^{-\frac{1}{2}} \int_{-\infty}^{\infty} \exp(-a^2 p^2 / 4\alpha) \exp(iapx) \, dp$$

$$= \left(\frac{a}{2}\right)(\alpha\pi)^{-\frac{1}{2}} \exp(-\alpha x^2) \int_{-\infty}^{\infty} \exp\left[-\left(\frac{ap}{2\sqrt{\alpha}} - i\sqrt{\alpha}\,x\right)^2\right] dp$$

$$= \pi^{-\frac{1}{2}} \exp(-\alpha x^2) \int_{-\infty}^{\infty} \exp(-y^2) \, dy$$

$$= \exp(-\alpha x^2) = \psi(x),$$

where

$$y \equiv \frac{ap}{2\sqrt{\alpha}} - i\sqrt{\alpha}\,x$$

and we have invoked condition (c) above:

$$\int_{-\infty}^{\infty} \exp(-y^2) \, dy = \sqrt{\pi}.$$

As another example, suppose we have the complex-valued

$$\psi(x) = \begin{cases} \exp(ikx) & |x| < L \\ 0 & |x| > L \end{cases} \tag{A.5}$$

where k is a positive real number. The Fourier transform of this function is

$$\varphi(p) = \left(\frac{2a}{\pi}\right)^{\frac{1}{2}} \frac{\sin[(k - ap)L]}{(k - ap)} \qquad (-\infty < p < \infty) \tag{A.6}$$

[2] For a listing of Fourier transforms, see Appendix C of the book by Sneddon, cited in the previous footnote.

The original function is then retrieved by

$$\frac{a}{\pi} \int_{-\infty}^{\infty} \frac{\sin [(k - ap)L]}{(k - ap)} \exp (iapx) \, dp = \frac{1}{\pi} \exp (ikx) \int_{-\infty}^{\infty} \frac{\sin yL}{y} \exp (-ixy) \, dy$$

$$= \begin{cases} \exp (ikx) & |x| < L \\ 0 & |x| > L \end{cases} = \psi(x),$$

where here

$$y \equiv k - ap$$

and we have used [2]

$$\mathscr{F} \frac{\sin yL}{y} = \left(\frac{1}{2\pi}\right)^{1/2} \int_{-\infty}^{\infty} \frac{\sin yL}{y} \exp (-ixy) \, dy$$

$$= \begin{cases} (\pi/2)^{1/2} & |x| < L \\ 0 & |x| > L \end{cases} \tag{A.7}$$

with a taken to be unity.

There are several properties of Fourier transforms which are of value in quantum theory.[3] The first of these which we shall mention is *Parseval's theorem*, which is derived in section 3.1. This theorem is expressed

$$\int_{-\infty}^{\infty} |\varphi(p)|^2 \, dp = \int_{-\infty}^{\infty} |\psi(x)|^2 \, dx \tag{A.8}$$

and is our way of expressing the equivalence of $\varphi(p)$ and $\psi(x)$. Another important property is provided by the *differentiation theorem*,

$$\mathscr{F}\psi^{(n)}(x) = (-iap)^n \mathscr{F}\psi(x) \qquad (n = 0, 1, \ldots) \tag{A.9}$$

where

$$\psi^{(n)}(x) \equiv \frac{d^n \psi}{dx^n}$$

tends to zero as $|x| \to \infty$ for all n. For $n = 0$ the theorem is just a statement of equation (A.1). For $n = 1$ we can prove the theorem by a partial integration:

$$\left(\frac{a}{2\pi}\right)^{1/2} \int_{-\infty}^{\infty} \frac{d\psi}{dx} \exp (-iapx) \, dx = \left(\frac{a}{2\pi}\right)^{1/2} \psi(x) \exp (-iapx) \Big|_{-\infty}^{\infty}$$

$$- iap\left(\frac{a}{2\pi}\right)^{1/2} \int_{-\infty}^{\infty} \psi(x) \exp (-iapx) \, dx$$

$$= -iap\mathscr{F}\psi(x).$$

For $n > 1$, the proof is by iteration of this procedure.

[3] For a good discussion of these properties, see Chapter IV of L. Schwartz, *Mathematics for the Physical Sciences*, Addison-Wesley Publ. Co., Reading, Mass., 1966.

Finally we mention the *convolution theorem*. If $\psi(x)$ and $\chi(x)$ possess invertible Fourier transforms $\varphi(p)$ and $\omega(p)$, respectively, then

$$\mathscr{F}^{-1}\varphi(p)\omega(p) = \left(\frac{a}{2\pi}\right)^{\frac{1}{2}} \int_{-\infty}^{\infty} \chi(y)\psi(x-y)\,dy$$

$$\equiv \psi*\chi \qquad \qquad (A.10)$$

To see this, we write

$$\int_{-\infty}^{\infty} \varphi(p)\omega(p) \exp{(iapx)}\,dp$$

$$= \left(\frac{a}{2\pi}\right)^{\frac{1}{2}} \int_{-\infty}^{\infty} \varphi(p) \exp{(iapx)} \int_{-\infty}^{\infty} \chi(y) \exp{(-iapy)}\,dy\,dp$$

$$= \int_{-\infty}^{\infty} \chi(y)\left(\frac{a}{2\pi}\right)^{\frac{1}{2}} \int_{-\infty}^{\infty} \varphi(p) \exp{[iap(x-y)]}\,dp\,dy$$

$$= \int_{-\infty}^{\infty} \chi(y)\psi(x-y)\,dy,$$

upon interchanging the order of integration. (The interchange is permitted so long as the operator $\mathscr{F}$ has an inverse.) In precisely the same way we can show that the convolution operation is symmetric under the operator $\mathscr{F}$. That is,

$$\mathscr{F}\psi(x)\chi(x) = \varphi*\omega$$

$$\equiv \left(\frac{a}{2\pi}\right)^{\frac{1}{2}} \int_{-\infty}^{\infty} \omega(p)\varphi(p-p')\,dp'. \qquad (A.11)$$

Equation (A.11) is used in Chapter 3 to compute the Fourier transform of the potential energy operator.

A.2. THE DELTA-"FUNCTION"

The formalism of quantum mechanics encounters difficulties when Hermitian operators possessing a continuous spectrum must be considered. The problem is that the "eigenvectors" corresponding to the "eigenvalues" have infinite lengths and thus cannot be normalized to unity. It follows by definition that these "wavefunctions" are not vectors in Hilbert space and that no physical interpretation by means of the probability postulate is possible. Obviously, we should like to surmount this difficulty in some reasonable way, but not at the expense of too much mathematical rigor. Although we cannot expect to get the eigenvectors back into Hilbert space,[4] at least we must provide a rigorous foundation for their use in the quantum theory. The theory of distributions, developed by Laurent Schwartz,[5] does just this.

Consider a set of functions $f(x)$ that are continuous, with continuous derivatives of all orders, and that vanish identically outside a finite domain. We shall call these functions *testing functions* and require that they form a linear space. A linear functional, called a *distribution*, is defined in terms of $f(x)$ by

$$F_s(f) \equiv \int_{-\infty}^{\infty} s(x)f(x)\, dx \qquad (A.12)$$

where s is a symbol which may or may not have values (that is, which may or may not be a function). The operations which can be defined for distributions are something like those usually encountered when dealing with ordinary functionals on a linear space. For example,

$$F_s(a_1 f_1 + a_2 f_2) = a_1 F_s(f_1) + a_2 F_s(f_2),$$

where the a's are scalars, defines the linearity property. However, the *product* of distributions does not always exist.[5]

The quantity $s(x)$ in equation (A.12) is called a *symbolic function*. Every integrable function is a symbolic function, but certain symbolic functions are neither integrable nor functions. As an example of the latter, consider the Dirac distribution

$$F_\delta(f) \equiv f(0) \equiv \int_{-\infty}^{\infty} \delta(x)f(x)\, dx$$

where $\delta(x)$ is a symbolic function. The Dirac distribution defines the delta-"function" $\delta(x)$ and permits us to treat the latter as an ordinary function with exceptional properties. For instance, if $f(x) = 1$ over some finite domain which includes the origin, and vanishes otherwise, we have immediately the result

$$1 = \int_{-\infty}^{\infty} \delta(x)\, dx$$

that often appears (unrigorously) as the definition of the delta-"function".
The derivative of a distribution is defined by

$$\frac{d}{dx} F_s(f) \equiv -F_s\left(\frac{df}{dx}\right). \qquad (A.13)$$

---

[4] That is, we cannot get them back into the kind of Hilbert space we have considered so far. The scalar product can be redefined for these eigenfunctions so that they will have finite magnitudes and span a Hilbert space, but the Hilbert space will have several properties which are quite different from those of the one we are dealing with. To avoid this new problem, physicists have chosen to use the delta-"function" where necessary.
[5] For a rigorous discussion of distributions, see L. Schwartz, *Mathematics for the Physical Sciences*, Addison-Wesley, Reading, Mass., 1966.

Thus *any* distribution, whether $s(x)$ be a function or not, possesses a derivative. In the case of the Dirac distribution we have

$$\frac{d}{dx} F_\delta(f) = -F_\delta\left(\frac{df}{dx}\right) = -\left(\frac{df}{dx}\right)_{x=0} \qquad (A.14)$$

If we *define*

$$F_{\delta^{(1)}}(f) \equiv \frac{d}{dx} F_\delta(f),$$

where $\delta^{(1)}(x)$ means $d\delta/dx$, we have

$$\int_{-\infty}^{\infty} \delta^{(1)}(x)f(x)\,dx = -\left(\frac{df}{dx}\right)_{x=0} \qquad (A.15)$$

which gives a meaning to the derivative of the delta-"function"! This result is easily generalized to

$$\int_{-\infty}^{\infty} \delta^{(n)}(x)f(x)\,dx = (-1)^n\left(\frac{d^n f}{dx^n}\right)_{x=0}.$$

Notice that the delta-"function" has no rigorous meaning unless it appears under an integral sign. It has no "value" independent of the operation of multiplication by a testing function and integration over the domain $(-\infty, \infty)$. However, at times it is convenient when doing quantum theory to write $\delta(x)$ without the integral; the underlying rigor is always understood in these cases. In particular, some properties of $\delta(x)$ derivable from the Dirac distribution are

$$\left.\begin{aligned}
\delta(x) &= \delta(-x) \\
\delta(ax) &= \frac{\delta(x)}{|a|} \quad (a \neq 0) \\
x\,\delta(x) &= 0 \\
f(x)\,\delta(x - x_0) &= f(x_0)\,\delta(x - x_0) \\
\int_{-\infty}^{\infty} \delta(x - t)\,\delta(t - \tau)\,dt &= \delta(x - \tau).
\end{aligned}\right\} \qquad (A.16)$$

The analogs of equations (A.16) for the nth derivative of the delta-"function" are

$$\left.\begin{aligned}
\delta^{(n)}(x) &= (-1)^n\,\delta^{(n)}(-x) \\
x^{n+1}\,\delta^{(n)}(x) &= 0 \\
\int_{-\infty}^{\infty} \delta^{(m)}(x - t)\,\delta^{(n)}(t - \tau)\,dt &= \delta^{(m+n)}(x - \tau)
\end{aligned}\right\} \qquad (A.17)$$

where $\delta^{(n)}(x)$ means $d^n\delta/dx^n$.

A.3. EQUATIONS OF THE STURM-LIOUVILLE TYPE

The Hamiltonian operator for a physical system will often be reducible to three operators, each of the general form

$$\mathscr{L}u(x) \equiv \frac{d}{dx}\left(p(x)\frac{du}{dx}\right) + q(x)u(x) \qquad (a \leqslant x \leqslant b),$$

where $\mathscr{L}$ is the one-dimensional operator, $u(x)$ is a one-dimensional eigenfunction, $p(x)$ is a function which possesses continuous derivatives up to at least order two and which does not vanish anywhere in the domain of $u(x)$, and $q(x)$ is a continuous function of x anywhere in its domain. The operator $\mathscr{L}$ is called a *Sturm-Liouville operator* and the differential equation

$$\mathscr{L}u(x) + \lambda w(x)u(x) = 0 \qquad (a \leqslant x \leqslant b) \qquad (A.18)$$

is called *an equation of the Sturm-Liouville type*. The Sturm-Liouville equation is one example of an eigenvalue problem. The problem, however, is not complete until boundary conditions are put on. They are

$$\bar{v}(x)p(x)\frac{du}{dx}\bigg|_{x=a} = \bar{v}(x)p(x)\frac{du}{dx}\bigg|_{x=b} \qquad (A.19)$$

$$v(x)p(x)\frac{\overline{du}}{dx}\bigg|_{x=a} = v(x)p(x)\frac{\overline{du}}{dx}\bigg|_{x=b} \qquad (A.20)$$

where $v(x)$ and $u(x)$ are any two (complex) solutions of equation (A.18). [It is to be remembered, however, that $p(x)$ is a real function.]

The operator $\mathscr{L}$ possesses a very important property. To discover it, let us form the integral

$$\int_a^b \bar{v}(x)\mathscr{L}u(x)\,dx = \int_a^b \bar{v}(x)\frac{d}{dx}\left(p(x)\frac{du}{dx}\right)dx + \int_a^b \bar{v}(x)q(x)u(x)\,dx. \quad (A.21)$$

Now,

$$\int_a^b \bar{v}(x)\frac{d}{dx}\left(p(x)\frac{du}{dx}\right)dx = \bar{v}(x)p(x)\frac{du}{dx}\bigg|_a^b - \int_a^b \frac{\overline{dv}}{dx}p(x)\frac{du}{dx}\,dx$$

$$= -\int_a^b \frac{\overline{dv}}{dx}p(x)\frac{du}{dx}\,dx$$

$$= -\frac{\overline{dv}}{dx}p(x)u(x)\bigg|_a^b + \int_a^b u(x)\frac{d}{dx}\left(p(x)\frac{\overline{dv}}{dx}\right)dx$$

$$= \int_a^b u(x)\frac{d}{dx}\left(p(x)\frac{\overline{dv}}{dx}\right)dx$$

where we have integrated by-parts twice and applied, successively, the boundary conditions (A.19) and (A.20). (Since $u(x)$ and $v(x)$ are *any* two solutions of equation (A.18), we may interchange them in the boundary conditions. This was done in equation (A.20) to make

$$\frac{\overline{dv}}{dx}\,p(x)u(x)\,\bigg|_a^b$$

vanish identically.) Equation (A.21) may now be written

$$\int_a^b \bar{v}(x)\mathscr{L}u(x)\,dx = \int_a^b u(x)\frac{d}{dx}\left(p(x)\frac{\overline{dv}}{dx}\right)dx + \int_a^b \bar{v}(x)q(x)u(x)\,dx$$

which implies

$$\int_a^b \bar{v}(x)\mathscr{L}u(x)\,dx = \int_a^b (\mathscr{L}\bar{v}(x))u(x)\,dx. \tag{A.22}$$

Equation (A.22) shows that $\mathscr{L}$ is a *self-adjoint* operator, according to the definition given in equation (3.33). It follows that the solutions of equation (A.18) and their corresponding eigenvalues are subject to the conditions (a) through (g) given in section 3.2, except for the statements regarding continuous eigenvalues. In particular we may state that the eigenfunctions of $\mathscr{L}$ are orthonormal, or can be made so, and that they constitute an invariant subspace of $\mathscr{L}$. These properties are just those mentioned in section 4.4.

Bibliography

This is a list of books which may be used in seeking viewpoints or presentations different from those in the present volume, or may be consulted if you wish to read further on some point lightly touched upon. The list is certainly not exhaustive, but does claim some breadth in the array of books offered.

I. PRIMEVAL QUANTUM CONCEPTS

1. Anderson, D. L. *The Discovery of the Electron.* D. Van Nostrand Co., Princeton, 1964.

2. Born, M. *Atomic Physics.* Hafner Publ. Co., New York, 1962.

3. Eisberg, R. M. *Fundamentals of Modern Physics.* John Wiley and Sons, New York, 1961.

4. Feynman, R. *The Feynman Lectures.* Addison-Wesley Publ. Co., Reading, Mass., 1963–1965. Vol. I–III.

5. Tomonaga, S. *Quantum Mechanics.* North-Holland Publ. Co., Amsterdam, 1962. Vol. I.

6. Wichmann, E. *Quantum Physics.* McGraw-Hill Book Co., New York, 1967. (Preliminary edition.)

II. BASIC QUANTUM THEORY

1. Dicke, R. H., and J. P. Wittke. *Introduction to Quantum Mechanics.* Addison-Wesley Publ. Co., Reading, Mass., 1960.

2. Dirac, P. A. M. *The Principles of Quantum Mechanics.* Oxford Univ. Press, London, 1958.

3. Fermi, E. *Notes on Quantum Mechanics.* University of Chicago Press, Chicago, 1961.

4. Heisenberg, W. *The Physical Principles of the Quantum Theory.* Dover Publications, New York, 1930.

5. Heitler, W. *The Quantum Theory of Radiation.* Oxford Univ. Press, London, 1954.

6. Landau, L. D., and E. M. Lifshitz. *Quantum Mechanics.* Oxford Univ. Press, London, 1958.

7. Lawden, D. F. *The Mathematical Principles of Quantum Mechanics.* Methuen and Co., London, 1967.

8. Mandl, F. *Quantum Mechanics.* Academic Press, New York, 1957.

9. Messiah, A. *Quantum Mechanics.* John Wiley and Sons, New York, 1961, 1962. Vol. I and II.

10. Newing, R. A., and J. Cunningham. *Quantum Mechanics.* Oliver and Boyd, Edinburgh, 1967.

11. Pauling, L., and E. B. Wilson, Jr. *Introduction to Quantum Mechanics.* McGraw-Hill Book Co., New York, 1935.

12. Roman, P. *Advanced Quantum Theory.* Addison-Wesley Publ. Co., Reading, Mass., 1965.

13. Saxon, D. S. *Elementary Quantum Mechanics.* Holden-Day, San Francisco, 1968.

14. Schiff, L. *Quantum Mechanics.* McGraw-Hill Book Co., New York, 1968.

15. Tomonaga, S. *Quantum Mechanics.* North-Holland Publ. Co., Amsterdam, 1966. Vol. II.

16. von Neumann, J. *The Mathematical Foundations of Quantum Mechanics.* Princeton Univ. Press, Princeton, 1955.

III. ATOMIC AND MOLECULAR STRUCTURE

1. Alexandroff, P. S. *An Introduction to the Theory of Groups.* Hafner Publ. Co., New York, 1959.

2. Allen, H. C., and P. C. Cross. *Molecular Vib-Rotors.* John Wiley and Sons, New York, 1963.

3. Barrow, G. M. *Introduction to Molecular Spectroscopy.* McGraw-Hill Book Co., New York, 1962.

4. Candler, C. *Atomic Spectra and the Vector Model.* D. Van Nostrand Co., 1964.

5. Falicov, L. M. *Group Theory and its Physical Applications.* The University of Chicago Press, Chicago, 1966.

6. Hammermesh, M. *Group Theory and its Application to Physical Problems.* Addison-Wesley Publ. Co., Reading, Mass., 1962.

7. Hanna, M. W. *Quantum Mechanics in Chemistry.* W. A. Benjamin, New York, 1965.

8. Hartree, D. R. *The Calculation of Atomic Structures.* John Wiley and Sons, New York, 1957.

9. Herzberg, G. *Atomic Spectra and Atomic Structure.* Dover Publications, New York, 1944.

10. Herzberg, G. *Infrared and Raman Spectra of Polyatomic Molecules.* D. Van Nostrand Co., Princeton, 1945.

11. Herzberg, G. *Spectra of Diatomic Molecules.* D. Van Nostrand Co., Princeton, 1950.

12. Meijer, P. H. E., and E. Bauer. *Group Theory.* John Wiley and Sons, New York, 1962.

13. Moore, C. E. *Atomic Energy Levels.* U.S. Bureau of Standards Circ. 467. U.S. Government Printing Office, Washington D.C., 1949, 1952, 1958. Vol. I–III.

14. Pauling, L., and S. Goudsmit. *The Structure of Line Spectra.* McGraw-Hill Book Co., New York, 1930.

15. Pauling, L., and E. B. Wilson, Jr. *Introduction to Quantum Mechanics.* McGraw-Hill Book Co., New York, 1935.

16. Rose, M. E. *Elementary Theory of Angular Momentum.* John Wiley and Sons, New York, 1957.

17. Slater, J. C. *Quantum Theory of Atomic Structure.* McGraw-Hill Book Co., New York, 1960. Vol. I and II.

18. Slater, J. C. *Quantum Theory of Molecules and Solids.* McGraw-Hill Book Co., New York, 1963. Vol. I.

19. Stevenson, R. *Multiplet Structure of Atoms and Molecules.* W. B. Saunders Co., Philadelphia, 1965.

20. Weyl, H. *The Theory of Groups and Quantum Mechanics.* Dover Publications, New York.

21. Wigner, E. P. *Group Theory and its Application to the Quantum Mechanics of Atomic Spectra.* Academic Press, New York, 1959.

IV. HIGH ENERGY PHENOMENA

1. Bjorken, J. D., and S. D. Drell. *Relativistic Quantum Mechanics.* McGraw-Hill Book Co., New York, 1964.

2. Dirac, P. A. M. *The Principles of Quantum Mechanics.* Oxford Univ. Press, London, 1958.

3. Eisberg, R. M. *Fundamentals of Modern Physics.* John Wiley and Sons, New York, 1961.

4. Elton, L. R. B. *Introductory Nuclear Theory.* W. B. Saunders Co., Philadelphia, 1966.

5. Fermi, E. *Nuclear Physics.* University of Chicago Press, Chicago, 1950.

6. Green, A. E. S. *Nuclear Physics.* McGraw-Hill Book Co., New York, 1955.

7. Hofstadter, R. *Nuclear and Nucleon Structure.* W. A. Benjamin, New York, 1963.

8. Kaplan, I. *Nuclear Physics.* Addison-Wesley Publ. Co., Reading, Mass., 1963.

9. Landau, L. D., and E. M. Lifshitz. *Quantum Mechanics.* Addison-Wesley Publ. Co., Reading, Mass., 1965.

10. Livesey, D. L. *Atomic and Nuclear Physics.* Blaisdell Publ. Co., Waltham, Mass., 1966.

11. Mandl, F. *Quantum Mechanics.* Academic Press, New York, 1957.

12. Messiah, A. *Quantum Mechanics.* John Wiley and Sons, New York, 1961, 1962. Vol. I and II.

13. Mott, N. F., and H. S. W. Massey. *Theory of Atomic Collisions.* Oxford Univ. Press, London, 1965.

14. Preston, M. A. *Physics of the Nucleus.* Addison-Wesley Publ. Co., Reading, Mass., 1966.

15. Segré, E. *Nuclei and Particles.* W. A. Benjamin, New York, 1965.

16. Schiff, L. *Quantum Mechanics.* McGraw-Hill Book Co., New York, 1968.

17. Wu, C. S., and S. A. Moszkowski. *Beta Decay.* John Wiley and Sons, New York, 1966.

V. MATHEMATICS AND MATHEMATICAL PHYSICS

1. Arfken, G. *Mathematical Methods for Physicists.* Academic Press, New York, 1966.

2. Butkov, E. *Mathematical Physics.* Addison-Wesley Publ. Co., Reading, Mass., 1968.

3. Dennery, P., and A. Krzywicki. *Mathematics for Physicists.* Harper and Row, New York, 1967.

4. Friedman, B. *Principles and Techniques of Applied Mathematics.* John Wiley and Sons, New York, 1956.

5. Halmos, P. *Introduction to Hilbert Space.* Chelsea Publ. Co., New York, 1957.

6. Halperin, I. *Introduction to the Theory of Distributions.* University of Toronto Press, Toronto, 1952.

7. Ince, E. L. *Ordinary Differential Equations.* Dover Publications, New York, 1956.

8. Jackson, J. D. *Mathematics for Quantum Mechanics.* W. A. Benjamin, New York, 1962.

9. Jeffereys, H., and B. S. Jeffereys. *Methods of Mathematical Physics.* Cambridge Univ. Press, London, 1956.

10. Margenau, H., and G. M. Murphy. *The Mathematics of Physics and Chemistry.* D. Van Nostrand Co., Princeton, 1956. Vol. I.

11. Mathews, J., and R. L. Walker. *Mathematical Methods of Physics.* W. A. Benjamin, New York, 1965.

12. Schwartz, L. *Mathematics for the Physical Sciences.* Addison-Wesley Publ. Co., Reading, Mass., 1966.

13. Watson, G. N. *Theory of Bessel Functions.* Cambridge Univ. Press, London, 1944.

Index

415

Index

Index

Index

Index

422

Index

Index

Valence bond method, 290
Variance of an observable, 67, 178
Variational method, 162f
  applied to helium atom, 218
  applied to hydrogen atom, 166
  applied to Stark effect, 164
Variational principle, 160
  and Born-Oppenheimer theorem, 280
  and Hartree equation, 224
  and Hartree-Fock equation, 229
  relation to perturbation theory, 162
Vector addition, 51
Vector, angular momentum, 105
  orbital, 105
  spin, 204
  total, 205
Vector area, 11, 123
Vector, basis, 54
  as a state vector, 65
  cartesian, 54
  relation to eigenvalue problem, 62
  relation to orthogonality, 55
Vector, cartesian, 52
  in Hilbert space, 53
Vector coupling coefficients, 207, 262f
Vector model of the atom, 204
  relation to group theory, 262f
  relation to spectral fine structure, 210
  relation to Zeeman effect, 208
Vector potential, 156
Vector space, (see Linear space)
Virial theorem, 129
Virtual transitions, 156

Wave equation, electromagnetic, 156
Wave equation, Schrödinger, (see
  Schrödinger equation of motion)
Wavefunction, 69
  as vector in Hilbert space, 69
Wavenumber, 98
Wave packet, (see Principle of superposition)
Wave-particle dualism, 46
Wave quantum mechanics, (see Schrödinger
  picture)
Weight function, in Sturm-Liouville equa-
  tion, 113
Wien displacement law, 15, 20
Wigner coefficients, 207
Wilson-Sommerfeld quantization rule, 126
Work function, 33, 48

X-rays, 4, 344

Yukawa potential, 341

Zeeman effect, 122, 208
  anomalous, 122, 208
  for hydrogenic atoms, 121, 215
  for many-electron atom, 208
  for sharp series, 130
  in strong magnetic field, 216
  of mercury yellow line, 122
  selection rules and, 124f
Zero-point energy, 181
  in blackbody radiation, 20